Hyperbranched Epoxy Resins

Hyperbranched Epoxy Resins

Synthesis, Applications, and Recycling

Edited by Daohong Zhang, Yu Jiang, and Junheng Zhang

WILEY-VCH

Editors

Daohong Zhang
South-Central Minzu University
182# Minzu Road, Wuhan 430074
P.R. China

Yu Jiang
South-Central Minzu University
182# Minzu Road, Wuhan 430074
P.R. China

Junheng Zhang
South-Central Minzu University
182# Minzu Road, Wuhan 430074
P.R. China

Cover Design: Wiley
Cover Images: Courtesy of the Editors

Typesetting: Straive, Chennai, India

Library of Congress Card No.: applied for

British Library Cataloguing-in-Publication Data A catalogue record for this book is available from the British Library.

Bibliographic information published by the Deutsche Nationalbibliothek
The Deutsche Nationalbibliothek lists this publication in the Deutsche Nationalbibliografie; detailed bibliographic data are available on the Internet at <http://dnb.d-nb.de>.

Print ISBN: 978-3-527-35480-1
ePDF ISBN: 978-3-527-85081-5
ePub ISBN: 978-3-527-85080-8
oBook ISBN: 978-3-527-85082-2

Printed and bound by CPI Group (UK) Ltd, Croydon, CR0 4YY

C9783527354801_290726

Contents

Preface

Hyperbranched Epoxy Resins: Synthesis, Applications, and Recycling aims to provide a comprehensive and authoritative overview of the science and technology of hyperbranched epoxy resins (HERs), from their precise synthesis and unique topological structures to their groundbreaking applications in high-performance materials and their pathways toward circularity through recycling. This book emerges at a critical time when the inherent brittleness and environmental persistence of conventional epoxy resins increasingly limit their potential in advanced applications. HERs, with their distinctive three-dimensional architecture, offer a powerful and versatile solution to these long-standing challenges, enabling the creation of epoxy-based materials that are simultaneously stronger, tougher, and more sustainable.

The contents of this book are structured to systematically guide the reader through the entire field. The opening chapters lay the essential groundwork, introducing the fundamental concepts of HERs and detailing the sophisticated characterization methods for understanding their chemical and topological structures. The core of the book is dedicated to a thorough exploration of the diverse synthesis methodologies that empower researchers to precisely tailor HER architectures. Subsequent sections pivot to practical impact, examining the application of HERs in areas ranging from adhesives and composite matrices to flame-retardant materials. A particular focus is placed on the homogeneous reinforcing and toughening mechanism, a key advantage that allows HERs to enhance both the strength and toughness of epoxy resins. Finally, the concluding chapters address the vital topic of sustainability, presenting the latest advances in the degradation and recycling of HERs and their composites, thus closing the loop on the material life cycle.

Each chapter in this book is a well-documented and yet concise contribution written by noted experts and authorities in their field. We are extremely grateful to all of them for taking time to share their knowledge and popularize it in a way understandable to a broad readership. We are also indebted to all the reviewers whose comments and remarks helped us very much in our editing work. Finally, Wiley-VCH deserves our sincere acknowledgments for their professional support and for striving to keep the entire project on time.

We expect that *Hyperbranched Epoxy Resins: Synthesis, Application, and Recycling* will be an indispensable resource for specialists in materials science, polymer chemistry, and chemical engineering, providing them with the knowledge to update their research and innovate in industrial practice. Simultaneously, we believe it will serve as a valuable companion for graduate students and scientists from related disciplines who wish to understand and engage with the cutting-edge developments that are shaping the future of high-performance and sustainable polymer materials.

November 2025
Wuhan, China

Daohong Zhang
Yu Jiang
Junheng Zhang

1

Background and Basic Knowledge of Hyperbranched Epoxy Resins

Junheng Zhang and Daohong Zhang

South-Central Minzu University, Hubei R&D Center of Hyperbranched Polymers Synthesis and Applications, School of Chemistry and Materials Science, No.182, Minzu Ave. Hongshan District, Wuhan Hubei Province, 430074, P.R.China

1.1 Chemistry, Synthesis, Manufacture, and Characterization of Epoxy Resins

1.1.1 Introduction

The term "epoxy resin" (EP) refers to both prepolymers and cured resins. Prepolymers contain reactive epoxy groups, 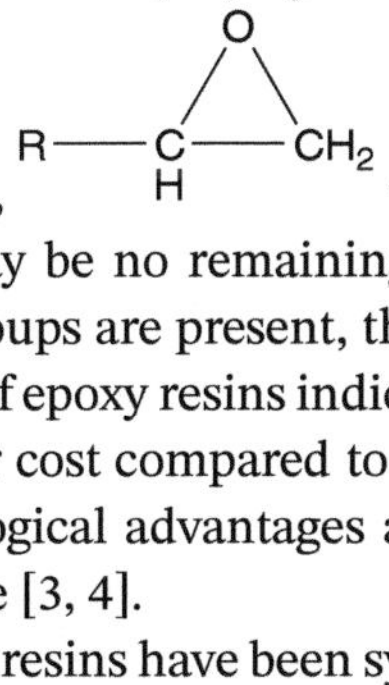

which give them their name. In cured resins, however, there may be no remaining reactive sites [1, 2]. Despite this, even when no more epoxy groups are present, these materials are still classified as epoxy resins. The market size of epoxy resins indicates their importance as industrial polymers. Due to their higher cost compared to other resins, they are typically used only when specific technological advantages are required. Many of these products offer significant added value [3, 4].

Although materials now known as epoxy resins have been synthesized since 1891, the first commercially available epoxy resin was not introduced until the 1940s and 1950s through the independent work of Swiss chemist Pierre Castan and American chemist Sylvan Greenlee, despite some patents dating back to the 1930s. The earliest commercially available epoxies were created by reacting bisphenol A (BPA) with epichlorohydrin, and this method remains the most common way of producing the majority of epoxies today, though other types of resins are also used [5–7].

Curing of an epoxy resin occurs when it reacts with a hardener, resulting in a rigid three-dimensional (3D) network. Hardeners used for curing typically have more than two reactive functional groups (i.e., $f > 2f > 2f > 2$). For common hardeners used with BPA-based resins, $f = 4f = 4f = 4$ (functionality), though the effective functionality is often $f = 2f = 2f = 2$. This effective functionality can be increased if the curing temperature is high enough to allow secondary hydroxy groups to participate in the curing process [8, 9].

Initially, when the epoxy resin mixes with the hardener, it forms slightly larger molecules [10, 11]. As curing progresses, these molecules grow longer. However, it's important to note that when half of the reactive sites have reacted, the resulting average molecular size is still relatively small. As the molecules continue to grow, highly branched structures begin to form, eventually leading to an extensively branched network. The key event in this process is when branching starts to occur, which marks the gelation point.

Before reaching the gel point, the sample can still dissolve in at least some solvents. Once the gel point is reached, however, the network becomes insoluble and begins to swell as it absorbs the solvent. At the gel point, both small branched molecules and solvable components are still present, meaning the curing sample has both sol and gel phases. The first gel formed is relatively soft and can be easily broken up. To create a stable structural material, curing must continue until most of the sample has transitioned into a stable 3D network.

During the curing process, the glass transition temperature (T_g) of the resin increases [12, 13]. When the cure temperature is well above T_g, the reaction between the epoxy and the hardener is chemically controlled, meaning the rate of the reaction determines its progress. In cases where complete reaction of every epoxy group is desired, the resin can be reheated or "post-cured" at a higher temperature to ensure full curing.

There are several methods for forming epoxy rings. These methods are among the most important processes leading to the development of epoxy compounds. The processes mainly involve the reaction of a halohydrin with a hydroxyl compound and the formation of epoxy products from unsaturated compounds using organic peracids. In industrial applications, epoxy resins require the formation of a 3D network through a reaction with a polyfunctional hardener [14–16].

Curing reactions are heavily influenced by the more reactive nature of the epoxy ring compared to "normal" noncyclic ethers (R–O–R′). Unlike typical ethers, which do not react with bases, ammonia, or amines, epoxy resins can react with some aliphatic amines at room temperature. This increased reactivity is attributed to the strain in the epoxy ring. The reactions most crucial for the formation and curing of epoxy resins involve either electrophilic attack on oxygen or nucleophilic attack on one of the ring carbons. In the case of unsymmetrically substituted epoxies, which are common in most epoxy resins 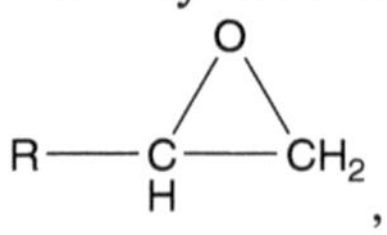, the ring opening is influenced by several factors: (i) the nature of the reagent or catalyst, which can be either electrophilic or nucleophilic; (ii) the nature of the substituent and the steric hindrance at the two carbon atoms; and (iii) the specific reagent used, which can lead to two possible outcomes in terms of which carbon undergoes ring opening.

1.1.2 The Synthesis of Epoxy Resins

Ethylene oxide can be manufactured by the direct oxidation of ethylene. However, these methods are less effective for higher olefins. The two primary methods for producing epoxy resins are (i) dehydrohalogenation of halohydrins and (ii) epoxidation of alkenes using peracids or their esters [17]. The formation of an epoxy group begins with the synthesis of a chlorohydrin. In the production of epichlorohydrin, propylene is first chlorinated. Next, allyl chloride undergoes reaction with hypochlorous acid to form dichlorohydrin. The dichlorohydrin is then dehydrochlorinated with a stoichiometric quantity of NaOH to produce epichlorohydrin.

Industrially, epichlorohydrin is typically reacted with sodium hydroxide to produce glycerol or is stripped with steam and purified via distillation. Another method involves the oxidation of propane to produce acrolein. Acrolein is then chlorinated to form 2,3-dichloropropionaldehyde, which is subsequently reduced to yield glycerol β,γ-dichlorohydrin.

Epichlorohydrin can also be synthesized from allyl chloride via epoxidation using a peracid. Epichlorohydrin is used to produce a wide range of epoxy resins, as the epoxy group readily reacts with hydroxyl-containing substances in the presence of an alkali catalyst, MOH. This reaction leads to the formation of a new epoxy ring through dehydrochlorination.

1.1.3 Properties of Epoxy Resins

For uncured resins, infrared (IR) spectroscopy is a simple and effective identification method. While there are compilations of spectra available, it is still useful to have a set of reference spectra recorded under consistent conditions. Spot tests are quick and convenient but should be verified for accuracy. For resin–hardener mixtures and compounded resins, a combination of IR and nuclear magnetic resonance (NMR) spectroscopy, along with chromatography, may be required for more comprehensive analysis [18, 19].

1.1.3.1 Epoxy Content

This is the analysis which is most commonly applied to uncured epoxy resins. This method, which was previously noted, is fairly intuitive for counting epoxy groups per unit mass. Published methods primarily focus on the completeness and quantitativeness of the reaction, but obtaining exact numbers can be challenging. Precision is crucial, and a keen attention to the required procedures and conditions is necessary for accurate results.

1.1.3.2 Determination of Hydroxyl Groups

When using an epoxy resin for which n is greater than or equal to 1, a secondary alcohol group will typically be present, unless the resin has reacted with something like BPA, which would lead to the formation of a branched molecule. Therefore, the resin must contain a secondary hydroxyl group, which needs to be accurately determined. Hydroxyl groups are also reactive toward curing agents, and knowing their quantity is essential for calculating the resin–hardener stoichiometry. These hydroxyl groups also participate in the reaction with the epoxy groups during curing, especially in reactions with amine hardeners.

Other types of hydroxyl groups may also be present, such as α-glycols formed when epoxy groups undergo hydrolysis, or phenolic hydroxyl groups resulting from unreacted phenol used in resin synthesis – particularly when large amounts of BPA are involved. One method to quantify hydroxyl groups is the use of lithium aluminum hydride, which selectively reacts with active hydrogen. This can be measured volumetrically or via gas–liquid chromatography. However, this method is difficult to use quantitatively and requires correction for the presence of water or other hydrogen-containing compounds. A more practical method is acetylation, where both hydroxy and epoxy groups react with acetic anhydride.

1.1.3.3 Chlorine Content

Pure DGEBA (diglycidyl ether of bisphenol A) does not contain chlorine, but commercial resins can have up to about 1% chlorine. Self-extinguishing resins can contain as much as 30% chlorine, serving as organic binders in specific applications. Additionally, bromated and fluorinated resins are also available. However, resins with high chloride content can have reduced stability, particularly when cured with amine hardeners. Flame retardancy is often a specialized feature, but resin manufacturers can develop formulations that meet the specific requirements of various applications.

For BPA-based epoxy resins, the presence of chlorine-like other polar impurities negatively affects their electrical properties, as well as their color and reactivity. Active chlorine can interfere with weaker base catalysts, such as tert-amine, by blocking the reaction sites. Therefore, the amount of active chlorine is critical. Furthermore, since the chlorine is organically bound, the resin may exhibit lower functionality, resulting in a less cross-linked network. The total chlorine content can be measured using the classical Parr bomb method. Organic chloride in resins exists in two forms: (i) hydrolyzable chlorine and (ii) inactive hydrolyzable chlorine (also known as saponifiable chlorine). The term “inactive hydrolyzable chlorine” refers to chlorine that is partially dehydrochlorinated. In contrast, inactive chloride is formed as a result of reactions between epichlorohydrin and secondary alcohol groups, or it may arise as an unintended side product during the addition of phenolic hydroxyl groups.

1.1.3.4 Molecular Structure

Both IR and NMR spectroscopy can be utilized to identify the type of epoxy resin and determine the structure of its repeat unit. These techniques are also useful for detecting hardeners and other additives in mixed epoxy resin–hardeners. The theory and practical applications of IR and NMR spectroscopy are well-documented in many texts, with Rabek offering a thorough discussion of their use in polymer chemistry.

1.1.3.5 Physical Properties

DGEBA, in its pure form, tends to crystallize; however, when mixed with higher molecular weight compounds, it loses its tendency to crystallize. In general, most epoxy resins are amorphous and do not crystallize. Upon cooling, they form a glassy, amorphous solid with a glass transition temperature (T_g), which has been measured using broad-line NMR spectroscopy. The most common method for characterizing resin softening is by measuring its softening point. Since resins soften gradually with temperature and do not exhibit a sharp melting point, the softening temperature is not precisely defined and can vary depending on the testing method. Ball-and-ring method: In this method, a resin sample is placed in a ring with a steel ball positioned on top. The temperature is gradually increased, and the softening point is reached when the ball sinks 2.5 cm (1 in.) under its own weight due to gravity. Durran's Mercury method: A fixed mass of resin is heated in a standard test tube, then cooled and hardened. Clean mercury is poured on top, and the temperature is increased. The softening point is recorded when the first drop of mercury falls through the resin to the bottom of the tube.

Good control over epoxy resins is essential, which requires comprehensive knowledge of their properties. Schoff provides a detailed discussion of various rheological measures. High-viscosity liquid resins can be challenging to mix with hardeners, making it difficult to achieve uniform molecular blending. Some hardeners may begin reacting at room temperature, causing the resin to thicken prematurely before it is fully mixed. This can lead to micro-gelation, where parts of the epoxy–hardener mixture behave differently than intended.

Even though hardeners may require high temperatures, preheating the resin is often necessary to reduce its viscosity and facilitate proper molecular-level blending. Inadequate mixing between the hardener and the epoxy prepolymer can lead to morphological irregularities in the cured epoxy. This is particularly important for applications such as casting, nondiluted adhesives, and lamination resins, where maintaining a low viscosity is crucial during fabrication. If viscosity increases before all the entrapped air has escaped, large air bubbles can remain in the final product.

Epoxy resin viscosity is influenced by factors such as molecular weight, molecular weight distribution, molecular structure, and temperature. For "compounded" resins, which are liquid at room temperature, viscosities can be measured directly without difficulty. However, for solid resins, a 40 wt% solution in diethylene glycol monobutyl ether is typically prepared, and its viscosity is measured using various methods. The most common methods include a bubble viscometer, such as the Gardner–Holdt tube, a capillary, or an instrument with a spindle. When measuring viscosity, it is essential to control both the solution concentration and the temperature.

For liquid resins, the viscosity may also be measured using a Hoeppler viscometer. This technique records the descent time of a ball in a resin-filled tube set at a specific tilt. The ball size, tube diameter, and tilt angle must meet the standardized specifications. Even with these measures in place, the viscosity results obtained using the Hoeppler viscometer may differ by around 15% when compared to those obtained through other methods like capillary flow or rotational viscosity tests.

The molecular structure of epoxy resin significantly influences its viscosity. Aromatic bisphenol-based epoxy resins, for example, have higher viscosities than cyclic aliphatics, which, in turn, have higher viscosities than linear types, as mentioned earlier [20, 21]. Other properties that can be determined for uncured resins include density and refractive index, the latter being particularly important as it depends

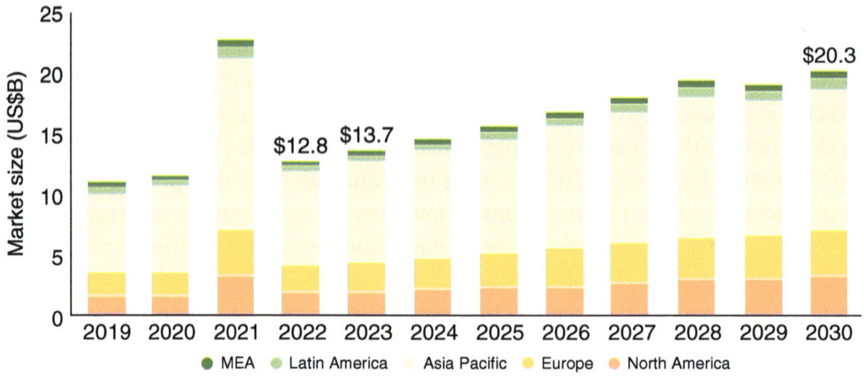

Figure 1.1 Epoxy resin market from 2019 to 2030.

critically on the resin's molecular structure and purity. Color is typically evaluated against established color standards. Additionally, the vapor pressure, especially for resins, is an important property, as is the flash point, which is crucial for safety considerations.

1.1.3.6 Epoxy Resin Market

The global epoxy resin market size was estimated at US$11.25 billion in 2023 and is anticipated to reach US$16.87 billion by 2030, growing at a compound annual growth rate (CAGR) of 6.3% from 2024 to 2030. The following are the leading companies in the epoxy resin market. These companies collectively hold the largest market share and dictate industry trends: 3M, Aditya Birla Management Corporation Pvt. Ltd., Atul Ltd., BASF SE, Solvay, Huntsman International LLC, KUKDO CHEMICAL CO., LTD., Olin Corporation, Sika AG, NAN YA PLASTICS CORPORATION, Jiangsu Sanmu Group Co., Ltd., Jubail Chemical Industries LLC, China Petrochemical & Chemical Corporation (SINOPEC), Hexion, Kolon Industries, Inc., Techstorm, and NAGASE & CO., LTD (Figure 1.1).

1.2 Chemistry, Synthesis, Manufacture, and Characterization of Hyperbranched Epoxy Resins

Hyperbranched polymers (HBPs) are the class of highly branched macromolecules. Compared with common linear polymers and other types of branched polymers, HBPs have the following advantages: nonentanglement or low entanglement, low viscosity, high solubility, compact molecular size, and the presence of numerous functional end groups [22–24]. In recent years, significant progress has been made in the synthesis and characterization of HBPs, leading to the development of many novel structures and functions. Owing to the convenience of nontoxic one-pot reaction and the potential of large-scale production, HBPs have already attracted much attention and have begun to be widely used in biomedical materials, analytical and detection chemistry, polymer coatings and processing aids, optoelectronic materials and devices, and polymer-derived ceramics [25–27].

HBPs are considered promising candidates for future bioimaging probes because of their tree-like architectures and the ease of their synthesis. They can be covalently or noncovalently coupled with, or encapsulated by, predesigned targeting sequences on their outer surface. Unlike other HBPs being less, epoxy-functionalized hyperbranched polymers (EHPs) have been used more widely than other ones as they would provide both reinforcement and toughening to conventional EPs due to their thermosetting property [28–31]. Table 1.1 shows properties of some typical liquid epoxy resins. The viscosities of some commercially available liquid epoxies containing linear chains exceed 2000 mPa·s, such as diglycidyl ether of bisphenol-A (DGEBA), diglycidyl ether of bisphenol-F (DGEBF), and *N,N,N′,N′*-tetraglycidyl-4,4′-diaminodiphenylmethane (TGDDM). In contrast, resins such as hyperbranched poly(trimellitic anhydride-diethylene glycol) ester epoxy resin (HTDE) with a polyester chain, hyperbranched poly (trimellitic anhydride ethylene glycol) epoxy (HTME) with a polyester chain, and hyperbranched epoxy resins with a silicone skeleton (HERSS) with a silicone skeleton chain—representing EHPs with relatively low epoxy equivalent weights—exhibit much lower viscosities than DGEBA or DGEBF. However, Boltorn E, 1, which is hyperbranched, has high viscosity. The main reason for the high viscosity of Boltorn

Table 1.1 The history of epoxy resin.

Years	Processes
1934	Schlack of I.G. Farbenindustrie AG in Germany filed a patent application for the preparation of reaction products of amines with epoxies, including one epoxy based on bisphenol A and epichlorohydrin.
1936	Pierre Castan of DeTrey Freres Co. produced a low-melting epoxy resin from bisphenol A and epichlorohydrin that gave a thermoset composition with phthalic anhydride.
1946	Ciba AG of Basel, Switzerland, first reported epoxy adhesive, and samples of casting resin were offered to the electrical industry.
1950s	The peracetic acid epoxidation of olefins was developed by Union Carbide in the United States and by Ciba AG in Europe for cycloaliphatic structures.
1960s	A number of multifunctional epoxy resins were developed for higher temperature applications. Ciba Products Co. manufactured and marketed o-cresol epoxy novolac resins, which had been developed by Koppers Co. Ciba Products marketed cycloaliphatic epoxy resins in 1963 and licensed several multifunctional resins from Union Carbide in 1965.
1970s	Epoxy resins became known for excellent chemical resistance properties, which led to the development and commercialization of epoxy vinyl ester resins.
1980s	Dow Chemical introduced a number of new, high-performance products such as hydrocarbon epoxy novolacs based on dicyclopentadiene.
1990s	Dow Chemical introduced a new epoxy-based thermoplastic resin, BLOX*, for gas barrier, adhesives, and coatings applications.

Table 1.2 Properties of liquid epoxy resins.

Type	Epoxy equivalent weight (g/eq)	Viscosity (25 °C, mPa·s)	Molecular weight
DGEBA (E44)	217–238	91,000	420–475
DGEBA (E54)	180–190	8000–11,000	360–380
DGEBA (E51)	184–194	12,000–15,000	370–390
DGEBF	160–180	2000–5000	330–360
TGDDM	110–120	3000–6000	440–480
EHPs (Boltorn E1)	875	Solid	10,500
EHPs (HTDE)	313–526	350–700	1700–3900
EHPs (HTME)	350–467	1000–1250	1300–2000
EHPs (HERSS)	417–588	104–697	1800–6600

E, 1 is its high epoxy equivalent weight of about 875 g/eq and the hydrogen bond interactions of a large number of hydroxyl groups [32–35] (Table 1.2).

Another major shortcoming of epoxy resins (EPs) is their low toughness and inherent brittleness, which limits their applications in low-shrinkage and high-toughness materials [36, 37]. Therefore, toughening modification of EPs is essential in practical applications. Compared with many other modifiers such as rubbers, epoxidized oleic esters, thermoplastic polyetherimide (PEI), core–shell particles, and graphene oxide, functional HBPs are considered the most effective additives for improving the performance of EPs [38–40]. Hydroxyl-functionalized HBPs (Boltorn H30, H50) and epoxide-functionalized HBPs (Boltorn E1, Boltorn E2), hydroxyl/epoxy-functionalized hyperbranched block copolycarbonate and polyethers with core/shell structure, and amine ended ones can increase the fracture toughness of DGEBA at least tens of times by only adding 5–20 wt% HBPs. However, their T_g and flexural strength and tensile strength are usually reduced to some degrees because of the low reactivity and low cross-linking density [41–43]. EHPs, carrying numerous reactive epoxy groups, can cross-link with hardeners and thus improve the tensile and flexural strength of DGEBA. For this reason, extensive research efforts have been devoted to synthesizing EHPs and modifying EPs.

There are then quite common types of strategy regarding the spacers used in the synthesis of EHPs, depending on the different mechanisms employed for the precursor or the epoxy monomer. One approach involves grafting epoxy groups onto precursors through etherification, esterification, hydrosilylation, or thiol-ene reactions (including Michael addition and coupling) [44, 45]. Another method is that the epoxy monomers are polymerized by proton/group transfer polymerization [46]. The third method involves the oxidation of double bonds of precursors.

This book wants to summarize the current situation concerning EHPs. This book will describe how EHPs were prepared, how EHPs are used, and what are the advantages and disadvantages. Finally, it will provide perspectives on the use of EHPs for reinforcing and toughening epoxy resins.

Acknowledgments

The authors gratefully acknowledge financial support from the National Natural Science Foundation of China (U23A20691), Natural Science Foundation of Hubei Province (2024AFB800), and Innovation Group of National Ethnic Affairs Commission of China (MZR20006).

References

1 Jin, F.L., Li, X., and Park, S.J. (2015). Synthesis and application of epoxy resins: a review. *J. Ind. Eng. Chem.* 29: 1–11.

2 Rakotomalala, M., Wagner, S., and Döring, M. (2010). Recent developments in halogen free flame retardants for epoxy resins for electrical and electronic applications. *Materials* 3: 4300–4327.

3 Liu, S., Chevali, V.S., Xu, Z.G. et al. (2018). A review of extending performance of epoxy resins using carbon nanomaterials. *Compos. Part B Eng.* 136: 197–214.

4 Chruściel, J.J. and Leśniak, E. (2015). Modification of epoxy resins with functional silanes, polysiloxanes, silsesquioxanes, silica and silicates. *Prog. Polym. Sci.* 41: 67–121.

5 Mohan, P. (2013). A critical review: the modification, properties, and applications of epoxy resins. *Polym. Plast. Technol. Eng.* 52: 107–125.

6 Saba, N., Jawaid, M., Alothman, O.Y. et al. (2016). Recent advances in epoxy resin, natural fiber-reinforced epoxy composites and their applications. *J. Reinf. Plast. Compos.* 35: 447–470.

7 Baroncini, E.A., Kumar Yadav, S., Palmese, G.R., and Stanzione, J.F. III (2016). Recent advances in bio-based epoxy resins and bio-based epoxy curing agents. *J. Appl. Polym. Sci.* 133: 44103.

8 Paluvai, N.R., Mohanty, S., and Nayak, S.K. (2014). Synthesis and modifications of epoxy resins and their composites: a review. *Polym. Plast. Technol. Eng.* 53: 1723–1758.

9 Kumar, S., Krishnan, S., Samal, S.K. et al. (2018). Toughening of petroleum based (DGEBA) epoxy resins with various renewable resources based flexible chains for high performance applications: a review. *Ind. Eng. Chem. Res.* 57: 2711–2726.

10 Chen, J., Chu, N., Zhao, M. et al. (2020). Synthesis and application of thermal latent initiators of epoxy resins: a review. *J. Appl. Polym. Sci.* 137: 49592.

11 Peerzada, M., Abbasi, S., Lau, K.T., and Hameed, N. (2020). Additive manufacturing of epoxy resins: materials, methods, and latest trends. *Ind. Eng. Chem. Res.* 59: 6375–6390.

12 Müller-Pabel, M., Rodríguez Agudo, J.A., and Gude, M. (2022). Measuring and understanding cure-dependent viscoelastic properties of epoxy resin: a review. *Polym. Test.* 114: 107701.

13 Rafique, I., Kausar, A., Anwar, Z., and Muhammad, B. (2016). Exploration of epoxy resins, hardening systems, and epoxy/carbon nanotube composite designed for high performance materials: a review. *Polym. Plast. Technol. Eng.* 55: 312–333.

14 Mi, X.Q., Liang, N., Xu, H.F. et al. (2022). Toughness and its mechanisms in epoxy resins. *Prog. Mater. Sci.* 130: 100977.

15 Ou, B.L., Wang, Y.W., and Lu, Y. (2021). A review on fundamentals and strategy of epoxy-resin-based anticorrosive coating materials. *Polym. Plast. Technol. Mater.* 60: 601–625.

16 Gonçalves, F.A.M.M., Santos, M., Cernadas, T. et al. (2022). Advances in the development of biobased epoxy resins: insight into more sustainable materials and future applications. *Int. Mater. Rev.* 67: 119–149.

17 Jia, P.Y., Ma, Y.F., Kong, Q.Y. et al. (2020). Progress in development of epoxy resin systems based on biomass resources. *Green Mater.* 8: 6–23.

18 Capricho, J.C., Fox, B., and Hameed, N. (2020). Multifunctionality in epoxy resins. *Polym. Rev.* 60: 1–41.

19 Verma, C., Olasunkanmi, L.O., Akpan, E.D. et al. (2020). Epoxy resins as anticorrosive polymeric materials: a review. *React. Funct. Polym.* 156: 104741.

20 Kumar, S., Krishnan, S., Mohanty, S., and Nayak, S.K. (2018). Synthesis and characterization of petroleum and biobased epoxy resins: a review. *Polym. Int.* 67: 815–839.

21 Kumar, S., Samal, S.K., Mohanty, S., and Nayak, S.K. (2018). Recent development of biobased epoxy resins: a review. *Polym. Plast. Technol. Eng.* 57: 133–155.

22 Tang, Y., Zhang, X.Y., Liu, X.Y. et al. (2024). Facile synthesis of hyperbranched epoxy resin for combined flame retarding and toughening modification of epoxy thermosets. *Chem. Eng. J.* 482: 148992.

23 Zhang, J.H., Jiang, C., Deng, G.Y. et al. (2024). Closed-loop recycling of tough epoxy supramolecular thermosets constructed with hyperbranched topological structure. *Nat. Commun.* 15: 4869.

24 Fan, H.Y., Sun, T., Zhang, S.J., and Wu, Z.J. (2025). High strength and toughness epoxy nanocomposites reinforced with epoxy-terminated hyperbranched polymer. *J. Appl. Polym. Sci.* 142: e57071.

25 Liu, H.N., Wang, C.J., Liu, S.Y. et al. (2025). Enhancing interfacial properties of epoxy coatings via hyperbranched modification of SiC fillers: experimental and simulation insights. *Chem. Eng. J.* 511: 161841.

26 Zhang, Y.B., Yan, H.X., Yu, R.Z. et al. (2024). Hyperbranched dynamic crosslinking networks enable degradable, reconfigurable, and multifunctional epoxy vitrimer. *Adv. Sci.* 11: 2306350.

27 Qiao, H.W., Chen, M.F., Chen, B. et al. (2023). Understanding the curing kinetics of boron-based hyperbranched polysiloxane reinforced and toughened epoxy resin by rheology. *Chem. Eng. J.* 467: 143542.

28 Zhang, Y.B., Yuan, J.S., Hu, J.Z. et al. (2024). Toward understanding the cross-linking from molecular chains to aggregates by engineering terminals of supramolecular hyperbranched polysiloxane. *Aggregate* 5: e404.

29 Miao, C.F., Zhou, X., Huang, X.Y. et al. (2023). Effectively synthesized functional Si-doped carbon dots with the applications in tyrosinase detection and lysosomal imaging. *Anal. Chim. Acta* 1279: 341789.

30 Wang, R.Z., Cao, L., Wang, W. et al. (2024). Construction of smart coatings containing core-shell nanofibers with self-healing and active corrosion protection. *ACS Appl. Mater. Interfaces* 16: 42748–42761.

31 Jiang, Y.Q., Yuan, X.Y., Wang, P.R. et al. (2025). Dihydroxy functionalized binuclear carbonate based poly(ionic liquid)s for highly efficient conversion of low-concentration CO_2 into cyclic carbonate. *J. Environ. Chem. Eng.* 13: 116247.

32 Xu, W.M., Zhou, X.Z., Fang, Y.L. et al. (2024). Water-assisted reprocessing and shape programming of epoxy vitrimer. *Adv. Funct. Mater.* 34: 2403681.

33 Wang, C., Huo, S.Q., Ye, G.F. et al. (2024). Strong self-healing close-loop recyclable vitrimers via complementary dynamic covalent/non-covalent bonding. *Chem. Eng. J.* 500: 157418.

34 Zeng, X.Q., Wang, L.G., Cao, Y. et al. (2024). Solvent-free and recyclable polyether/polyphenol supramolecular adhesives with robust, water- and low-temperature-resistant properties. *ACS Appl. Polym. Mater.* 6: 1336–1347.

35 Cai, L., Li, Y., Zhang, X. et al. (2025). A mussel protein-inspired biomimetic soy protein adhesive with strong moisturizing properties. *Chem. Eng. J.* 515: 163875.

36 Wang, M.L., Liu, Z.Y., Wang, T.C. et al. (2025). The UV aging resistance of strong and ductile epoxy resin achieved by UV-induced self-compensated crosslinking points. *Chem. Eng. J.* 510: 161661.

37 Shen, Q.B., Yu, B.B., Chen, D. et al. (2025). Novel furfuryl-based phosphaphenanthrene containing N/P flame retardant: towards epoxy resin with excellent flame retardancy, enhanced toughness and low dielectric constant. *Chem. Eng. J.* 519: 165034.

38 Wang, C.J., Xin, M.H., Gao, M. et al. (2025). A rigid and flexible combination 3D network with hyperbranched materials as bridge and the enhancement of mass transfer resistance in coatings. *Compos. Part B Eng.* 304: 112688.

39 Zhou, Y.X., Huang, Q.B., Wang, L., and Wang, X.H. (2024). Enhancing cross-linking network for superior wet strength of paper by sustainable hyperbranched polyimines. *ACS Appl. Mater. Interfaces* 16: 15383–15393.

40 Son, W., Saha, S., Tran, D.T. et al. (2024). Rational design of semi-interpenetrating network based on hyperbranched polyglycerol grafted MXene/polyurethane-epoxy for compressed hydrogen storage. *Carbon* 228: 119427.

41 Shao, J.W., Li, X.Y., Liu, T. et al. (2024). Multiple function hyperbranched polysiloxane nanoclusters for controlling a cross-linking structure to convert soy meal into a strong, tough, and multifunctional adhesive. *ACS Sustain. Chem. Eng.* 12: 3966–3976.

42 You, T.L., Guo, G.X., and Li, W.B. (2024). Waterborne liquid-like coatings with high transparency, superior scratch resistance, and antismudge properties. *ACS Appl. Mater. Interfaces* 16: 43907–43917.
43 Tang, Y., Zhang, X.Y., Ge, A.X. et al. (2024). Novel hyperbranched phosphaphenanthrene-terminaled silicon-nitrogen cocuring agent towards high-performance epoxy thermosets. *Prog. Org. Coat.* 197: 108828.
44 Cai, J.J., Zhang, X.R., Wang, Z.M. et al. (2024). A bio-based antibacterial epoxy resin coating with outstanding flame retardant and mechanical properties. *Prog. Org. Coat.* 190: 108369.
45 Khan, A. (2025). Thiol-epoxy 'click' reaction in polymer synthesis. *Prog. Polym. Sci.* 169: 102022.
46 Lv, X., Zhang, H.P., Zhang, H. et al. (2025). Exploring the properties and performance of Lewis base organocatalysts in epoxy/polyester hybrid low-cure powder coatings. *Prog. Org. Coat.* 200: 109053.

2

Properties and Characterization Methods of Hyperbranched Epoxy Resins

Kai Cheng and Xian Wang

South-Central Minzu University, Hubei R&D Center of Hyperbranched Polymers Synthesis and Applications, Key Laboratory of Catalysis and Energy Materials Chemistry of Ministry of Education & Hubei Key Laboratory of Catalysis and Materials Science, 182 Minzu Ave. Hongshan District, Wuhan, Hubei Province 430074, P.R. China

2.1 Introduction

Dendrimers are uniformly branched macromolecules with a globular-shaped structure, which show a degree of branching (DB) equal to 1. Dendrimers have tunable functional end groups, which render them with excellent physical and chemical properties. Dendrimers have a well-defined structure with uniform molar mass. However, the synthesis processes for dendrimers are laborious, the purifications after each step are complex, scaling-up is difficult, and yields are very low. Hyperbranched polymers (HBPs) are highly branched macromolecules with three-dimensional ellipsoid-like structures. HBPs emerge as the fourth major class of new polymers after linear polymers, branched polymers, and cross-linked polymers. HBPs are usually prepared in a one-pot synthesis. The synthesis process of HBPs is relatively facile, and the scaling-up is relatively easy. However, the synthesis methods of HBPs limit control over molar mass and branching accuracy, which leads to heterogeneous polymers with a wide molar mass distribution and branching dispersion. As the in-between analogs of linear polymers and dendrimers, HBPs show great superiority in properties and performance. Unlike dendrimers, HBPs have a DB less than 1 and show potential for branching in each repeating unit. The chemical composition, topological structure, and functional group of polymers play an important role in determining the material properties. HBPs possess an ellipsoid-like three-dimensional structure, a random branching structure with $DB < 1.0$ (usually, 0.4–0.6), a wide distribution of molar mass (usually, polydispersity index [PDI] > 1.0), and large amounts of functional groups located at both linear and terminal units. Compared with dendrimers, the facile synthesis process and low cost of HBPs make them more widely used in various applications such as processing additives, catalysts, encapsulation, drug delivery, and composites. Both dendrimers and HBPs show excellent physical and chemical

Table 2.1 Comparison of HBPs with dendrimers and linear polymers.

Polymer	Linear	HBP	Dendrimer
Synthesis	One-step, easy	One-step, easy	Multi-step, hard
Topology	1D, linear	3D, ellipsoid	3D, globular
Scaling-up	Easy	Easy	Difficult
DB	0	0.4–0.6	1
Polydispersity index	>1.1	>1.1	1.0
Viscosity	High	Low	Low
Solubility	Low	High	High
Entanglement	Strong	Weak	Weak
Location of functional group	Two ends	Linear and terminal units	Terminal units

properties when compared with linear polymers, such as better solubility, low molecular entanglement, lower melt and solution viscosity, large amount of terminal functional groups, self-assembly behaviors, and usually high mechanical strength. The differences between HBPs, dendrimers, and linear polymers are listed in Table 2.1.

Many HBPs have been synthesized such as polyesters, polycarbonates, polyether, polyurea, polyurethanes, and polyglycerols. HBPs are also used as novel tougheners for epoxy resins due to their branched ellipsoid topology, low viscosity, and large amount of terminal functional groups. The unique topological structure endows HBPs with large free volume, which improves the toughness of epoxy thermosets. Compared with other HBPs, epoxy-ended hyperbranched polymers (EHPs) with good thermosetting properties can simultaneously strengthen and toughen epoxy resins, which have been extensively used in adhesives, coatings, painting, and electrical insulating materials. EHPs are a unique class of highly branched, globular macromolecules that have gained significant attention in various fields due to their distinct properties and potential applications. EHPs possess a highly branched tree-like ellipsoid architecture with epoxy resin-ended group.

Compared with the linear polymers of similar molecular weight (MW), the topological structure endows epoxy resin with lower solution viscosity, lower molecular entanglement, and better solubility. The viscosity of EHPs first decreases and then increases with an increase in the MW, and the lowest viscosity appears at the MW of approximately 2000–5000 g/mol (Figure 2.1), which shows a unique rheology property. The EHPs often exhibit good solubility in a range of organic solvents due to their highly branched structure and the nature of the epoxy end groups. The presence of numerous terminal epoxy groups in EHPs also provides a high density of reactive sites for further functionalization or cross-linking reactions. EHPs also show high thermal stability, making them suitable for high-temperature applications. The thermal stability of EHPs can be further tuned by adjusting the chemical composition and specific functional group. EHPs also exhibit enhanced mechanical properties such as high mechanical strength, enhanced

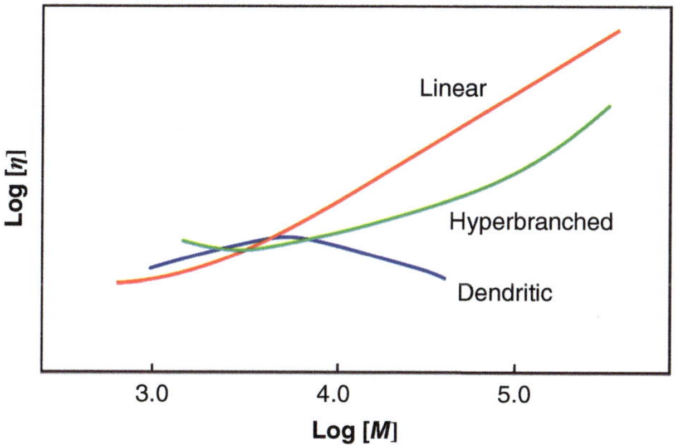

Figure 2.1 The difference of intrinsic viscosity versus polymers with various structures.

impact resistance, flexibility, and elongation. These mechanical properties can be fine-tuned through the control of the DB, cross-linking density, and MW to meet specific application requirements.

The epoxy end groups in EHPs can undergo curing reactions with a variety of hardeners such as amines and anhydrides, leading to the formation of highly cross-linked networks. This cross-linking capability is particularly valuable in developing thermosetting materials with tailored mechanical and thermal properties. The large free volume of EHPs endows it with the property to form nanocomposite. The EHPs can be effectively used as compatibilizers or interfacial agents in nanocomposites, which enhances the dispersion and interaction between nanofillers and the EHPs matrix. Some EHPs can be designed to be biodegradable depending on the synthetic route and choice of monomers, which offers environmentally friendly alternatives in certain applications.

EHPs are often used as precursors of functional materials and multifunctional additives for toughening the bisphenol A diglycidyl ether (DGEBA). Compared with other additives, EHPs show the best comprehensive performance in modifying DGEBA. The synthesis and properties of EHPs can be finely tuned to meet specific requirements, making them an exciting area of research and development in polymer science.

2.2 Properties and Characterization Methods of Hyperbranched Epoxy Resin

There are enormous synthetic developments in the field of hyperbranched epoxy resin (HER), which contribute to the unique characteristics and material properties of HER. Great efforts have been made to characterize the properties of HBPs to understand the relationship between their structure and properties. There are many excellent reviews that describe the characterization of HBP by different groups [1–5]. These characterization methods of HBPs can also be used to analyze the properties of

HERs. The results of these works show that HERs possess very complex multidimensional structures with broad distribution in molar mass, DB, hydrodynamic radius, and chemical composition. HERs have unique structure and rheological properties that are different from linear polymers, less branched polymers, and dendrimers. In this section, the most important properties of HERs are discussed. An overview of the characterization methods for HER DB, MW and its distribution, topology, and rheological properties is also integrated.

2.2.1 Degree of Branching

DB is one of the most important properties of HERs, as highly branching is an important characteristic of HERs that is different from the linear polymers and cross-linked networks. DB is a widely used structural parameter to quantitatively characterize the branching features of HERs. Understanding and controlling the DB is crucial in HER synthesis and processing to tailor the final material's properties for specific applications. Compared to the perfectly branched dendrimers with DB equals 1, the HERs have additional linear units. DB is defined as the ratio of the molar fraction of branched and terminal units relative to the total number of branching sites. The HERs made from AB_2-type monomers contain dendritic (*D*), linear units (*L*), and terminal units (*T*) as shown in Figure 2.2. DB can be calculated from Eq. (2.1) [6–8]:

$$\mathrm{DB} = \frac{D+T}{D+T+L} \tag{2.1}$$

In the case of high degree of polymerization (DP), the number of *T* is approximately that of *D*. The equation can be simplified as Eq. (2.2), which is more suitable for the calculation of DB as suggested by Yan et al. [7] and Frey and coworkers [8].

$$\mathrm{DB} = \frac{2D}{2D+L} = \frac{1}{1+L/2D} \tag{2.2}$$

However, for the HERs made from A_2B_3-type monomers, the compositions are more complex than those made from AB_2-type monomers as shown in Figure 2.3.

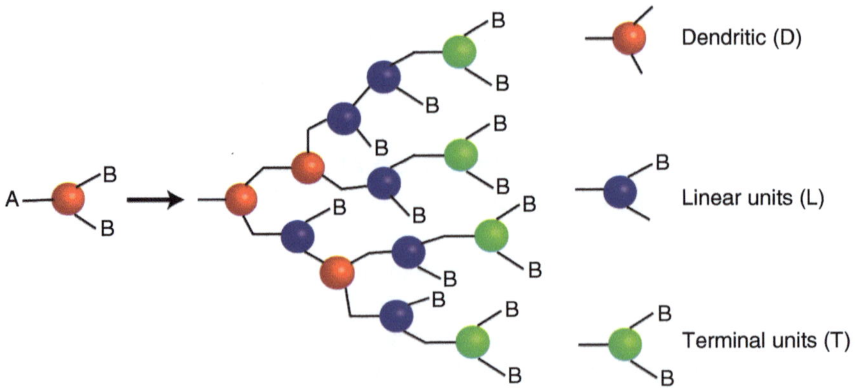

Figure 2.2 Different units in HERs made from AB_2-type monomer.

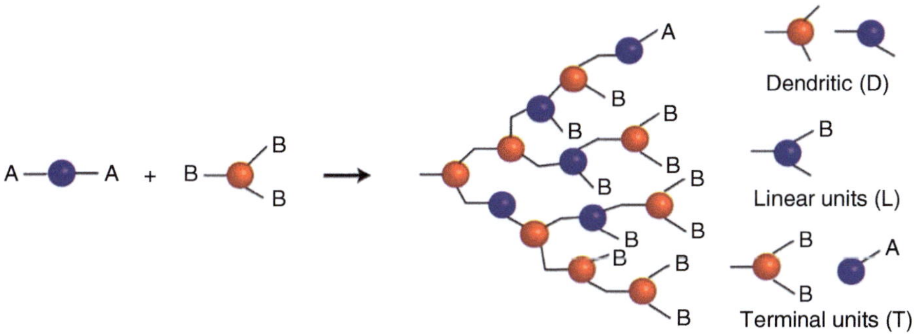

Figure 2.3 Different units in HERs made from A_2B_3-type monomer.

The characterization of DB of these HERs is difficult as more complex units need to be considered.

Both direct and indirect methods have been used to determine the DB of HERs. The direct measurement methods include nuclear magnetic resonance (NMR) and the degradation of HERs, followed by chromatographic analysis. The indirect method includes the viscometry approach.

2.2.1.1 NMR Spectroscopy

NMR refers to the physical process in which the nucleus of an atom with a nonzero spin quantum number (with an odd number of protons or neutrons) undergoes Zeeman splitting under the action of the external static magnetic field, forming different energy levels, absorbing the radio frequency radiation of a specific frequency, and transiting from a low-energy state to a high-energy state. In 1946, this NMR phenomenon was discovered independently by Professor Bloch of Stanford University and Professor Purcell of Harvard University, who shared the Nobel Prize in Physics in 1952. The Zeeman splitting of nucleus is related to the surrounding environment of nucleus, which can be used to provide atomic-level resolution information about the structure of molecules. NMR spectroscopy is a powerful analytical technique used in chemistry and biochemistry to determine the structure, dynamics, and composition of molecules. NMR spectroscopy is also a very important tool for the structural determination of HERs.

NMR is the most powerful technique to determine the DB of HERs that include dendritic (*D*), linear units (*L*), and terminal units (*T*) using NMR spectroscopy. Both 1D and 2D NMR spectroscopy have been employed to measure the DB of HERs. 1D ^{1}H NMR spectroscopy has long been used to differentiate the *D*, *L*, and *T* units to calculate the DB of HBPs because of its high sensitivity. Huber et al. used ^{1}H NMR spectroscopy to differentiate the hydrogen in secondary amide of different structural units (*D*, *L*, *T*) for the DB measurement of hyperbranched poly(etheramide) (Figure 2.4a) [9]. Behera and Ramakrishnan utilized ^{1}H NMR spectroscopy to discriminate between the benzylic protons belonging to different structural units and measure the DB of hyperbranched polyethers (Figure 2.4b) [10]. Shpotya et al. used ^{1}H NMR spectroscopy to quantitatively characterize the methyl group in structural units and study the DB of hyperbranched polyester (Figure 2.4c) [11].

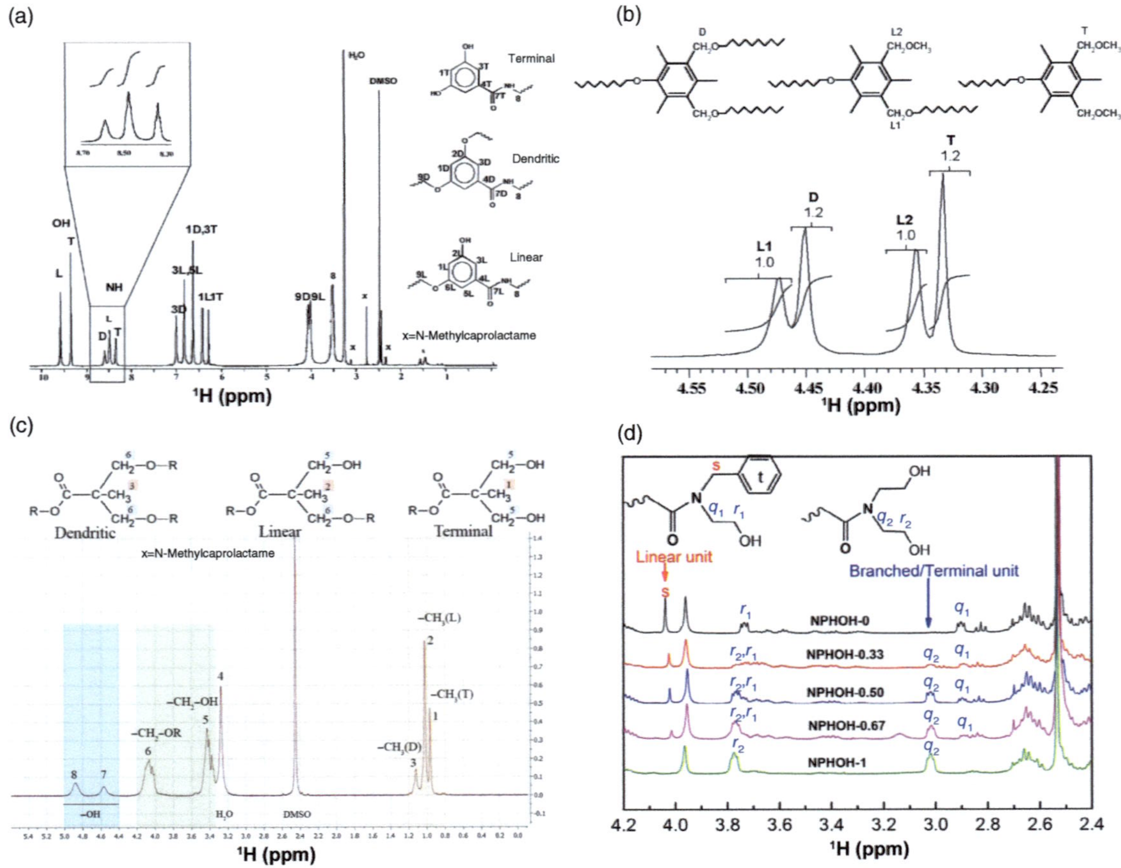

Figure 2.4 ^{1}H NMR spectroscopy for the determination of dendritic, linear units, and terminal units of hyperbranched poly(etheramide). *Source:* [9] / with permission of John Wiley & Sons (a), hyperbranched polyethers (b). *Source:* [10] / with permission of John Wiley & Sons, hyperbranched polyester. *Source:* [11] / with permission of Springer Nature (c), and hydroxyl-ended hyperbranched polymers. *Source:* [12] / with permission of Elsevier (d).

Chen et al. utilized ^{1}H NMR spectroscopy to calculate the DB of hydroxyl-ended HBPs (Figure 2.4d) [12]. However, for complex HERs, the accurate determination of DB using ^{1}H NMR spectroscopy is not enough due to the low resolution of ^{1}H NMR spectroscopy, which may contribute to severe peak overlapping.

However, due to severe peak overlapping in ^{1}H NMR spectroscopy, the resolution of ^{1}H NMR is sometimes too poor to distinguish the different units of HBPs. ^{13}C, ^{15}N, ^{19}F, ^{29}Si, and ^{31}P NMR spectroscopies and 2D ^{13}C—^{1}H Heteronuclear Single Quantum Coherence (HSQC), ^{15}N—^{1}H HSQC NMR spectroscopy can also be used to measure the DB of heteroatom-containing HBPs (Figure 2.5) [13–21]. De and Karak measured the DB of HER using the ^{13}C NMR spectroscopy. The monosubstituted (*T*), disubstituted (*L*) and trisubstituted (*D*) CH_2 carbon of triethanol amine of HERs can be differentiated and quantified by ^{13}C NMR spectroscopy (Figure 2.5a) [13]. Somwangthanaroj and coworkers calculated the DB of hyperbranched BPA/PEG epoxy resins using the ^{13}C NMR spectra, which can be used to discriminate the four units of central carbon atoms of pentaerythritol (Figure 2.5b) [14]. Hawker and Chu used 1D ^{19}F NMR to differentiate the linear and terminal units of the fluoro-terminated hyperbranched poly(ether ketone) and calculate the DB (Figure 2.5c) [15]. Yoon and Son utilized 1D ^{29}Si NMR to measure the DB of hyperbranched poly(carbosilarylenes) (Figure 2.5d) [16]. Holycross and Chai employed ^{1}H—^{13}C Heteronuclear Multiple-Quantum Correlation (HMQC) NMR spectrum to study the DB of hyperbranched poly(ethylene imine) (Figure 2.5e) [17]. Zhu et al. applied ^{1}H—^{13}C HSQC NMR spectrum to measure the DB of complicated hyperbranched poly(sulfone-amine) (Figure 2.5f) [18]. Although the resolutions of ^{13}C, ^{15}N, ^{19}F, ^{29}Si, and ^{31}P NMR spectroscopies and 2D ^{13}C—^{1}H HSQC, ^{15}N—^{1}H HSQC NMR spectroscopy are much higher than the traditional ^{1}H NMR spectrum, quantitative analysis of DB is more difficult due to the relatively low signal-to-noise ratio. Besides, NMR is the most powerful and in situ analysis method for the accurate determination of the DB of HERs.

2.2.1.2 The Degradable Method and Indirect Viscometry Method

For some HBPs, the resolutions of 1D ^{1}H—^{13}C NMR spectra and 2D ^{1}H—^{13}C HSQC spectra are insufficient to discriminate the *D*, *L*, and *T* units of HBPs. The DB can't be determined directly by NMR spectroscopy. Under this condition, the degradable approach could be used to measure the DB of HBPs. The HBPs can be degraded into *L*, *D*, and *T* units, which could be quantitatively analyzed by chromatographic and NMR analysis. Kambouris and Hawker degraded the methyl ether-terminated hyperbranched polyester into *D*, *L*, and *T* subunits and then analyzed the hydrolysis products using high-performance liquid chromatography (HPLC) and capillary gas chromatography to determine the DB of HBPs [22]. Bolton and Wooley determined the DB of hyperbranched polycarbonates using a reductive degradation procedure followed by HPLC and ^{1}H NMR analysis [23]. The degradable method required the HBPs to be completely degradable to facilitate the accurate determination of DB.

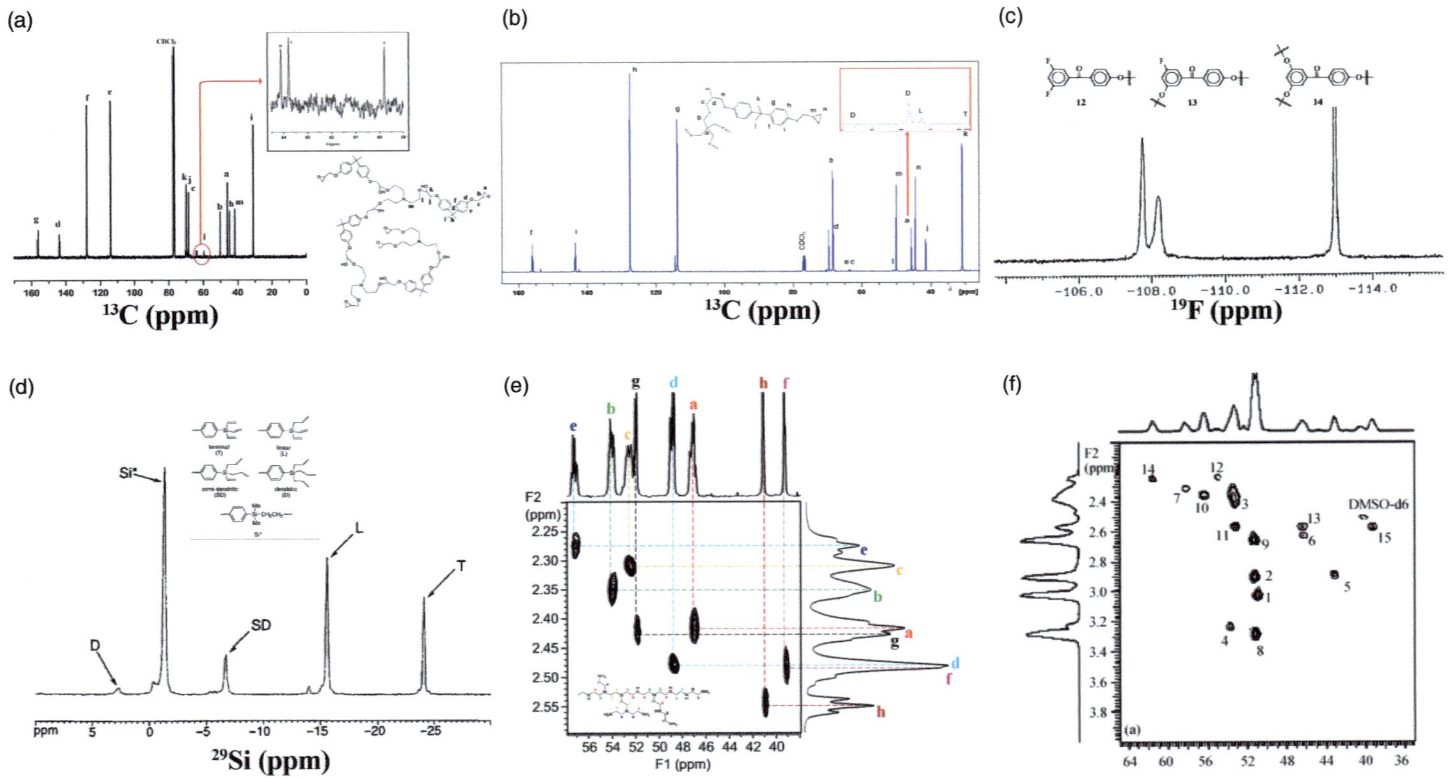

Figure 2.5 NMR spectroscopy for the determination of dendritic, linear units, and terminal units of hyperbranched polymers. (a) ^{13}C NMR spectrum of the hyperbranched epoxy. *Source:* [13] / with permission of The Royal Society of Chemistry, (b) ^{13}C NMR spectrum of HER. *Source:* [14]/MDPI/CC BY 4.0. (c) 1D ^{19}F NMR spectrum of the fluoro-terminated hyperbranched poly(ether ketone). *Source:* [15] / with permission of American Chemical Society, (d) 1D ^{29}Si NMR spectrum of hyperbranched poly(carbosilarylenes). *Source:* [16] / with permission of American Chemical Society, (e) 2D ^{1}H–^{13}C HMQC spectrum of hyperbranched poly(ethylene imine). *Source:* [17] / with permission of American Chemical Society, and (f) 2D ^{1}H–^{13}C HSQC spectrum hyperbranched poly(sulfone-amine). *Source:* [18] / with permission of Springer Nature.

The viscometry approach is a more conventional and less sophisticated method to indirectly determine the extent of branching of HBPs compared with NMR spectroscopy and the degradable method. HBPs exhibit lower viscosity than linear polymers, and the relationship between intrinsic viscosity ($[\eta]$) and molecular weight (M) follows the Mark–Houwink equation:

$$[\eta] = KM^{\alpha} \tag{2.3}$$

where K and α are the Mark–Houwink constants. The α for HBPs with a globular structure is less than 0.5, whereas the α for randomly coiled linear polymers is about 1.0. The lowering of the intrinsic viscosity value for HBPs is caused by less entanglements and more compact structure in HBPs, which indicates a large DB value. The larger value in DB may contribute to a compact structure with less chain entanglements, which led to a lower intrinsic viscosity.

As one of the most important parameters for HBPs, DB has a strong influence on the solution and mechanism properties of HBPs such as free volume, chain entanglement, hydrodynamic radius, mechanical strength, viscosity, glass transition temperature, and degree of crystallization. The accurate determination of DB may help establish the structure–property relationship of HBPs.

2.2.2 Molecular Weight and Molecular Weight Distribution

MW and its distribution are the other two important parameters for HBPs. HBPs show a broad MW distribution than linear polymers. The MW distribution can be described by the PDI, which is used as a measure of the breadth of the MW distribution. PDI is defined as the ratio of the mass-weighted molecular weight (Mw) to the number-weighted molecular weight (Mn). The MW distribution usually becomes broad with an increase in the DP. The increase in DP makes the HBPs become large and confers more functional group at the terminal ends, which then react with monomer to give a wider MW distribution [24]. The PDI can be narrowed mainly by the slow addition of monomers into the system.

Gel permeation chromatography (GPC), light scattering techniques, matrix-assisted laser desorption-ionization time-of-flight mass spectroscopy (MALDI-TOF MS) techniques, and diffusion-ordered spectroscopy (DOSY) have been used to determine the MW and its distribution.

2.2.2.1 GPC and Light Scattering Techniques

MW and PDI of HBPs are generally determined by GPC. GPC, also called size-exclusion chromatography (SEC), is one of the most popular methods for the separation of different kinds of polymers and macromolecules [25–27]. The GPC method is cheap, fast, simple, and reproducible. Online detectors such as light scattering detectors, viscometers, and mass spectrometers have been developed for GPC methods [28–30]. As a chromatography method, the separation in GPC is solely based on the size of molecules that share the partial exclusion from the pores of the stationary phase. Larger molecules are too large to enter some of the pores, pass

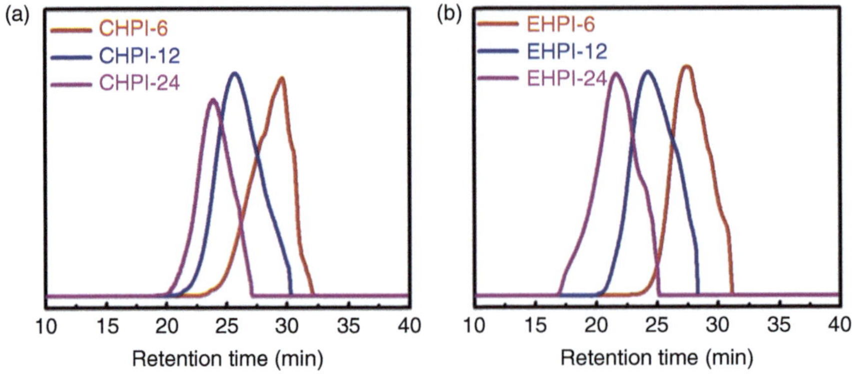

Figure 2.6 GPC traces of HEPs CHPI-*n* and EHPI-*n*. *Source:* [31] / with permission of Elsevier.

through the column faster than smaller molecules, and elute before the solvent. Smaller molecules can enter the pores and elute at the end of the chromatography, which is close to the solvent. GPC with differential refractive index detection and UV detection, followed by calibration with a standard polymer, is the most frequently used method to determine MW and PDI. HBPs with a large PDI usually show a broad peak in the GPC chromatogram. The Mw and Mn of HBPs can be calculated from the width and shape of the GPC peak by a calibration procedure and the use of an absolute detection technique. Yu et al. used GPC to determine the Mw, Mn, and PDI of HEPs (Figure 2.6) [31].

Light scattering techniques can also be used to study the MW of HBPs. Multiangle laser light scattering (MALLS) and low-angle laser light scattering (LALLS) are the two main types of light scattering techniques that give information about the MW of HBPs. MALLS is less sensitive to the impurities of polymers and can give detailed information about the MW and PDI of HBPs simultaneously compared with LALLS. However, due to the aggregates in the sample solution, the MALLS technique alone is not sufficient to measure the MW and PDI of HBPs. The MALLS are used in combination with GPC, termed GPC-DRI-MALLS, to give knowledge about the MW distribution of HBPs.

2.2.2.2 MALDI-TOF MS

The newly developed MALDI-TOF MS techniquehas been used to determine the MW of HBPs and plays an essential role in the analysis of polymers. The 2002 Nobel Prize in Chemistry was awarded to John Fenn and Koichi Tanaka for the contribution to the development of electrospray and MALDI. MALDI-TOF MS is a powerful analytical technique for high-molecular-weight and labile macromolecules such as synthetic polymers. As a soft ionization, MALDI utilizes a laser to strike a large excess of organic matrix that is mixed with a small quantity of analyte to make the analyte molecules enter the gas phase without fragmenting or decomposing them. A major advantage of MALDI is that it predominantly produces singly charged ions, in which case m/z and mass scales are identical, facilitating the analysis of mixtures.

The matrix of MALDI absorbs photon energy from the laser beam, transferring the energy into the excitation energy of the solid system and diluting the analyte molecules to avoid intermolecular interaction and aggregation. The choice of the right matrix is key to the soft ionization of HBPs. For MALDI-TOF MS, α-cyano-4-hydroxycinnamic acid (CHCA), 2,5-dihydroxybenzoic acid (DHB), dithranol (DIT), *trans*-2-[3-(4-tert-butylphenyl)-2-methyl-2-propenylidene] malononitrile) (DCTB), and *trans*-3-indoleacrylic acid (IAA) are the most frequently used matrixes. Each matrix exhibits distinct ionization behavior depending on analyte chemistry. Several strategies exist for the addition of matrix and analyte to the MALDI sample plate, although the dried droplet and sandwich methods are the most popular. Excess matrix can enhance signal, but beyond saturation, it may suppress analyte ionization, and therefore, optimal polymer-to-matrix volume ratios are required.

MALDI-TOF MS is a powerful method for the detection of the MW of monodisperse HBPs. For HBPs with broad MW distributions, the lower MW species are preferably activated and desorbed, and the higher MW species are harder to activate and desorb. The lower MW can be detected, while the higher MW species are not detected due to low intensities in the MS. Therefore, HBPs with large PDI needed to be fractionated by GPC to monodisperse fractions and then measured by MALDI-TOF MS. The combination of SEC and MALDI-TOF MS has been used to measure the MW and PDI of HBPs with a broad MW distribution [32, 33]. Gooden et al. utilized MALDI-TOF MS to investigate the cyclization in HBP synthesis (Figure 2.7a) [34]. Burgath et al. also used MALDI-TOF MS to study the cyclization in the synthesis of hyperbranched aliphatic polyesters (Figure 2.7b) [35]. Jayakannan et al. applied MALDI-TOF MS to characterize the fractionation of hyperbranched polyether (Figure 2.7c) [33]. Chikh L et al. used MALDI-TOF MS to determine the absolute average molar masses and PDI of hyperbranched polyesters (Figure 2.7d) [32].

A newer type of instrument, MALDI-ToF/ToF, is also applied to hyperbranched polyesters, consisting of a linear ToF coupled to a reflectron ToF analyzer. Here, post source decay (PSD) takes place in the linear ToF and fragment mass analysis in the reflectron ToF device [36]. These instruments are equipped with collision cells in the linear ToF region to enhance the fragmentation extent by collision-induced dissociation CID (also known as collisionally activated dissociation, CAD). The extra collision may, however, promote minor, high-energy fragmentation pathways that are structurally diagnostic, yet undetectable without collisional activation [37].

2.2.2.3 Diffusion-ordered Spectroscopy and NMR Spectroscopy

The detailed description of DOSY is discussed in Section 2.2.3.2. Diffusion NMR can be used to measure the self-diffusion coefficient, D_τ, which is caused by the Brownian motion of molecules in liquids. As a well-established NMR method, DOSY can be utilized to measure the diffusion coefficients for individual resonances in NMR spectra to analyze the mixtures of polymers with different MWs. ^{1}H DOSY NMR can be applied to predict the MWs of narrowly dispersed polymers based on the linear correlation between the log D_τ and log M_w and the external calibration curves [38–40]. Compared with other characterization techniques, such as GPC and

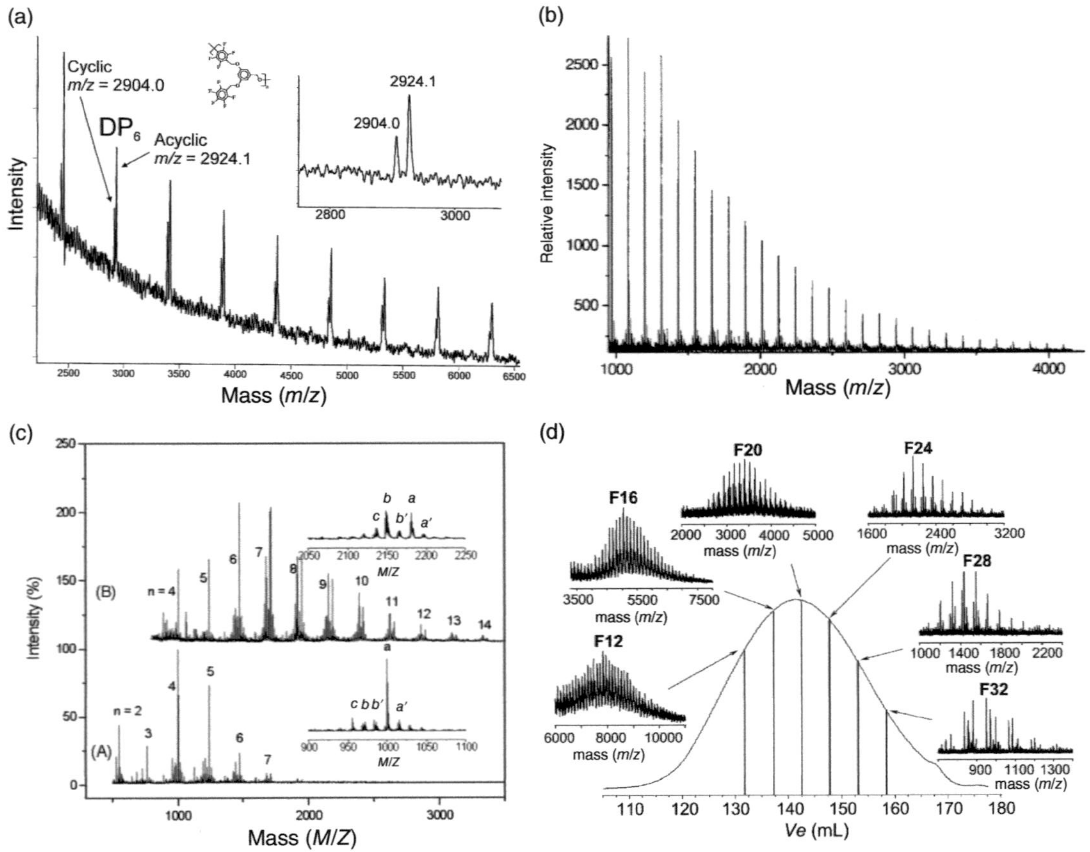

Figure 2.7 MALDI-TOF mass spectroscopy of HBPs [32–35]. (a) Hyperbranched cyclophane polymer. *Source:* [34] / with permission of American Chemical Society, (b) hyperbranched aliphatic polyesters. *Source:* [35] / with permission of John Wiley & Sons, (c) fractions of hyperbranched polyether. *Source:* [33] / with permission of John Wiley & Sons, and (d) fractions of hyperbranched polyester P4-PE. *Source:* [32] / with permission of American Chemical Society.

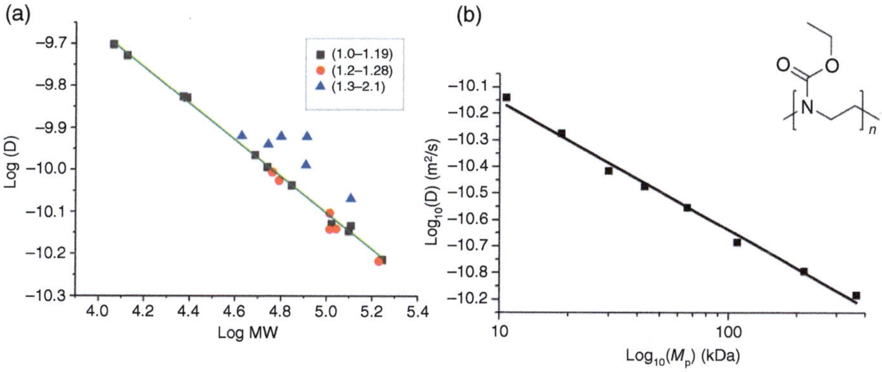

Figure 2.8 Diffusion coefficients for PMMA samples with broad dispersities (a) *Source:* [41] / with permission of American Chemical Society and for poly(2-ethyl-2-oxazoline). (b) *Source:* [42] / with permission of The Royal Society of Chemistry.

MALDI-TOF MS, DOSY is faster and doesn't require the purification of polymer sample. In this way, the linear correlation between log D_{τ} and log M_{w} of a series of standard polymer samples with the same conformations can be determined as the external calibration curves. However, DOSY is not frequently used to study HBPs with large PDI. Ruzicka et al. utilized ^{1}H DOSY NMR to determine the reliability of the calibration curves by increasing the dispersity of polymethyl methacrylate (PMMA) and showed that when the dispersity is greater than 1.3, the calibration curves exhibit significant deviations from that of the samples with lower dispersity (Figure 2.8a) [41]. Monnery et al. applied ^{1}H DOSY NMR to a series of narrow-distribution Polymers and their results showed that the apparent hydrodynamic radii scaled with MW as expected (Figure 2.8b) [42].

2.2.3 Topology

The topology of HBP, including size and shape, can significantly influence the properties of HBPs. Long-chain linear HBPs and hyperbranched–hyperbranched polymers are the two major complex topologies of HBPs. The unique branching topology of HBPs confers better solubility, low molecular entanglement, lower melt and solution viscosity, large amount of terminal functional groups, self-assembly behaviors, and weak mechanical strength. The topological microstructure of HBPs such as size (hydrodynamic radius, mean square radii gyration) and shape factor have significant effects on the mechanical and interfacial performance of HBPs. It's of great importance to build the structure–property relationship of HBPs to improve the mechanical and interfacial performance of HBPs. Dynamic light scattering (DLS), DOSY, and GPC are utilized to study the size and shape of HBPs.

2.2.3.1 Dynamic Light Scattering

DLS, which is also called photon correlation spectroscopy (PCS) or quasi-elastic light scattering (QLS), is a very powerful tool for studying and measuring the diffusion behavior and size distribution of particles (polymers, proteins, DNA, RNA) in solution. In the static light scattering (SLS) experiments, a monochromatic beam of light encounters the solution containing the polymers, and light scatters in all directions as a function of the size and shape of the polymers. For SLS experiments, the intensity of scattered light is analyzed as time-averaged intensity. The average intensity provides valuable information on the MW and radius of gyration of polymers. For DLS experiments, the monochromatic incident light impinges onto a solution with polymer particles in Brownian motion. The intensity fluctuations of scattered light due to the Brownian motion of polymer in solution are analyzed. The diffusion coefficient (D_τ) that correlates with the hydrodynamic size of polymers can be determined. DLS is utilized to measure the Brownian motion of polymer in solution, which relates the motion to the hydrodynamic radius (R_h) and diffusion coefficient (D_τ) of polymer particles. The Brownian motion of polymer particles depends on the temperature, their size, and solvent viscosity.

In the DLS experiment, the intensity correlation function $G_2(\tau)$ that describes the motion of polymer under investigation can be measured and expressed as follows:

$$g_2(\tau) = 1 + \beta e^{-2D_\tau q^2 \tau} \tag{2.4}$$

where β is the coherence factor that depends on detector area, q is the Bragg wave vector that is proportional to solvent refractive index n, τ is the duration of the experiment, and D_τ is the translational diffusion coefficient.

The D_τ is valuable to determine the hydrodynamic parameters. The hydrodynamic radius (R_h) is defined as the radius of a hypothetical sphere that diffuses at the same rate as the particle under investigation and can be determined using the Stokes–Einstein equation:

$$D_\tau = \frac{k_B T}{6\pi\eta R_h} \tag{2.5}$$

For the measurements of D_τ by DLS experiments, the concentration of polymer particles under investigation is of great importance. Lower concentration of particles leads to weak signals, and higher concentration of particles makes the multiple scattering phenomenon more severe. By using DLS, Zhang and coworkers obtained the average particle size and particle spread of hyperbranched epoxy and DGEBA blends (Figure 9a) [43]. They determined the average particle size and particle size distributions of HERs NPEHP-*n* from DLS (Figure 9b) [44]. They also measured the particle size distributions of EFTH-*n* using DLS (Figure 9c) [45] (Figure 2.9).

2.2.3.2 Diffusion-ordered Spectroscopy

DOSY NMR is a powerful tool to measure the diffusion coefficients of molecules that are related to their MWs and hydrodynamic radius. DOSY NMR can also be used to characterize the MWs of polymer using the calibration curve from monodisperse reference standards, which utilizes the linear correlation of log D to log MW [41, 46, 47]. Compared with other characterization techniques, such as

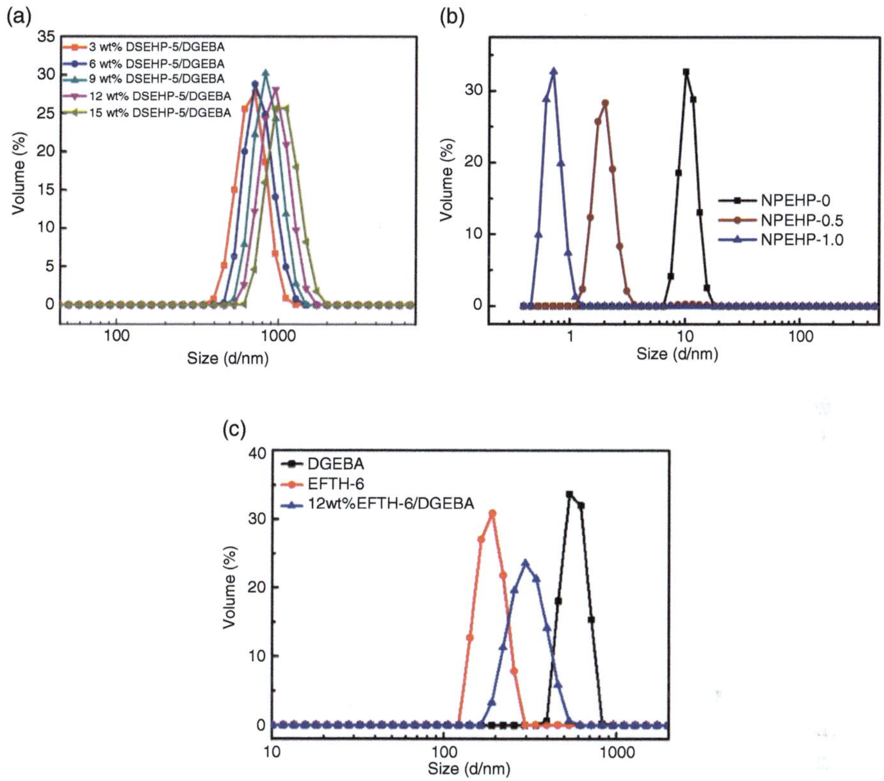

Figure 2.9 Particle size distributions of HERs determined by DLS. (a) DSEHP-5/DGEBA. *Source:* [43] / with permission of Elsevier, (b) NPEHP-*n*. *Source:* [44] / with permission of Elsevier, and (c) EFTH-6/DGEBA. *Source:* Adapted from [45].

GPC and DLS, DOSY is faster and does not require the separation and purification of polymers. GPC and DLS techniques have many inherent drawbacks such as being time-consuming, requiring relatively pure samples, and retaining a 10–20% systematic error even with calibration [25].

The DOSY experiment yields a pseudo-2D spectrum with chemical shifts on the *x*-axis and self-diffusion coefficients on the *y*-axis, which means the diffusion spectrum at each chemical shift can be gained and the diffusion coefficient of the polymer can be measured. DOSY NMR can be used to determine the self-diffusion coefficients of all compounds in the mixture without purification. DOSY NMR experiment utilizes the theme of refocusing magnetization while utilizing gradient pulses to apply a spatially dependent phase to the NMR signal. The basic pulse sequence for measuring the diffusion using the pulsed field gradient (PFG) spin–echo is shown in Figure 2.10.

The first part of the PFG spin–echo sequence labels the nuclear spin with a phase factor, which depends on the position of spin in the NMR tube. Then the diffusion occurs during the delay of Δ. A second gradient afterward refocuses the spatial dependence of the first gradient pulse. As the molecule diffuses away from the

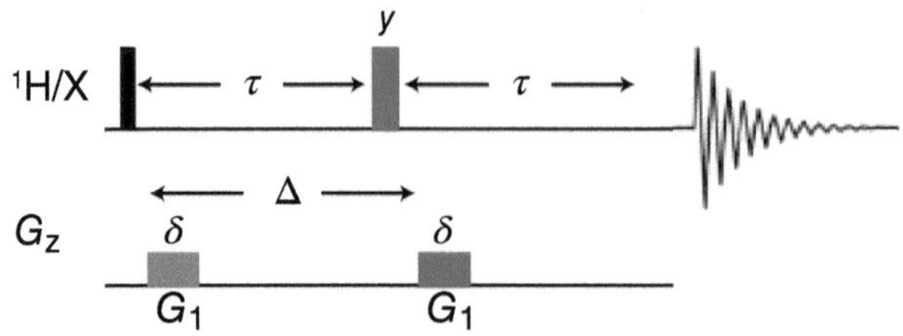

Figure 2.10 The PFG spin–echo sequence for diffusion measurement. The diffusion during the period Δ is characterized by a series of measurements with increasing gradient pulse strengths (G_1) of fixed duration δ.

initial position along the z-axis during the diffusion time Δ, the signal can hardly be completely refocused and will be attenuated. The attenuated amplitudes of signals that refocused at the end of the sequence depend on the translational diffusion rate of the molecules. The larger the diffusion, the lower the NMR signal intensity. The following Stejskal–Tanner equation is used to quantitatively describe the NMR signal intensity of DOSY experiment.

$$I_G = I_0 \exp\left[-(\gamma\delta G)^2 D\left(\Delta - \frac{\delta}{3}\right)\right] \tag{2.6}$$

where γ is the gyromagnetic ratio of the nuclear, δ is the gradient pulse and is also called "little delta," G is the gradient strength, I_0 is the signal intensity at a gradient strength of zero, Δ is the diffusion time and is also called "big delta," and D is the translational self-diffusion coefficient. The diffusion coefficient can be measured by progressively increasing the gradient strength (G) while keeping the diffusion time (Δ) and the length of the gradient pulses (δ) constant and fitting the peak intensity to a series of gradient strengths, as given in Eq. (2.6). The diffusion coefficient can be used to calculate the hydrodynamic radius using Stokes–Einstein equation. However, Eq. (2.6) can only be used to quantify the diffusion coefficient of monodisperse polymer. For polydisperse polymer, Eq. (2.6) should be adjusted. The signal decay in the PSG-NMR diffusion measurement can be quantified as follows [48]:

$$\ln\left(\frac{I_G}{I_0}\right) = -\left[(\gamma\delta G)^2 D\left(\Delta - \frac{\delta}{3}\right)\right]^{\beta} \tag{2.7}$$

where β is the stretch parameter that is used to quantify the influence of polydispersity of polymer on the signal decay in the diffusion measurement. The β decreases as the polydispersity increases. For monodisperse polymers, the β equals 1. For polydisperse polymers, the β ranges from 0.5 to 1.

Wang et al. utilized ^{1}H PFG diffusion NMR to measure the diffusion coefficients of hyperbranched polyethylenimine HPEI-C4-C2 and HPEI-C2-C4 [49]. Their results showed that after phase transition, HPEI-C4-C2 showed a bigger diffusion coefficient than HPEI-C2-C4, which indicates that HPEI-C4-C2 possessed a denser packing than HPEI-C2-C4 (Figure 2.11a). Baille et al. applied PFG NMR spectroscopy to study the self-diffusion of hyperbranched polyglycidols (HBPG-1, HBPG-2, HBPG-3, HBPG-4) in polyvinyl alcohol (PVA) solutions and gels to investigate the shape effect on the self-diffusion of HBPs (Figure 2.11b) [50]. They plotted the self-diffusion coefficient and the effective size of HBPG as a function of the MW to verify the shape

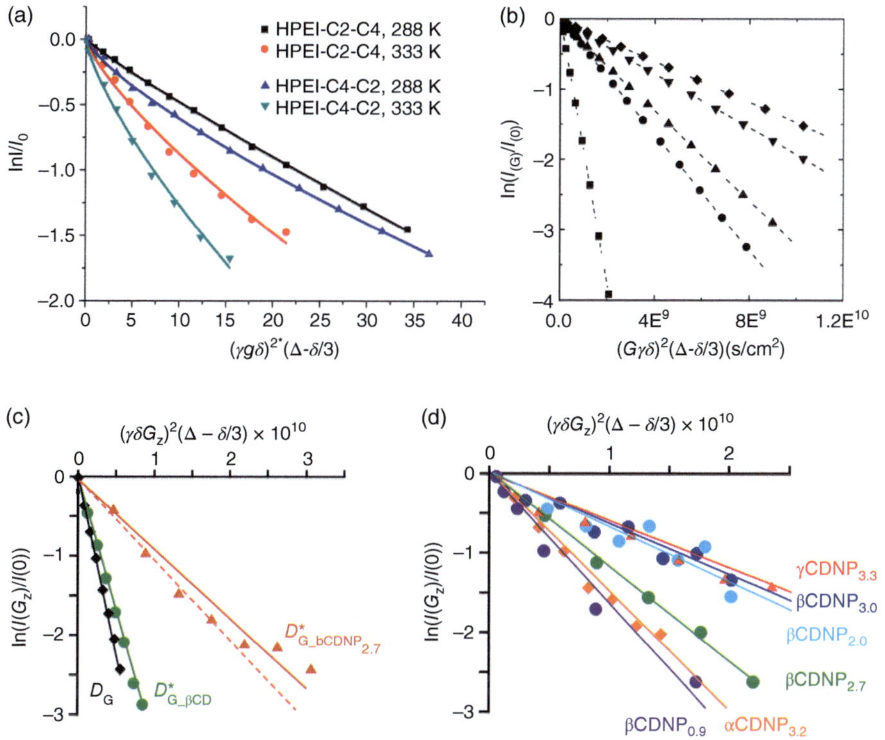

Figure 2.11 The plots of DOSY signal intensity versus $[g^2\gamma^2\delta^2(\Delta - \delta/3)]$ of HBPs. (a) The plots of CH_3 signal intensity of C2 group versus $[g^2\gamma^2\delta^2(\Delta - \delta/3)]$ of the dilute hyperbranched polyethylenimine HPEI-C4-C2 and HPEI-C2-C4 solutions. *Source:* [49] / with permission of American Chemical Society. (b) The plots of NMR signals versus $[g^2\gamma^2\delta^2(\Delta - \delta/3)]$ of hyperbranched polyglycidols (HBPG) HBPG-1 (circles), HBPG-2 (up triangles), HBPG-3 (down triangles), and HBPG-4 (diamonds). *Source:* [50] / with permission of American Chemical Society. (c, d) The plots of NMR signals versus $[g^2\gamma^2\delta^2(\Delta - \delta/3)]$ of cyclodextrin hyperbranched polymers (CDNP). *Source:* [51] / with permission of American Chemical Society.

of HBPG molecules qualitatively. Their results showed that HBPG conformation is like an expanded coil. Doan et al. explored the binding constant of cyclodextrin (CD) HBPs with vanillin using DOSY-NMR (Figure 2.11c,d) [51]. Their results showed that CD HBPs dramatically increased the binding constant of CD to vanillin.

2.2.4 Rheological Properties

HBPs with a high DB and ellipsoid-like or globular-shaped structures show distinctive rheological properties, which make them quite different from linear polymers. The rheological properties such as the storage (G'), the loss modulus (G''), viscosity, and entanglements open a wide range of application fields for HBPs. Due to the lack of entanglements, the HBPs behave as Newtonian fluids [52]. HBPs are used as rheology modifiers, coatings, nanotechnology, and drug delivery. The DB and topology of HBPs make the rheological properties of HBPs deviate a lot from that of linear polymer. Understanding and manipulating HBPs rheological properties through molecular design and processing conditions are crucial for optimizing their performance in specific applications. The characterization of the rheological properties of HBPs is of great importance to understand the influence of DB, topology, and chemical composition on the rheological properties of HBPs. The viscosity, the storage modulus (G'), and the loss modulus (G'') can be measured by rheometer and dynamic mechanical analyzer (DMA).

2.2.4.1 Rheometer and DMA

A rheometer is an instrument used to measure the flow or deformation properties of different materials, which can also be referred to as strain or strain rate, respectively. The rheometer can be used by applying different stress or strain to the materials and measuring the corresponding strain rate or stress response to analyze the rheological properties of materials. The rheometer can give a continuous speed (or shear rate) scanning process for a complete rheological curve. For an advanced rotational rheometer that operates in dynamic oscillation mode; the viscosity; viscoelastic properties (G'', G', tan δ) with respect to time, temperature, frequency, and stress–strain; and transient response.

Ye et al. carried out dynamic oscillations and steady shear measurements to study the rheological behavior of highly branched polymers and linear polymers using a Stress Tech HR rheometer [53]. For HBPs, the G'' was always dominant over the G' in the whole frequency and no cross between the two curves is observed due to a lack of chain entanglements (Figure 2.12a). For linear polymers, crossovers between G' and G'' were observed, which suggests significant entanglements in linear polymers (Figure 2.12b).

Magnusson et al. synthesized a series of hyperbranched aliphatic polyethers with varying DB to study the effects of DB on the rheological properties of HBPs (Figure 2.13) [54]. Their result shows that DB has an important effect on the rheological behavior of the HBPs. The most highly branched polymers are amorphous materials with nearly no chain entanglements and are unable to crystallize.

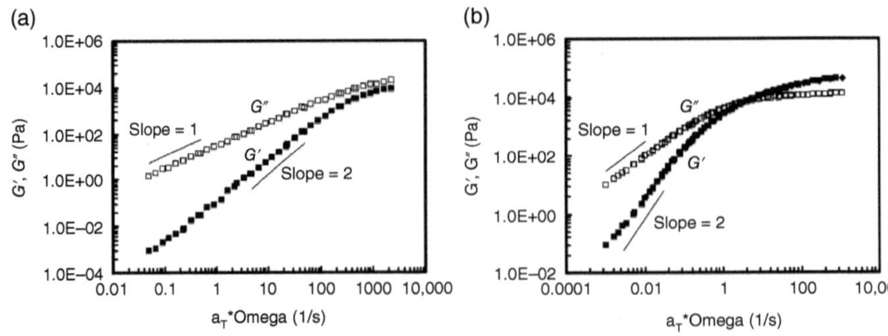

Figure 2.12 Storage modulus (G') and loss modulus (G'') versus angular frequency at 60 °C for polymers with highly branched topology (a) and nearly linear chain topology (b). *Source:* [53] / with permission of John Wiley & Sons.

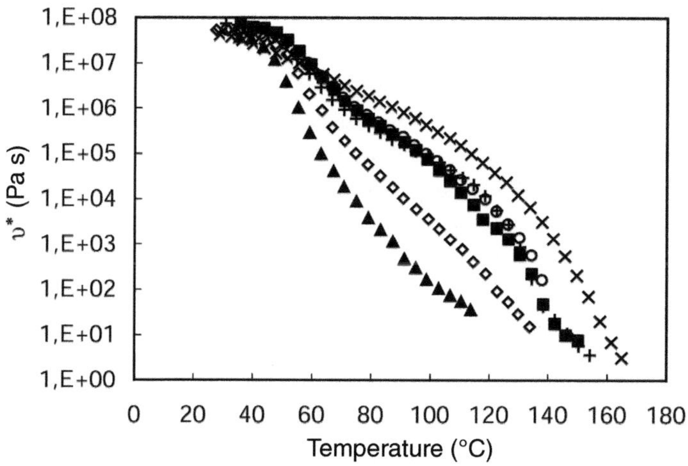

Figure 2.13 Complex dynamic viscosity (η^*) versus temperature for hyperbranched polyethers with different DB: 0.43 (▲), 0.36 (+), 0.21 (■), 0.38 (◇), and 0.30 (○). *Source:* [54] / with permission of Elsevier.

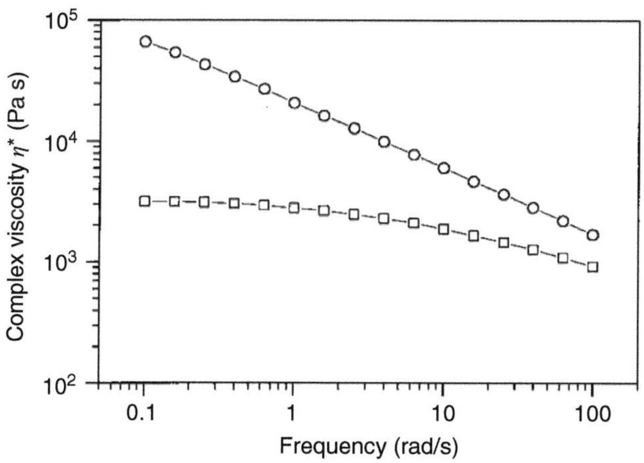

Figure 2.14 Complex viscosity versus frequency for OH- (○) and C12-terminated (□) aromatic HB polyester. *Source:* [55] / with permission of John Wiley & Sons.

Whereas the less branched polymers have a distinct rubbery plateau and show a significantly higher viscosity than the highly branched polymers and can be partially crystallized.

Besides the branched topology, the terminal groups also affect the rheological properties of HBPs. Schmaljohann et al. studied the effect of alkyl chain modifications of the terminal OH group on the rheological properties of HBPs (Figure 2.14) [55]. Nonentangled elastic behavior can be observed for the OH-terminated polymer. For C12-terminated polymers, typical behavior for a viscous melt with a plateau at low frequencies was observed. These results were also confirmed by the analysis of viscoelastic properties (G'', G', tan δ) of both polymers.

2.3 Conclusion and Prospect

In this chapter, we first summarized the properties of HBPs and HERs. HERs possess a branched ellipsoid topology with low viscosity and a large amount of terminal functional groups. HBPs are usually prepared in a one-pot synthesis, and the scaling-up process is relatively easy. The property differences between HBPs, dendrimers, and linear polymers have also been discussed. Compared with linear polymers and dendrimers, HBPs show great superiority in properties and performance. HBPs usually show excellent physical and chemical properties when compared with linear polymers, such as better solubility, low molecular entanglement, lower melt and solution viscosity, large amount of terminal functional groups, self-assembly behaviors, and high mechanical strength. The structure–property relationship of HBPs is also summarized. The chemical composition, topological structure, and functional group of polymers play an important role in determining the material properties. HBPs possess an ellipsoid-like three-dimensional structure, a random branching structure with DB < 1.0 (usually, 0.4–0.6), a wide distribution of molar mass (usually, PDI > 1.0), and large amounts of functional groups located at both linear and terminal units. The structures of HBPs such as DB, MW, size, shape, and shape factor offer them unique mechanical performances. The characterization of HERs is a critical component of research in this area, and this chapter focuses on the characterization of solution and bulk properties.

DB is the most important parameter for HBPs. As a widely used structural parameter to quantitatively characterize the branching features of HERs, DB is the key factor to determine the solution properties, thermal properties, rheological properties, and performances of HBPs. In the past two decades, NMR, the degradation method followed by chromatographic analysis, and the viscometry approach have been used to evaluate the DB of HERs. NMR is the most frequently used method to measure the DB of HERs. Both 1D and 2D NMR spectroscopy are utilized to measure the DB of HERs. The main challenge for DB measurement by NMR is the signal assignments of different units. For 1D NMR spectroscopy, ^{1}H NMR spectroscopy has long been utilized to differentiate the structural units (*D*, *L*, and *T*) to measure the DB because of its high sensitivity. For complex HBPs, the resolution of ^{1}H NMR is insufficient to differentiate the structural units of polymers due to the severe peaks overlapping. ^{13}C NMR can also be used to address this problem and discriminate between the different units of polymer owing to its high resolution and can be applied to quantitatively investigate the DB. The major problem for ^{13}C NMR is the relatively low sensitivity. Quantitative ^{13}C NMR spectroscopy is usually time-consuming and requires large amounts of samples. For heteroatom-containing HBPs, ^{15}N, ^{19}F, ^{29}Si, and ^{31}P NMR spectroscopy can also be used to discriminate the structural units and measure the DB. Besides ^{1}H and ^{13}C NMR spectroscopy, 2D ^{13}C—^{1}H correlation spectroscopy and ^{15}N—^{1}H correlation spectroscopy (HSQC and HMQC) can also be applied to differentiate the different units of complex HBPs. Recently developed pure shift NMR (broadband homonuclear-decoupled NMR) that offers a large improvement in spectral resolution by suppressing the homonuclear couplings can be used to simplify the 1D and 2D spectra of HERs and contribute to the measurements of

DB [56]. For more complex HERs, the resolution of NMR spectra is insufficient to discriminate among the different structural units and thus cannot measure the DB. Under this condition, the degradable method can be applied to degrade HERs into small units, which then can be quantitatively analyzed by chromatographic analysis and NMR analysis. However, the degradable approach required the HERs to be purified before degradation and completely degraded to facilitate the accurate determination of DB. Another indirect approach is the viscometry method, which is based on the relationship between intrinsic viscosity and the MW that follows the Mark–Houwink equation. Large value in DB may contribute to a compact structure with less chain entanglement and lower intrinsic viscosity. Higher intrinsic viscosity value for HERs can be caused by more entanglements and less compact structure in HERs, which indicates a small DB value. As one of the most important parameters for HERs, DB has a great influence on the solution and mechanism property of HERs, such as viscosity, glass transition temperature, degree of crystallization, free volume, chain entanglement, hydrodynamic radius, and mechanical strength. The accurate measurement of DB can help establish the structure–property relationship of HERs. However, due to the broad distribution in MW and topology, the measured DB is the average value, which also shows a distribution in value between 0.5 and 1.0. How to measure the accurate distribution of DB can also be very important for the deep understanding of structure–property relationship.

MW and its distribution are also important parameters for HERs. GPC, light scattering techniques, MALDI-TOF MS techniques, and DOSY NMR have all been used to determine the MW and its distribution. GPC with differential refractive index detection and UV detection, followed by calibration with a standard polymer (polystyrene), is the most frequently used method to determine the MW and PDI of HERs. The Mw, Mn, and PDI can be calculated from the width and shape of the GPC peak through a calibration procedure. However, the principle of separation in GPC is based on the polymer hydrodynamic volume and not the MW, which makes inherent drawbacks for GPC as a relative method with respect to column, solvent, molecular shape, and temperature. The raw data from GPC can be converted into an MW based on the calibration procedure under the same conditions. As a powerful method for the detection of MW and PDI of HERs, MALDI-TOF MS is preferable for the monodisperse HERs over the polydisperse HERs as lower MW can be detected and higher MW are hard to detect due to the low intensities in the MS. Therefore, the combination of SEC and MALDI-TOF MS has been used to measure the MW and PDI of HERs with a broad MW distribution. DOSY NMR can also be utilized to predict the MW of narrowly dispersed HERs based on the linear correlation between the logarithm of diffusion coefficient and the MW. DOSY NMR doesn't require the purification of HERs.

The unique branching topology of HERs such as size and shape confers lower melt and solution viscosity, better solubility, low molecular entanglement, and special mechanical strength. DLS, DOSY NMR, and GPC are utilized to study the size and shape of HERs. The size and shape of HERs are related to the Brownian motion of HERs in solution. DLS and DOSY NMR can be utilized to study the diffusion behavior of HERs in solutions, and the hydrodynamic radius can be determined using Stokes–Einstein equation.

Many new HER classes cannot readily be characterized with the well-established methods due to the complex structure of HERs. HERs possess very complex 3D ellipsoid structures with multidimensional, broad distributions in molar mass, DB, and chemical structure. HERs have a unique behavior in solution that differs strongly from that of linear and less branched polymers. This behavior is influenced by dense branching, MW, the nature of end groups, and the chemistry of the repeating units. Great care needs to be taken to adapt characterization methods and to elucidate the structure before any structure–property relationships can be established.

Acknowledgments

The authors gratefully acknowledge financial support from the National Natural Science Foundation of China (22276221) and the Fundamental Research Funds for the Central Universities of South-Central Minzu University (CZQ25009).

References

1 Voit, B.I. and Lederer, A. (2009). Hyperbranched and highly branched polymer architectures–synthetic strategies and major characterization aspects. *Chem. Rev.* 109 (11): 5924–5973.

2 Jiang, W.F., Zhou, Y.F., and Yan, D.Y. (2015). Hyperbranched polymer vesicles: from self-assembly, characterization, mechanisms, and properties to applications. *Chem. Soc. Rev.* 44 (12): 3874–3889.

3 McKee, M.G., Unal, S., Wilkes, G.L., and Long, T.E. (2005). Branched polyesters: recent advances in synthesis and performance. *Prog. Polym. Sci.* 30 (5): 507–539.

4 Mu, B., Liu, T.T., and Tian, W. (2019). Long-chain hyperbranched polymers: synthesis, properties, and applications. *Macromol. Rapid Commun.* 40 (17): 1800471.

5 Hao, D., Guo, X., Zhu, X. et al. (2024). Progress in synthesis, modification, characterization and applications of hyperbranched polyphosphate polyesters. *Des. Monomers Polym.* 27 (1): 62–86.

6 Hawker, C.J., Lee, R., and Frechet, J.M.J. (1991). One-step synthesis of hyperbranched dendritic polyesters. *J. Am. Chem. Soc.* 113 (12): 4583–4588.

7 Yan, D., Müller, A.H.E., and Matyjaszewski, K. (1997). Molecular parameters of hyperbranched polymers made by self-condensing vinyl polymerization. 2. Degree of branching. *Macromolecules* 30 (23): 7024–7033.

8 Hölter, D., Burgath, A., and Frey, H. (2003). Degree of branching in hyperbranched polymers. *Acta Polym.* 48 (1–2): 30–35.

9 Huber, T., Böhme, F., Komber, H. et al. (1999). New hyperbranched poly(ether amide)s via nucleophilic ring opening of 2-oxazoline-containing monomers. *Macromol. Chem. Phys.* 200 (1): 126–133.

10 Behera, G.C. and Ramakrishnan, S. (2007). Dramatic effect of solvent on the NMR spectral resolution of hyperbranched polyethers. *J. Polym. Sci. A Polym. Chem.* 45 (8): 1474–1480.

11 Shpotya, V., Perepukhov, A., Maksimychev, A. et al. (2022). Structure determination of hyperbranched polyester BOLTORN H40 by 1D- and 2D-NMR spectroscopy. *Polym. Bull.* 80 (4): 4523–4534.

12 Chen, S.F., Xu, Z.J., and Zhang, D.H. (2018). Synthesis and application of epoxy-ended hyperbranched polymers. *Chem. Eng. J.* 343: 283–302.

13 De, B. and Karak, N. (2013). Novel high performance tough hyperbranched epoxy by an A_2+ B_3 polycondensation reaction. *J. Mater. Chem. A* 1 (2): 348–353.

14 Boonlert-Uthai, T., Samthong, C., and Somwangthanaroj, A. (2019). Synthesis, thermal properties and curing kinetics of hyperbranched BPA/PEG epoxy resin. *Polymers (Basel)* 11 (10): 1545.

15 Hawker, C.J. and Chu, F. (1996). Hyperbranched poly(ether ketones): manipulation of structure and physical properties. *Macromolecules* 29 (12): 4370–4380.

16 Yoon, K. and Son, D.Y. (1999). Syntheses of hyperbranched poly(carbosilarylenes). *Macromolecules* 32 (16): 5210–5216.

17 Holycross, D.R. and Chai, M.H. (2013). Comprehensive NMR studies of the structures and properties of PEI polymers. *Macromolecules* 46 (17): 6891–6897.

18 Zhu, X.Y., Chen, L., Chen, Y. et al. (2008). Using 2D NMR to determine the degree of branching of complicated hyperbranched polymers. *Sci. China Ser. B* 51 (11): 1057–1065.

19 Maier, G., Zech, C., Voit, B., and Komber, H. (1998). An approach to hyperbranched polymers with a degree of branching of 100%. *Macromol. Chem. Phys.* 199 (12): 2655–2664.

20 Lois, J.H. and Kenwright, A.M. (1997). A simple 'one pot' route to the hyperbranched analogues of Tomalia's poly(amidoamine) dendrimers. *Chem. Commun.* 19: 1877.

21 Merino, S., Brauge, L., Caminade, A.M. et al. (2001). Synthesis and characterization of linear, hyperbranched, and dendrimer-like polymers constituted of the same repeating unit. *Chemistry* 7 (14): 3095–3105.

22 Kambouris, P. and Hawker, C.J. (1993). A versatile new method for structure determination in hyperbranched macromolecules. *J. Chem. Soc., Perkin Trans. 1* (22): 2717–2721.

23 Bolton, D.H. and Wooley, K.L. (2002). Hyperbranched aryl polycarbonates derived from A_2B monomers versus AB_2 monomers. *J. Polym. Sci. A Polym. Chem.* 40 (7): 823–835.

24 Yan, D. and Zhou, Z. (1999). Molecular weight distribution of hyperbranched polymers generated from polycondensation of AB_2 type monomers in the presence of multifunctional core moieties. *Macromolecules* 32 (3): 819–824.

25 Berek, D. (2010). Size exclusion chromatography–a blessing and a curse of science and technology of synthetic polymers. *J. Sep. Sci.* 33 (3): 315–335.

26 Gaborieau, M. and Castignolles, P. (2011). Size-exclusion chromatography (SEC) of branched polymers and polysaccharides. *Anal. Bioanal. Chem.* 399 (4): 1413–1423.

27 Uliyanchenko, E. (2014). Size-exclusion chromatography-from high-performance to ultra-performance. *Anal. Bioanal. Chem.* 406 (25): 6087–6094.

28 Gruendling, T., Guilhaus, M., and Barner-Kowollik, C. (2009). Fast and accurate determination of absolute individual molecular weight distributions from mixtures of polymers via size exclusion chromatography–electrospray ionization mass spectrometry. *Macromolecules* 42 (17): 6366–6374.

29 Wyatt, P.J. (1993). Light scattering and the absolute characterization of macromolecules. *Anal. Chim. Acta* 272 (1): 1–40.

30 Lecacheux, D., Lesec, J., and Quivoron, C. (2003). High-temperature coupling of high-speed GPC with continuous viscometry. I. Long-chain branching in polyethylene. *J. Appl. Polym. Sci.* 27 (12): 4867–4877.

31 Yu, C., Xu, Z., Ma, X. et al. (2021). Hyperbranched polymers containing epoxy and imide structure. *Prog. Org. Coat.* 151: 106031.

32 Chikh, L., Tessier, M., and Fradet, A. (2008). Polydispersity of hyperbranched polyesters based on 2,2-bis(hydroxymethyl)propanoic acid: SEC/MALDI–TOF MS and ^{13}C NMR/kinetic-recursive probability analysis. *Macromolecules* 41 (23): 9044–9050.

33 Jayakannan, M., Van Dongen, J.L.J., Behera, G.C. et al. (2002). SEC-MALDI-TOF mass spectral characterization of a hyperbranched polyether prepared via melt transetherification. *J. Polym. Sci. A Polym. Chem.* 40 (24): 4463–4476.

34 Gooden, J.K., Gross, M.L., Mueller, A. et al. (1998). Cyclization in hyperbranched polymer syntheses: characterization by MALDI-TOF mass spectrometry. *J. Am. Chem. Soc.* 120 (39): 10180–10186.

35 Burgath, A., Sunder, A., and Frey, H. (2000). Role of cyclization in the synthesis of hyperbranched aliphatic polyesters. *Macromol. Chem. Phys.* 201 (7): 782–791.

36 Scionti, V. and Wesdemiotis, C. (2011). Tandem mass spectrometry analysis of polymer structures and architectures. In: *Mass Spectrometry in Polymer Chemistry* (ed. C. Barner-Kowollik, T. Gruendling, J. Falkenhagen, and S. Weidner), 57–84. Wiley.

37 Town, J.S., Jones, G.R., Hancox, E. et al. (2019). Tandem mass spectrometry for polymeric structure analysis: a comparison of two common MALDI–ToF/ToF techniques. *Macromol. Rapid Commun.* 40 (13): 1900088.

38 Chen, A., Wu, D., and Johnson, C.S. (2002). Determination of molecular weight distributions for polymers by diffusion-ordered NMR. *J. Am. Chem. Soc.* 117 (30): 7965–7970.

39 Hou, J. and Pearce, E. (2021). Characterization of polymer molecular weight distribution by NMR diffusometry: experimental criteria and findings. *Anal. Chem.* 93 (22): 7958–7964.

40 Rosenboom, J.G., De Roo, J., Storti, G. et al. (2016). Diffusion (DOSY) ^{1}H NMR as an alternative method for molecular weight determination of poly(ethylene furanoate) (PEF) polyesters. *Macromol. Chem. Phys.* 218 (1): 1600436.

41 Ruzicka, E., Pellechia, P., and Benicewicz, B.C. (2023). Polymer molecular weights via DOSY NMR. *Anal. Chem.* 95 (20): 7849–7854.

42 Monnery, B.D., Jerca, V.V., Hoogenboom, R., and Swift, T. (2024). Polymer conformation determination by NMR spectroscopy: comparative diffusion ordered ^{1}H-NMR spectroscopy of poly(2-ethyl-2-oxazoline)s and poly(ethylene glycol) in D_2O. *Polym. Chem.* 15 (30): 3077–3085.

43 Ma, X., Guo, W., Xu, Z. et al. (2020). Synthesis of degradable hyperbranched epoxy resins with high tensile, elongation, modulus and low-temperature resistance. *Compos. Part B* 192.

44 Wang, Y., Chen, S., Guo, W. et al. (2019). The precise effect of degree of branching of epoxy-ended hyperbranched polymers on intrinsic property and performance. *Prog. Org. Coat.* 127: 157–167.

45 Chen, X., Chen, S., Xu, Z. et al. (2020). Degradable and recyclable bio-based thermoset epoxy resins. *Green Chem.* 22 (13): 4187–4198.

46 Viel, S., Capitani, D., Mannina, L., and Segre, A. (2003). Diffusion-ordered NMR spectroscopy: a versatile tool for the molecular weight determination of uncharged polysaccharides. *Biomacromolecules* 4 (6): 1843–1847.

47 Nelson, T.F. and Ward, C.P. (2023). Diffusion-ordered spectroscopy for rapid and facile determination of consumer plastic molecular weight. *Anal. Chem.* 95 (22): 8560–8568.

48 Gong, X.L., Hansen, E.W., and Chen, Q. (2012). Molecular weight distribution characteristics (of a polymer) derived from a stretched-exponential PGSTE NMR response function—simulation. *Macromol. Chem. Phys.* 213 (3): 278–284.

49 Wang, B., Xiao, T., Fu, X.B. et al. (2017). Thermoresponsive hyperbranched polymers with spatially isomerized groups: NMR implication to their thermoresponsive behaviors. *Macromolecules* 50 (24): 9647–9655.

50 Baille, W.E., Zhu, X.X., and Fomine, S. (2004). Study of self-diffusion of hyperbranched polyglycidols in poly(vinyl alcohol) solutions and gels by pulsed-field gradient NMR spectroscopy. *Macromolecules* 37 (23): 8569–8576.

51 Doan, A.T.N., Doan, V.T.H., Katsuki, J. et al. (2022). Dramatically increased binding constant of water-soluble cyclodextrin hyperbranched polymers: explored with diffusion ordered NMR spectroscopy (DOSY). *ACS Omega* 7 (13): 10890–10900.

52 Farrington, P.J., Hawker, C.J., Frechet, J.M.J. et al. (1998). The melt viscosity of dendritic poly(benzyl ether) macromolecules. *Macromolecules* 31 (15): 5043–5050.

53 Ye, Z., AlObaidi, F., and Zhu, S. (2004). Melt rheological properties of branched polyethylenes produced with Pd- and Ni–diimine catalysts. *Macromol. Chem. Phys.* 205 (7): 897–906.

54 Magnusson, H., Malmström, E., Hult, A., and Johansson, M. (2002). The effect of degree of branching on the rheological and thermal properties of hyperbranched aliphatic polyethers. *Polymer* 43 (2): 301–306.

55 Schmaljohann, D., Häußler, L., Pötschke, P. et al. (2000). Modification with alkyl chains and the influence on thermal and mechanical properties of aromatic hyperbranched polyesters. *Macromol. Chem. Phys.* 201 (1): 49–57.

56 Foroozandeh, M., Adams, R.W., Meharry, N.J. et al. (2014). Ultrahigh-resolution NMR spectroscopy. *Angew. Chem. Int. Ed. Engl.* 53 (27): 6990–6992.

3

Synthesis Methods of Hyperbranched Epoxy Resins

Yu Jiang[1], Aolin Wang[1], Yufei Zhang[1], Mengting Du[1], and Sufang Chen[2]

[1]South-Central Minzu University, Hubei R&D Center of Hyperbranched Polymers Synthesis and Applications, Key Laboratory of Catalysis and Energy Materials Chemistry of Ministry of Education & Hubei Key Laboratory of Catalysis and Materials Science, No.182, Minzu Ave. Hongshan District, Wuhan, 430074, P.R. China
[2]Wuhan Institute of Technology, School of Chemical Engineering and Pharmacy, State Key Laboratory of Green and Efficient Development of Phosphorus Resources, No.206, Guanggu 1st road, Donghu New & High Technology Development Zone, Wuhan 430073, P.R. China

3.1 Introduction

Epoxy resin (EP), as one of the most widely consumed thermosetting resins, was synthesized by Prileschajew in 1939 [1]. This widespread commercial success is largely attributable to its ingenious molecular architecture. Its molecular structure features highly reactive epoxy groups (—C_2H_4O—) that provide exceptional adhesive properties and enable cross-linking reactions with diverse curing agents [2–6]. The incorporation of a benzene ring structure imparts high rigidity, heat resistance, and chemical stability, guaranteeing that cured epoxy resins maintain consistent mechanical properties even at high temperatures. Moreover, the inclusion of ether bonds (—O—) and hydroxyl groups (—OH) improves the resin's capacity to be wetted, hence promoting stronger chemical interaction with the surfaces of substrates [7, 8]. Hydroxyl groups facilitate hydrogen bonding with other molecules, hence enhancing intermolecular interactions and overall material performance. Taken together, these structural attributes play a decisive role in shaping the material's key performance characteristics. These include its remarkable versatility, highly flexible formulation design, broad applicability, outstanding adhesion, minimal curing shrinkage, superior mechanical strength, excellent electrical insulation, and strong resistance to chemical corrosion—all of which underpin its commercial success [9]. However, the highly cross-linked three-dimensional network of EPs after curing also leads to significant drawbacks: increased brittleness, low fracture strain, poor impact strength, and low resistance to crack propagation [10, 11]. This inherent brittleness severely limits epoxy's engineering applications, particularly in composite structures requiring toughness.

Hyperbranched epoxy resins (HERs), which were first synthesized by Sorensen in 1996 [12], have attracted great attention due to their unique molecular structural features, including a highly branched backbone, abundant functional terminal groups [13], intrinsically low viscosity [14], and topological conformation [15]. As a choice for next-generation EPs, they can be used as a prepolymer cured by a hardener or as a toughener to toughen other commercial EPs [16]. Previous studies have demonstrated that a homogeneous toughening effect can be achieved by modulating the topological and chemical structures of HERs, resulting in resins that exhibit not only satisfactory toughness but also well-balanced mechanical properties [17].

Due to the important role of HERs in synergistically enhancing and toughening epoxy resins, researchers are dedicated to developing new synthesis methods for HERs. To the best of our knowledge, two main synthetic strategies have been developed for the construction of HERs until now, including the one-step polymerization of epoxy monomers and post-polymerization modification (PPM) of hyperbranched polymers. In the following sections, the synthesis of HERs is introduced from these two aspects.

3.2 One-step Polymerization of Epoxy Monomers

The one-pot polymerization strategy enables the direct synthesis of HERs by leveraging the polymerization of epoxy-containing monomers [12, 18, 19]. This approach incorporates proton transfer polymerization, step-growth polycondensation, and step-growth polyaddition. The inherent epoxy groups in the monomers not only serve as reactive functional groups during the process but are also partially retained in the resulting polymer structure.

Compared to the multistep post-polymerization strategy, the one-pot method offers considerable advantages and demonstrates unique value in the field of HER synthesis. The most apparent advantage is the simplicity of a single reaction step. Isolation and purification of the intermediate polymer are excluded, leading to a dramatic reduction in processing time, energy consumption, and overall cost. Moreover, the polymerization chemistry can be designed to precisely control the number of epoxy groups through monomer stoichiometry, enabling excellent control over the degree of branching (DB) and the final epoxy equivalent weight (EEW). Furthermore, the one-pot method avoids the undesirable side reactions during the PPM step, which may effectively reduce the epoxy functionality and introduce structural defects.

3.2.1 Proton Transfer Polymerization

The concept of proton transfer polymerization was first illustrated by Frechet and coworkers in 1999, involving that protons (hydrogen ions) are transferred between different molecules or different atoms within a molecule during the reaction to initiate and drive the polymerization reaction [20]. For the synthesis of HERs, the polymerization is centered on the activation of nucleophilic reagents by proton transfer to drive selective ring-opening reactions of epoxy groups. Using this polymerization method, the preparation of HERs from AB_n and $A_2 + B_3$ monomers were all reported.

As shown in Figure 3.1b, the phenolic double epoxides, as an AB_2-type epoxy-contained monomer, have been used to prepare the HERs. Under alkaline conditions, the phenolic hydroxyl group ($pK_a \approx 10$) is deprotonated to form a phenoxide anion, which then attacks the less substituted carbon of the epoxide ring and

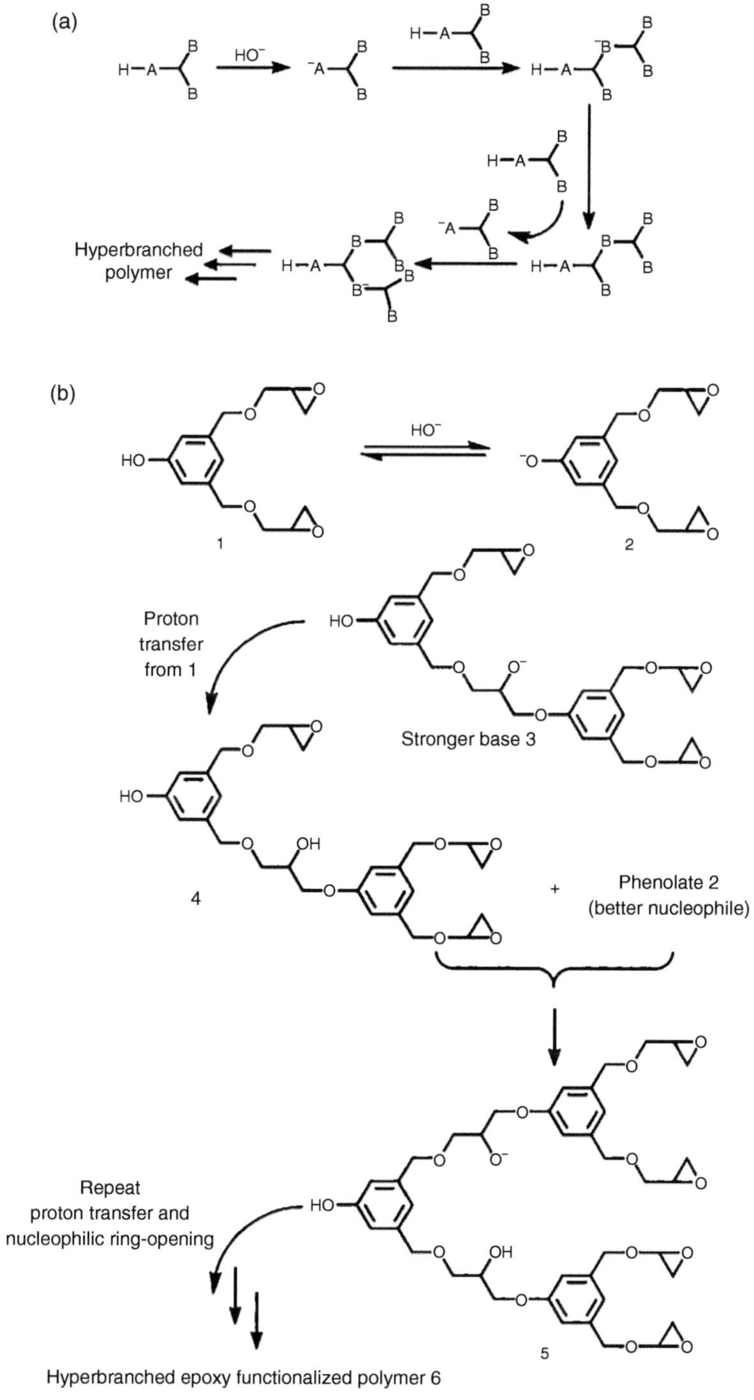

Figure 3.1 (a) Synthetic route of the proton transfer polymerization. (b) Proton transfer polymerization of AB_2 monomers containing phenolic and epoxide groups. *Source:* [20] / with permission of American Chemical Society.

results in an alkoxide intermediate. Given the higher basicity of the alkoxide anion ($pK_a \approx 17$), a rapid proton transfer occurs from the phenolic species, regenerating the phenoxide anion. This process establishes a cyclic mechanism that sustains chain propagation and branching. The proton transfer polymerization technique offers a straightforward route for synthesizing HERs, and the resulting resins show promising potential for applications in adhesives and coatings. Moreover, modification of their terminal epoxy groups can yield hydrolytically stable materials with diverse peripheral functionalities [20].

Gadwal et al. [21] reported a strategy to synthesize HERs by proton transfer polymerization of AB_2 monomers containing thiol and epoxy groups, resulting in the hyperbranched polymer scaffolds with dual reactivity (Figure 3.2), and to functionalize them for potential applications in gene delivery and other fields. Each

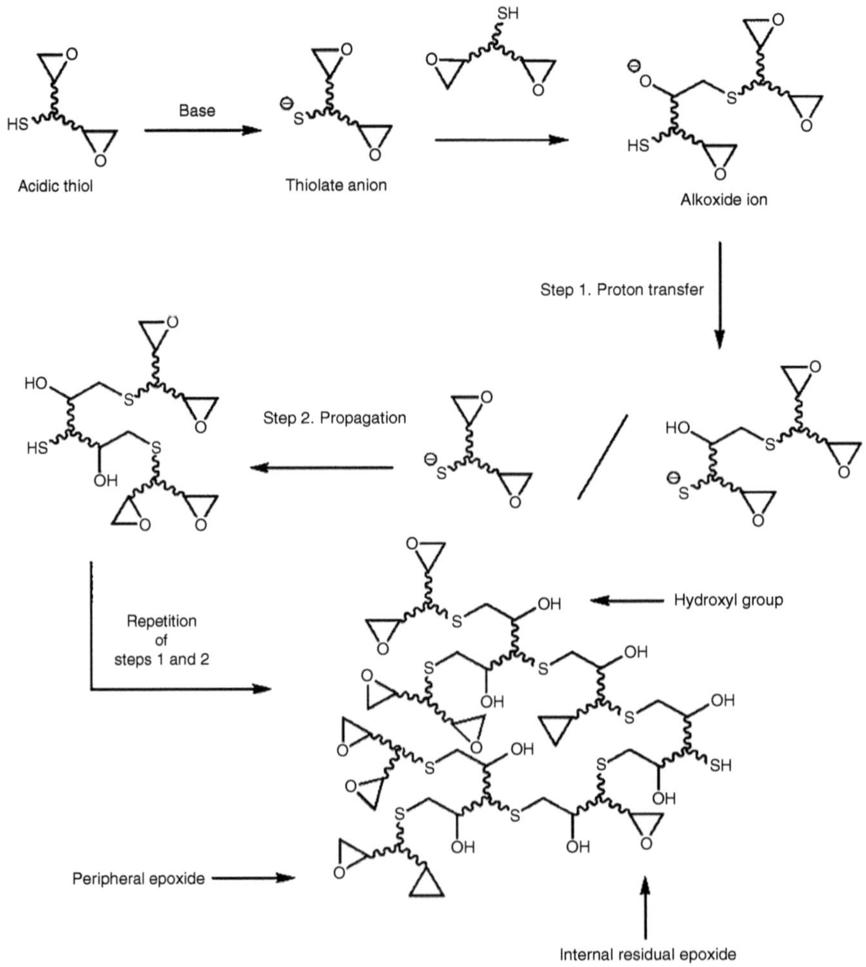

Figure 3.2 Synthetic route of the proton transfer polymerization of AB_2 monomers containing thiol and epoxide groups. *Source:* [21] / with permission of American Chemical Society.

AB_2 monomer contains two epoxy groups and one thiol group. The proton transfer polymerization involves a nucleophilic attack of thiolate anion to epoxy group under basic conditions, to form the hyperbranched polymers. HERs with a DB of 65–69% and approximately 2% disulfide-bonded structural defects can be obtained with various catalytic and reaction conditions. Both the epoxy and hydroxyl groups in the resulting HERs can be selectively functionalized with groups such as alkyl, aryl, ethylene oxide chains, and ammonium moieties. This study offers a novel strategy for synthesizing hyperbranched polymers and demonstrates the potential of the resulting bifunctionalized cationic polymers in gene delivery applications, thereby contributing to the advancement of related research fields.

Fréchet and coworkers [22] reported a preliminary study on the preparation of polyether-based HERs by the $A_2 + B_3$ type of proton transfer polymerization. As shown in Figure 3.3, the polymerization employs 1,2,7,8-diepoxyoctane (**1**) as

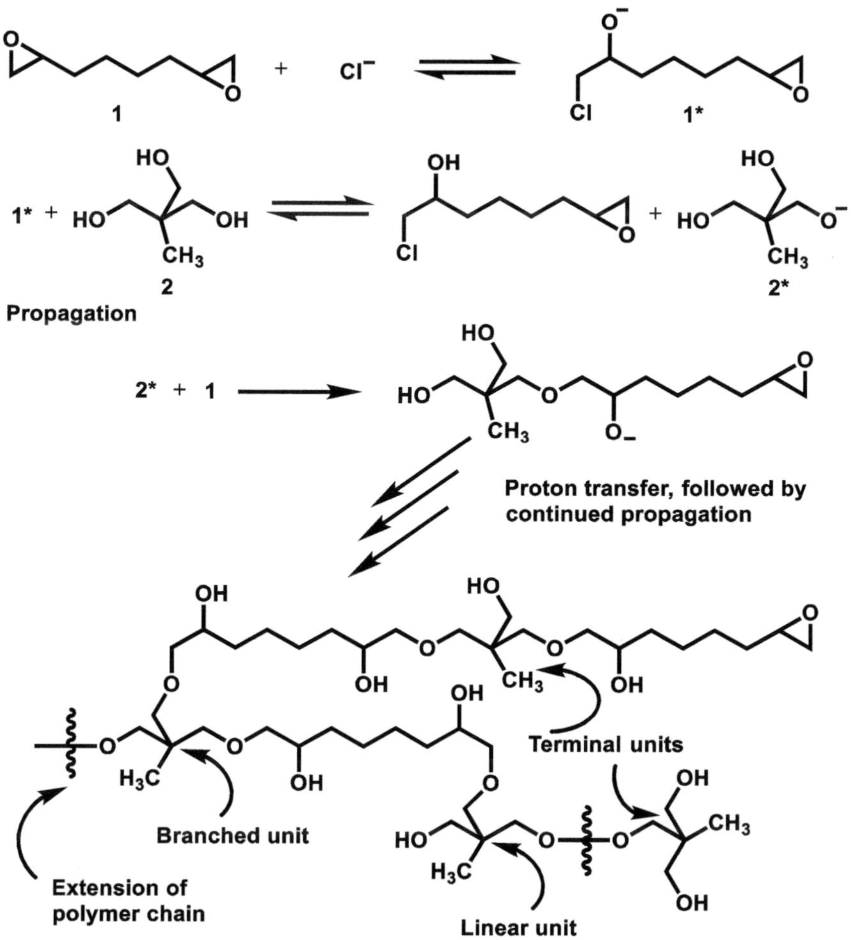

Figure 3.3 Synthetic route of the proton transfer polymerization of $A_2 + B_3$. *Source:* [22] / with permission of American Chemical Society.

the A_2 monomer, 1,1,1-tris(hydroxymethyl)ethane (**2**) as the B_3 monomer, and tetra-*n*-butylammonium chloride (5 mol% based on **1**) as the nucleophilic catalyst. Initiation occurs by nucleophilic attack of the chloride ion on the epoxide of **1** at the less-hindered, terminal carbon to give a secondary alkoxide. The equilibrium between primary and secondary alkoxides will be established through rapid proton exchange. The subsequent propagation proceeds by nucleophilic attack of primary alkoxides on epoxide rings to give the aliphatic polyether epoxy. The present study provides a method for preparing novel hyperbranched aliphatic polyethers, which hold promise for applications in low-viscosity epoxy resins and demonstrate potential uses in adhesives and coatings.

Although the polymerizations shown in Figures 3.1 and 3.2 produce HERs under basic conditions, the chloride ion-catalyzed proton transfer polymerization offers distinct advantages in controlling the molecular weight (MW) of the resulting HERs [23]. Under basic conditions, the polymerization relies on the much lower pK_a of a phenol relative to that of a secondary alcohol. This pK_a difference allows for the fast, thermodynamically favorable transfer of phenolic protons to secondary alkoxides that are created during epoxide ring opening. However, a rapid growth in the MW and ultimately gelation was observed if the polymerization was not carefully monitored and terminated at the appropriate time. This phenomenon results from the reactions of secondary alkoxides in the later stages of the polymerization, when the concentration of phenol is very low and proton transfer can no longer take place. In comparison, the use of chloride ions as a nucleophilic catalyst enables the polymerization to reach high MWs without cross-linking. As shown in Figure 3.4, an equilibrium exists between the epoxide ring opening by the chloride ion and base-induced ring closure of the resulting chlorohydrin. This mechanism constitutes a type of controlled polymerization in which the chlorohydrin group serves as a regulator capable of moving the polymer into a dormant phase by ring closure, as well as transferring the active site to another monomer or oligomer for further propagation. During the later stages of polymerization, consumption of the phenolic groups halts proton transfer, shifting the equilibrium toward chlorohydrin ring closure. As a result, the polymerization rate decreases so significantly that the reaction is effectively self-terminating.

Compared to HERs with aliphatic backbone, aromatic epoxide-based hyperbranched polymers possess a practical competitive advantage owing to their enhanced material properties and lower cost, which results from a simpler preparation process. By selecting suitable commercially available bifunctional and trifunctional monomers, proton transfer polymerization can be employed

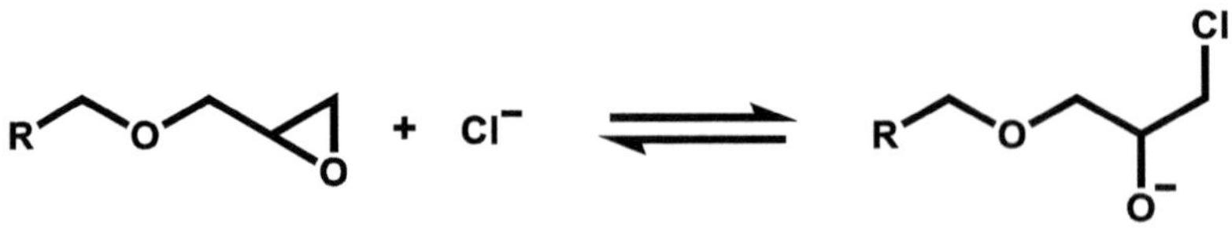

Figure 3.4 Equilibrium between the epoxide ring opening by chloride ion and base-induced ring closure of the resulting chlorohydrin. *Source:* [23] / John Wiley & Sons.

to synthesize HERs with structure most close to the conventional bisphenol A (BPA)-type epoxy resin.

Ma et al. [24] synthesized an epoxy-terminated hyperbranched polymer (EHP) via proton transfer polymerization, employing BPA as the A_2 monomer and trimethylolpropane triglycidyl ether (TMPTG) as the B_3 monomer (Figure 3.5). Tetrabutylammonium bromide was employed as the catalyst for the polymerization, which leveraged the pK_a difference between phenol and secondary alcohol, as well as the dynamic equilibrium between bromide-induced epoxide ring opening and bromohydrin ring closure. The resulting HERs exhibited a structure analogous to that of conventional BPA-based epoxy resin (E20), demonstrating good solubility in polar solvents, low solution viscosity, and high thermal stability. This successful synthesis of hyperbranched analogues to BPA-type epoxy resins not only offers a novel strategy for epoxy modification but also shows promising performance in coating applications, indicating potential for broadening the applicability of epoxy-based materials.

The polymerization process of the proton transfer polymerization is largely dependent upon the monomer concentration and the reaction temperature. Lv et al. [25] reported the preparation of hyperbranched polyether epoxy (HBPEE) by bulk proton transfer polymerization, employing hydroquinone (HQ) and TMPGE as A_2 and B_3 monomers, respectively (Figure 3.6). It was observed that the bulk polymerization carried out at higher temperature serves a better route for synthesizing HBPEE compared with that carried out in solvent. In the bulk polymerization, excess of B_3 is employed to ensure an epoxy-terminated product. In addition, the excess amount of liquid tri-epoxy monomer can act as a solvent to decrease the concentration of A_2 monomer to avoid gelation. The synthesized HBPEE is proven to be an effective toughener for diglycidyl ether of bisphenol A (DGEBA) in spite of its relatively rigid aromatic polyether structure, explained by the deformation of nano-sized free volume and the formation of oriented fibrils.

Luo et al. [26] synthesized an HBPEE with an aromatic backbone via one-step $A_2 + B_3$ proton transfer polymerization using 4,4′,4″-trihydroxytriphenylmethane (THTPM) and DGEBA as monomers (Figure 3.7). The resulting HER was employed to modify a conventional epoxy resin based on DGEBA. It was found that the modified system containing 5 wt% HER exhibited outstanding mechanical properties, an increased glass transition temperature, and no phase separation. This study presents an efficient hyperbranched modifier that simultaneously enhances the toughness and strength of epoxy resins without compromising thermal properties, demonstrating significant potential for applications requiring high-performance epoxy materials.

Overall, one-step proton transfer polymerization provides a facile way to synthesize HERs, which can be obtained in a one-step procedure without the isolation and purification of the intermediate polymer. Especially, the A_2 and B_3 monomers used for proton transfer polymerization are usually commercially available or can be easily prepared. Moreover, the polyether-based HERs were usually obtained from this polymerization method, which is chemically more stable and flexible than the HERs with polyester backbone. Therefore, these polymers are more desirable in places

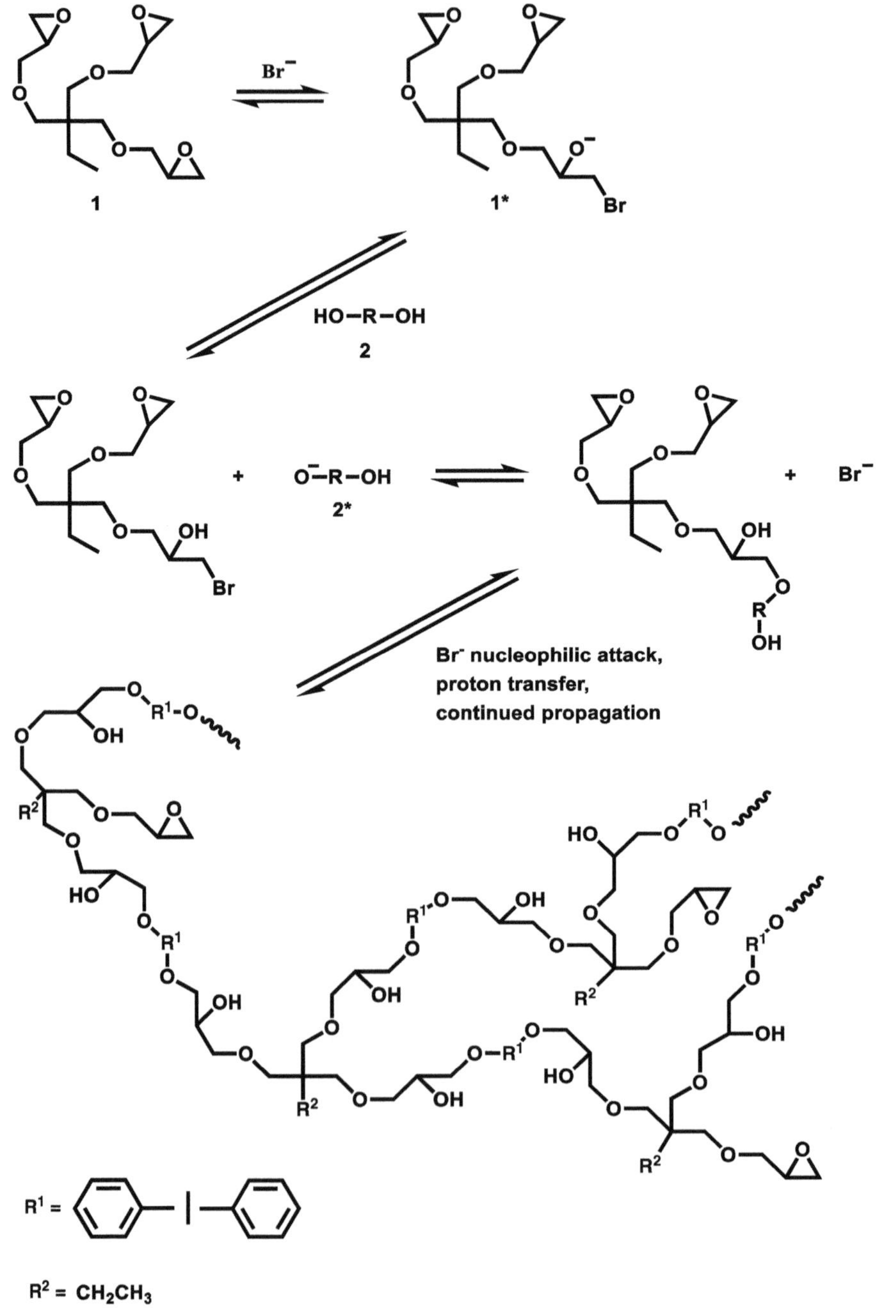

Figure 3.5 Synthetic route of the synthesis of aromatic epoxide-based HERs. *Source:* [24] / with permission of Springer Nature.

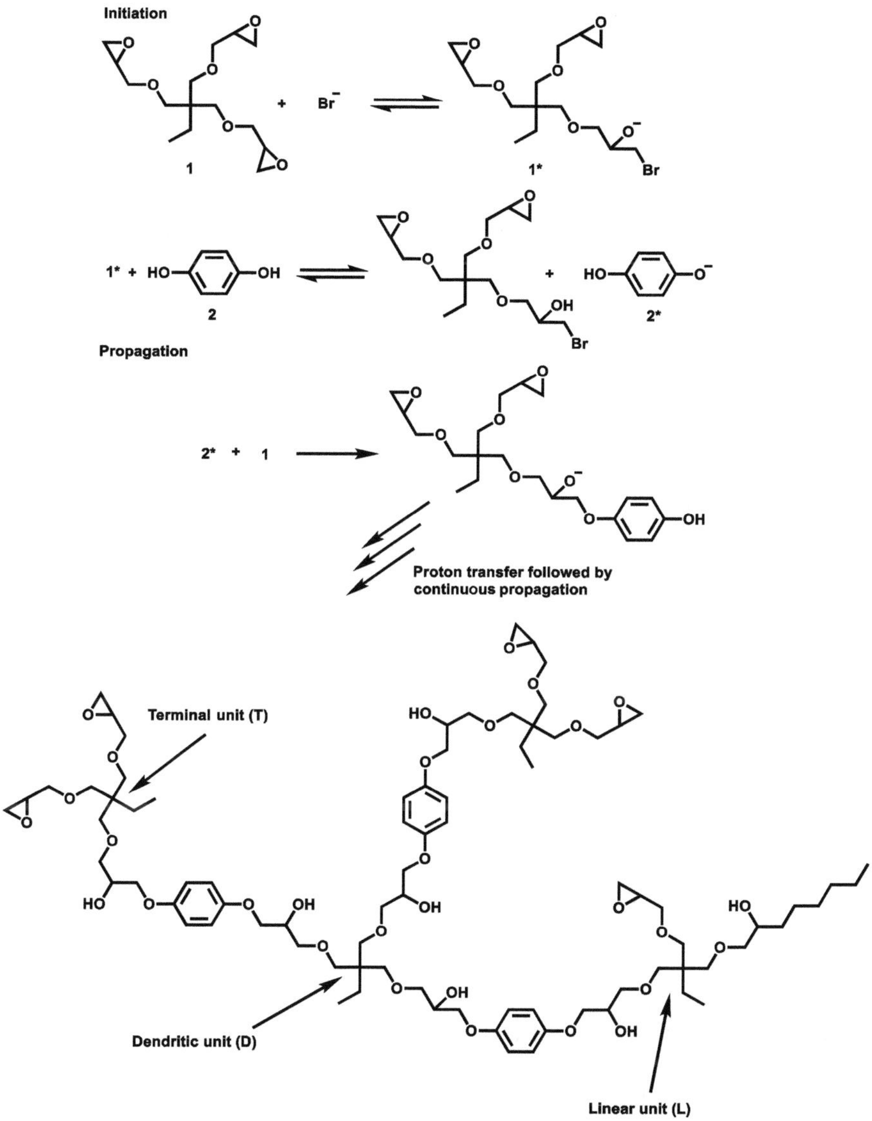

Figure 3.6 Synthetic route of hyperbranched polyether epoxy by bulk proton transfer polymerization. *Source:* [25] / with permission of Springer Nature.

where high chemical stability is demanded. Due to the presence of a large number of active terminal functional groups in its molecular structure, these functional groups provide abundant sites for the subsequent chemical reaction, resulting in a product with high chemical reaction activity.

However, the one-step proton transfer polymerization method for preparing HERs also presents certain limitations. Regarding raw materials, this approach demands both high purity and high reactivity. The presence of impurities in the feedstocks can reduce the reaction rate, promote side reactions, and adversely

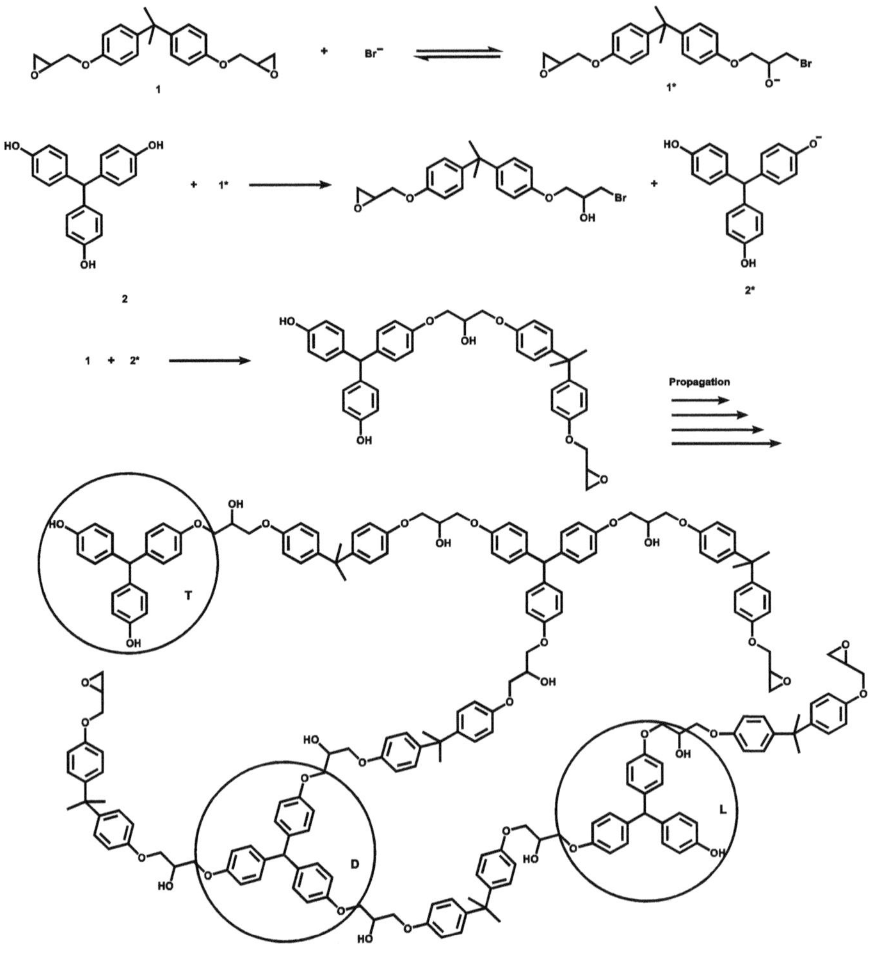

3

Figure 3.7 Synthetic route of hyperbranched polyether epoxy from DGEBA and THTPM. *Source:* [26] / with permission of John Wiley & Sons.

affect the structure and properties of the resulting polymer. Additionally, the reactivity of some monomers may be constrained by their molecular structures and physical characteristics, making it challenging to achieve the desired reaction selectivity and conversion efficiency. Furthermore, the ring-opening polymerization of epoxy groups may proceed too vigorously, leading to excessive chain growth and cross-linking, which can easily cause gelation and pose significant challenges for reaction control.

3.2.2 Step-growth Polycondensation

Step-growth polycondensation is one of the most common and versatile methods for synthesizing hyperbranched polymers. This process relies on repetitive reactions between multifunctional monomers, typically of the AB_n type (where $n \geq 2$), in which A and B represent complementary functional groups capable of reacting with each other [13, 27, 28]. Taking an AB_2 monomer as an example—which contains one A functional group and two B functional groups—the polymerization process can be described in the following steps:

1) Initiation: The process begins when two AB_2 molecules react. The A group of one monomer reacts with the B group of another, forming a dimer and a by-product (e.g., water, alcohol). Crucially, the dimer still retains one A group and multiple (three) B groups, meaning it remains highly functional and can continue to react.
2) Propagation: This reaction continues in a statistical manner. The remaining A and B groups on the growing oligomers and polymers react with each other. Because each monomer has a higher functionality ($f > 2$), branching is introduced at every step where an AB_2 monomer is incorporated.
3) Termination: The polymerization continues until the functional groups are exhausted or the viscosity becomes too high. Unlike chain-growth polymerization, there is no distinct initiation or termination step; molecules of all sizes are present throughout the reaction.

Notably, the generation of small by-products (e.g., water molecules) is an important feature of the step-growth polycondensation. The presence of these small-molecule by-products affects the equilibrium of the reaction and the properties of the products, so it is usually necessary to take some measures to remove the small-molecule by-products during the reaction process, in order to promote the reaction in the direction of generating hyperbranched polymers.

Considering that designing and synthesizing a pure and stable AB_n monomer is challenging and costly for HERs, the one-pot polymerization of A_n and B_m monomers via step-growth polycondensation emerges as a highly advantageous and commercially viable alternative route. This approach utilizes the step-growth copolymerization of two different monomers to directly form a hyperbranched structure with inherent epoxy functionality in a single pot [18, 19, 29, 30]. It is an efficient route that uses commercially available starting materials. For instance, in a typical $A_2 + B_3$ system, the A_2 monomer usually features epoxy groups as the functional groups, with DGEBA—a widely used industrial epoxy resin—serving as the most prevalent and commercially accessible example [31, 32]. Another trifunctional molecule containing three identical nucleophilic functional groups serves as the B_3 monomer, which can readily open the epoxy ring of the A monomer. Common B_3 examples include triamines, triphenols, tricarboxylic acids, and others [29, 33].

The polymerization mechanism can be illustrated using an $A_2 + B_3$ system as an example. A B_3 monomer can react with one epoxy group from an A_2 molecule, creating an adduct with the remaining unreacted A and B functional groups. Its functionality is greater than 2 (behaving as a "pseudo-AB_2" monomer), immediately in situ generating a branch point. Subsequently, this adduct can now react with other A_2 or B_3 monomers, or with other growing oligomers. The remaining functional groups continue to react via the same mechanism. The high functionality of the B_3 monomer ensures that branch points are incorporated during polymerization, leading to the formation of a randomly branched, hyperbranched macromolecular architecture. Notably, the A_2 monomers become permanent parts of the polymer backbone, with each unit contributing one unreacted epoxy group to the final structure. Therefore, as the hyperbranched polymer grows, epoxy groups are automatically and directly incorporated into its structure without the need for a separate functionalization step. With this strategy, the properties of the obtained HERs (e.g., MW, DB, epoxy value) can be tuned by altering the A_2-to-B_3 ratio and the choice of specific monomers.

Niu et al. [18] synthesized a novel silicone-containing hyperbranched epoxy resin (SHBEp) via a one-pot $A_2 + B_3$ polycondensation reaction between (3-glycidyloxypropyl)trimethoxysilane (A-187) and excess neopentyl glycol (NPG, Figure 3.8). It is worth noting that the reaction was carried out under solvent- and

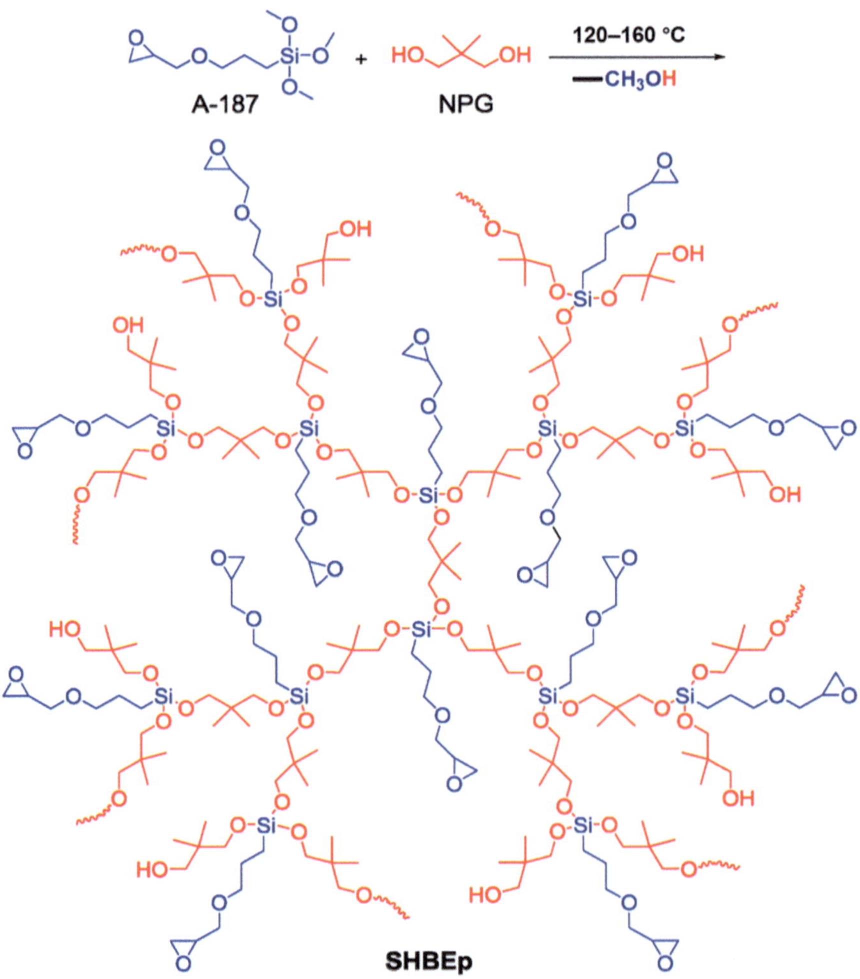

Figure 3.8 Synthetic route of SHBEp. *Source:* [18] / with permission of The Royal Society of Chemistry.

catalyst-free conditions without any catalyst; otherwise, the ring-opening reaction of the epoxide group will be triggered by —OH at high temperature, which is hard to control and easily leads to gelation. The resulting SHBEp can be scaled up easily to produce the commercial product at a low cost. Moreover, the as-prepared SHBEp could be used to toughen bismaleimide (BMI) resins, and it was found that the addition of 8 wt% SHBEp can make the thermoset exhibit significant improvement in the impact and flexural strengths compared to that of the neat BMI thermoset. This methodology furnishes great universality as a platform for the preparation of HERs.

Qiao et al. [34] reported that the preparation of a hyperbranched polysiloxane (PBPS) with Si–O–B backbone and epoxy ends is synthesized using a silicone via $A_2 + B_3$ polycondensation using phenylboronic acid as an A_2 monomer and 3-glycidoxypropyltrimethoxysilane (GPTMS) as a B_3 monomer (Figure 3.9). During the polymerization, the condensation reaction involving the hydroxyl groups of GPTMS and the boronic acid moieties predominates, while the ring-opening reaction of the epoxy group is negligible. This is likely due to the relatively low reaction temperature (80 °C). Moreover, the rapid consumption of hydroxyl groups during the condensation with PBA may further diminish the possibility of reaction between hydroxyl and epoxy groups, demonstrating the successful synthesis of HERs with PBPS structure. Notably, when used to toughen conventional thermoset epoxy resins, the boron of PBPS can act as a Lewis acid catalyst for epoxy ring-opening polymerization and efficiently reduce the induction time of curing of the epoxy-amine system. The resultant high cross-link density, combined with the flexible segments of PBPS, creates a synergistic effect that delivers excellent toughening performance.

BPA-based epoxy prepolymers are commercially available and contain the requisite epoxy functionality, making them ideal monomers for constructing HERs. However, the predominant synthetic strategy avoids their direct use as primary reactants. Instead, a more versatile approach involves a reversion to the fundamental monomers: BPA and epichlorohydrin (ECH). Usually, a DGEBA is formed "in situ" from the condensation of BPA with ECH, serving as the A_2 monomer. This approach offers superior control of the polymer's DB and MW over the polymerization process of the one-pot, step-growth polycondensation reactions. In contrast, starting with a BPA-based epoxy prepolymers imposes structural limitations and complicates the regulation of three-dimensional network architectures.

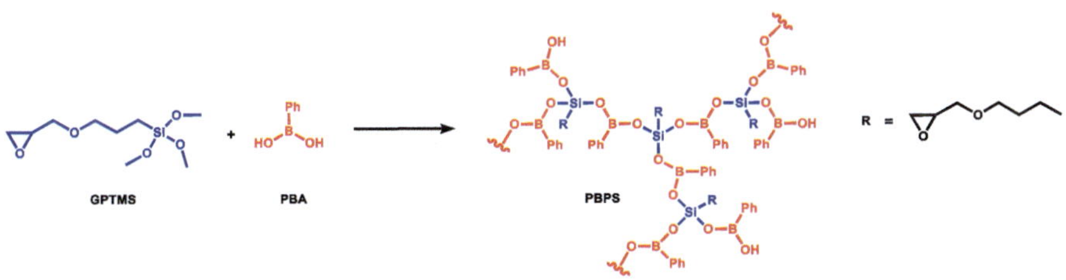

Figure 3.9 Synthetic route of PBPS. *Source:* [34] / with permission of Elsevier.

De and Karak [35] synthesized HERs (TAHE) via an $A_2 + B_3$ polycondensation approach using BPA, ECH, and triethanolamine (Figure 3.10). Due to the higher acidity of the phenolic protons, BPA exhibited greater reactivity toward ECH than triethanolamine. As a result, a DGEBA was first formed and served as the A_2 monomer, which subsequently reacted with triethanolamine (B_3) to yield the HERs. The DB, EEW, hydroxyl value, and viscosity of the HERs can be regulated by adjusting the wt% of the B_3 reactant. The TAHE20 with the lowest DB (DB = 0.79) was achieved at an optimal A_2:B_3 molar ratio of 3:1. This ratio promoted both a higher percentage of trisubstituted branching units and a greater substitution of chain ends to form terminal epoxides. In contrast, a lower ratio (2:1) yielded the fewest trisubstituted

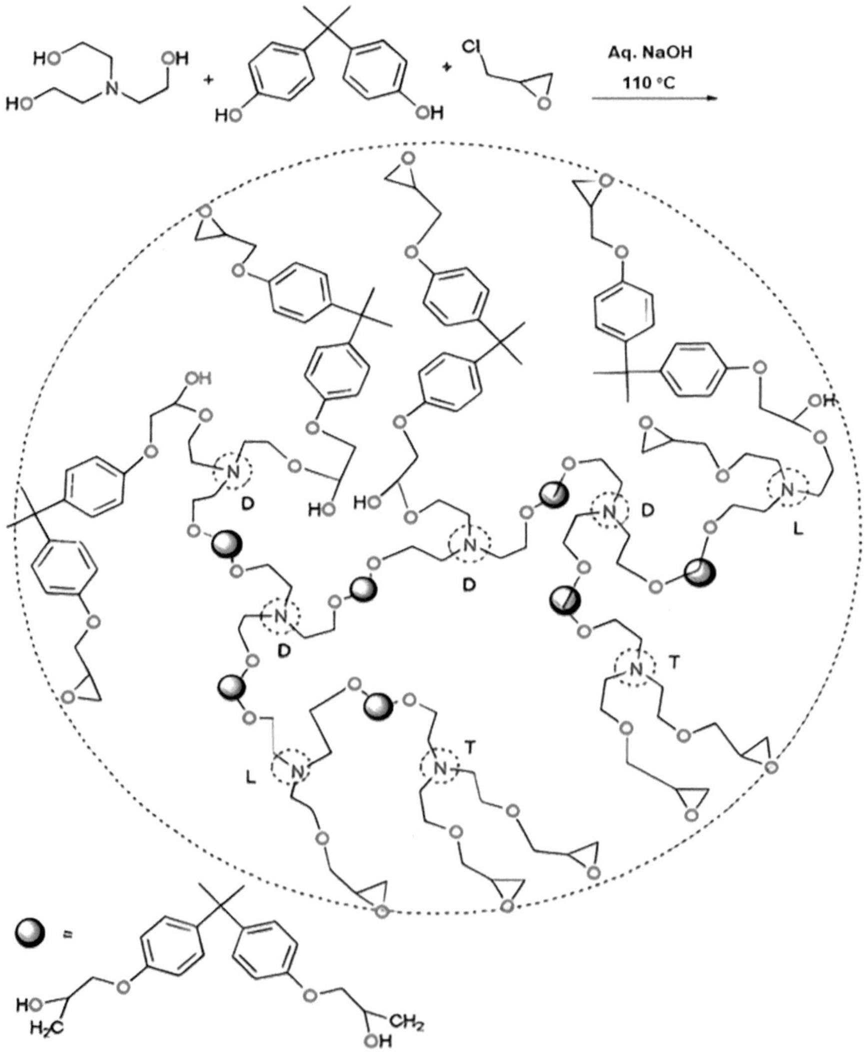

Figure 3.10 Synthetic route of TAHE. *Source:* [35] / The Royal Society of Chemistry.

branches, while a higher ratio (6:1) caused steric crowding around the B_3 monomer, which impeded the formation of branching units. Consequently, TAHE20 exhibits the lowest EEW, hydroxyl value, and viscosity among all the HERs synthesized at different A_2:B_3 molar ratios.

Boro and Karak [36] further synthesized a bio-based HER called tetraarylethylene (TAE) by the polycondensation of tannic acid, BPA, and ECH (Figure 3.11). Different wt% of tannic acid, which is the B_3 monomer, was used for the polymerization, and it was found that the use of more than 7.5 wt% tannic acid would cause gel formation. The prepared TAE showed a good solubility than that of DGEBA, due to the hyperbranched structure with a large number of polar groups in the former.

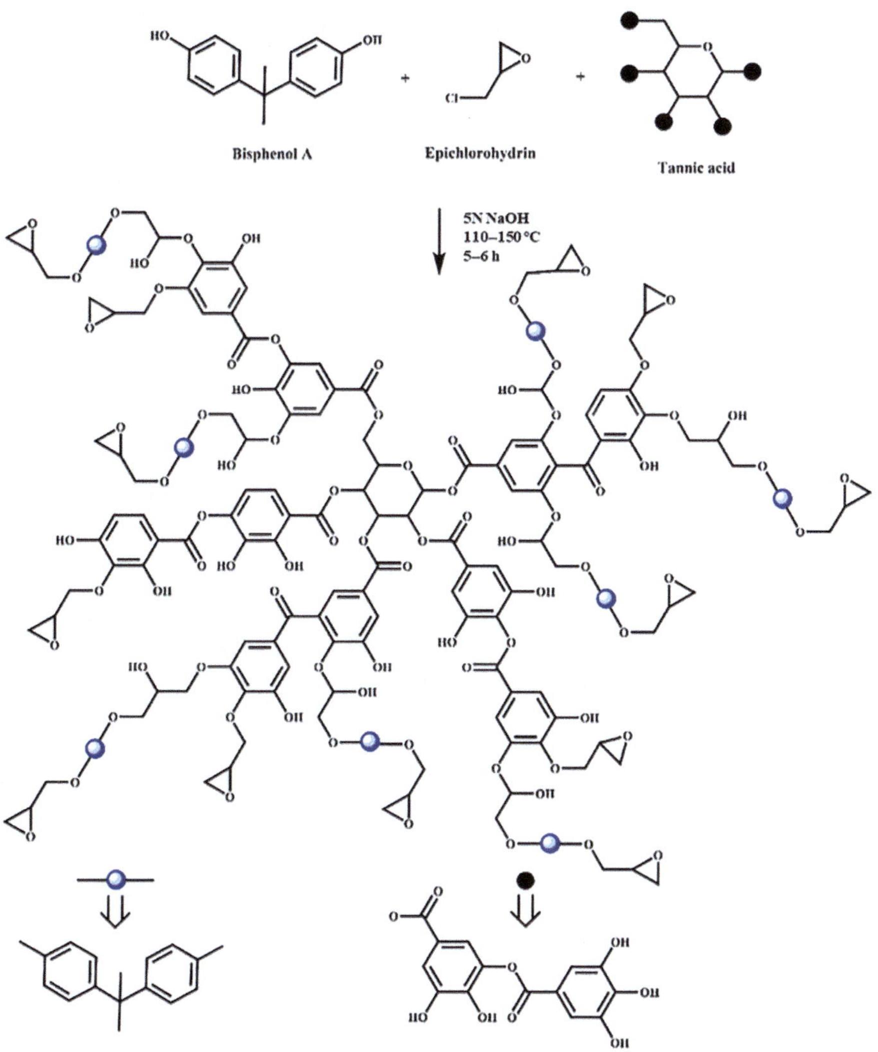

Figure 3.11 Synthetic route of TAE. *Source:* [36] / with permission of Elsevier.

TAE can be mixed with reduced graphene oxide (RGO) and then cured with a poly(amido-amine) hardener to prepare the epoxy resin composites. The tensile strength, scratch hardness, impact strength, initial degradation temperature, and gloss of the cured resins were all improved due to the presence of the hyperbranched structure.

Fei et al. [29] reported a similar approach for synthesizing biphenyl backboned hyperbranched epoxy resins (BPHBEs) through a one-pot $A_2 + B_3$ polycondensation of 2,2′,6,6′-tetramethyl-4,4′-biphenol (TBP), ECH, and triethanolamine (Figure 3.12). They prepared three BPHBEs with varying TEA contents, all of which exhibited high intrinsic viscosities—significantly greater than that of conventional epoxy resins like DGEBA. When incorporated into epoxy/anhydride and epoxy/jeffamine curing systems, these hyperbranched resins were found to improve the impact strength, tensile strength, and elongation at break of the cured EPs. This

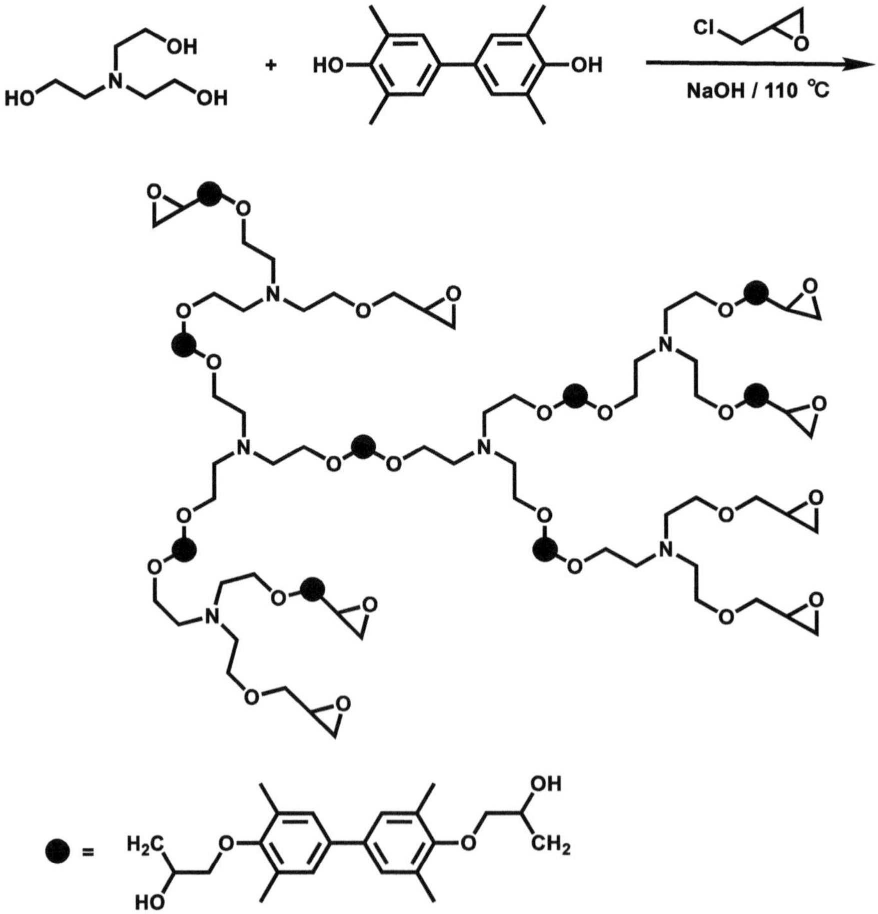

Figure 3.12 Synthetic route of BPHBE. *Source:* [29] / with permission of Springer Nature.

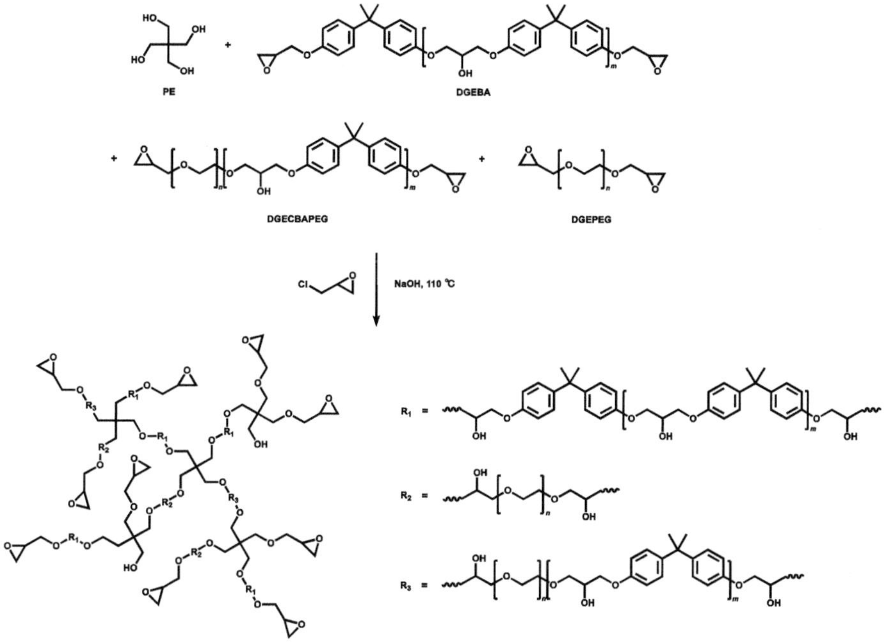

Figure 3.13 Synthetic route of HBE. *Source:* [30] / MDPI / CC BY 4.0.

work provides an effective approach for the synthesis of epoxy resin toughners and toughening EPs via a nonphase separation mechanism.

Boonlert-uthai et al. [30] reported the synthesis of HERs (HBE) via $A_2 + B_4$ polycondensation using BPA and polyethylene glycol (PEG) as reactants and pentaerythritol as branching point (Figure 3.13). The A_2 monomers—diglycidyl ether of polyethylene glycol (DGEPEG), DGEBA, and diglycidyl ether copolymer of bisphenol A and polyethylene glycol (DGECBAPEG)—were formed in situ through the polymerization of BPA, PEG, and ECH. The incorporation of PEG chains significantly increased the DB and EEW of the resulting HBEs. When cured by diethylenetriamine, HERs containing 5–10 wt% PEG exhibited decreases in the onset curing temperature, peak curing temperature, and glass transition temperature. In contrast, the sample with 15 wt% PEG showed an increase in these thermal properties, which was attributed to its lowest EEW. The curing kinetics were evaluated by fitting the experimental data of the curing behavior of all resins with the Šesták–Berggren equation. This report presents a strategy for developing hyperbranched BPA/PEG epoxy resins, which would be suitable for electronics applications.

The one-step, step-growth polycondensation method offers significant advantages for synthesizing HERs. It allows for precise molecular design by controlling monomer type and ratio, enabling the tailoring of material properties for specific applications. This structural control facilitates performance enhancement; for instance, incorporating biphenyl structures imparts excellent low dielectric constant, low water absorption, and flame retardancy, making these resins highly suitable for advanced fields like electronic packaging and aerospace.

Despite its benefits, this method has limitations. The reaction kinetics are inherently slow, often requiring long durations to achieve high conversion rates, which hampers industrial production efficiency. Furthermore, the process is susceptible to side reactions like degradation or cross-linking, potentially compromising product quality. Additional drawbacks include a relative scarcity of suitable monomer varieties and a lack of diversity in the fundamental reaction types available, which can restrict the structural diversity and property range of the resulting resins. Improvements could involve developing more efficient catalysts, optimizing reaction conditions, and exploring new monomer systems.

3.2.3 Step-growth Polyaddition

Step-growth polyaddition is also one of the most common and versatile methods for synthesizing HERs. Similar to the step-growth polycondensation, the typical AB_n or $A_n + B_m$ system is usually employed for the polymerization strategy [37–39]. Notably, it should be noticed that step-growth polyaddition offers a pathway for constructing hyperbranched architectures without the generation of small-molecule by-products, which is different from polycondensation. This method relies on the direct addition reactions between multifunctional monomers, such as the addition of amines to epoxide. The absence of condensation by-products simplifies the reaction process, as it eliminates the need for their continuous removal to drive the reaction equilibrium forward. Consequently, step-growth polyaddition often proceeds under milder conditions and with enhanced efficiency. While it similarly utilizes AB_n or $A_n + B_m$ system, the library of compatible functional groups is distinct, providing complementary opportunities for molecular design and expanding the scope of attainable polymer properties.

Srinivasa Rao and Samui [40] synthesized a series of photoactive liquid crystal HERs, which could be used for optical recording through the $A_2 + B_3$ polyaddition (Figure 3.14). Photoactive bis(benzylidene)alkanone diepoxides and trimesic acid were used for the A_2 and B_3 monomers, respectively. Tetra-*n*-butylammonium chloride (TBAC) was employed as a nucleophilic catalyst to initiate the addition reaction of carboxyl group to the epoxide. The obtained hyperbranched polymers exhibited a DB in the range of 0.46–0.49. Due to their compact structure, all hyperbranched polymers showed higher glass transition temperature (T_g) values compared to their linear analogues. This study provides a basis for the structural design of HERs and their applications in the field of photoactive liquid crystals.

Theis et al. [41] reported the synthesis of hyperbranched epoxy–amine adducts via a one-pot microwave-assisted AB_2 polyaddition (Figure 3.15). The AB_2 monomer, 4-(oxiran-2-ylmethoxy)benzonitrile, was generated in situ through Pd/C-catalyzed hydrogenation of a precursor compound, followed by epoxy–amine polyaddition. Mass spectrometry identified oligomeric homologues (up to six repeating units) and large aggregates, the latter resulting from Cu^{2+} complexation by the poly(amino alcohol) segments.

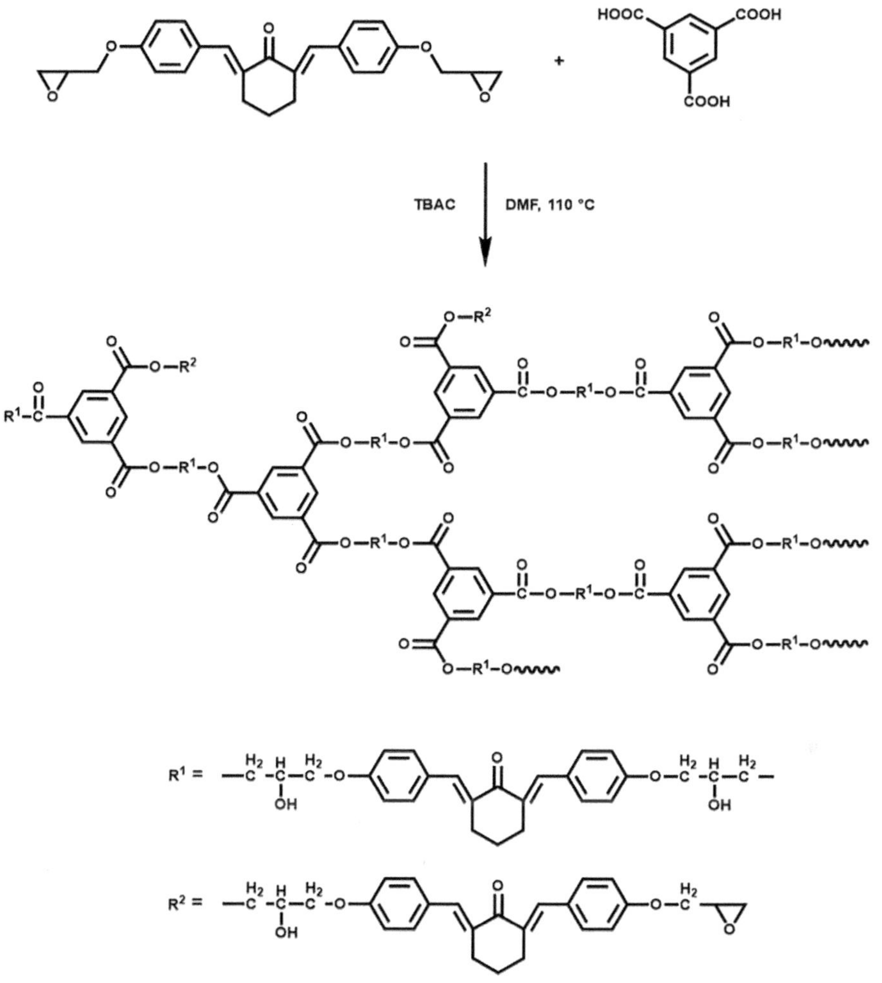

Figure 3.14 (a) Synthetic route of linear and hyperbranched polymer. (b) Structure of hyperbranched polymer. *Source:* [40] / with permission of John Wiley & Sons.

This report illustrates the typical approach for synthesizing HERs through step-growth polyaddition using the AB_2-type monomer. However, similar to step-growth polycondensation, the synthesis of such AB_n monomers is challenging due to reactivity conflicts between the different functional groups, leading to a limited diversity of available monomers. In comparison, the method utilizing readily available A_n and B_m monomers proves to be a superior alternative.

Zhong et al. [37] synthesized a vanillin-based diepoxide, which can react with trimethylolpropane (TMP) to give vanillin-based HERs (FEHBP) undergoing an $A_2 + B_3$-type polyaddition (Figure 3.16). The FEHBP was further mixed with the vanillin-based diepoxide to be cured by citric acid (CA), resulting in a hyperbranched epoxy vitrimer. The hyperbranched topological structure of FEHBP

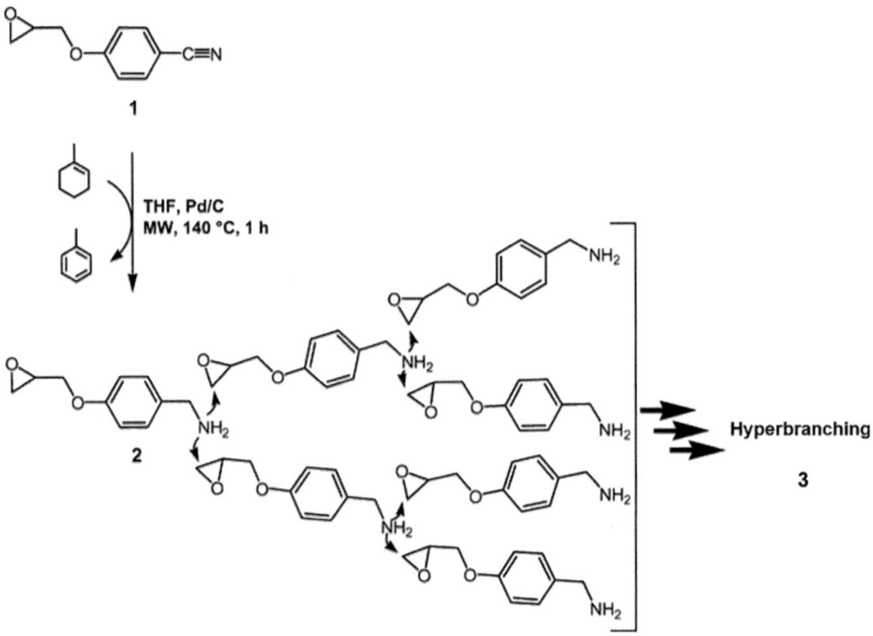

Figure 3.15 Synthetic route of epoxide–amine adducts by MW-assisted transfer hydrogenation. *Source:* [41] / with permission of John Wiley & Sons.

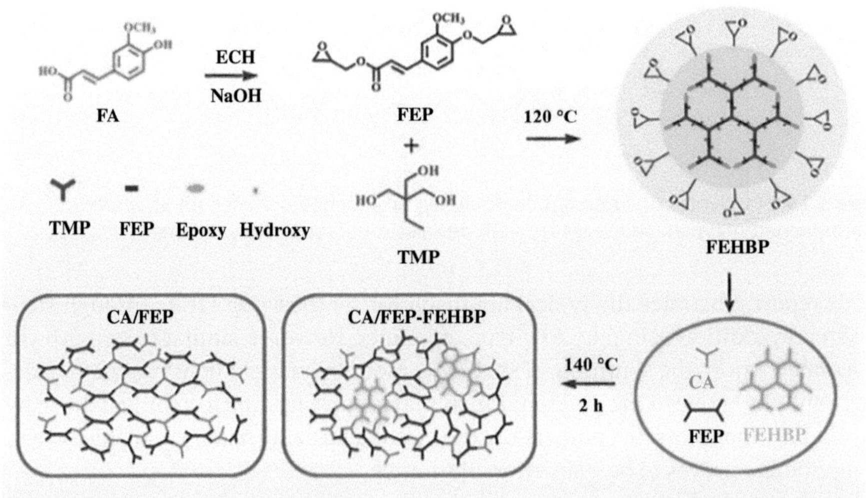

Figure 3.16 Synthetic route of FEHBP and the relevant hyperbranched epoxy vitrimers. *Source:* [37] / with permission of American Chemical Society.

increased the free volume and cross-link density of epoxy vitrimers, resulting in simultaneous enhancement of strength, modulus, and toughness of the vitrimers. Moreover, the hydroxyls of FEHBP catalyzed the dynamic transesterification, leading to excellent weldability, reparability, malleability, and reprocessability of the resultant epoxy vitrimers. Furthermore, the resulting epoxy vitrimers could be degraded into recycled ferulic acid and recycled CA-derived polyester, which can be used to prepare high-performance epoxy vitrimers again. This study developed an efficiency approach to prepare closed-loop recyclable HERs through $A_2 + B_3$-type polyaddition.

Similarly, vanillin-based HERs (VEHBP) were developed with the same polymerization strategy by Zhang's group (Figure 3.17) [38]. Vanillin-based diepoxide (VEP) was mixed with TMP to give hyperbranched VEHBP via $A_2 + B_3$-type polyaddition. VEHBP was then incorporated into the commercial DGEBA to be cured by 4-aminophenyl disulfide (4-AFD), resulting in a recyclable hyperbranched epoxy vitrimer. The introduction of hyperbranched topology significantly improved the tensile strength, modulus, toughness, and creep resistance of the resulting epoxy vitrimers.

The A_2 and B_3 monomers are readily synthesized or sourced, providing a direct pathway to incorporate desired functionalities into these monomers prior to polymerization. This inherent flexibility allows for the strategic design and selection of functionalized A_2 or B_3 monomers. For instance, phosphorus-based monomers have been widely developed for the $A_2 + B_3$ polymerization to produce flame-retardant epoxy resins (FREPs)—addressing the high flammability inherent in conventional epoxy.

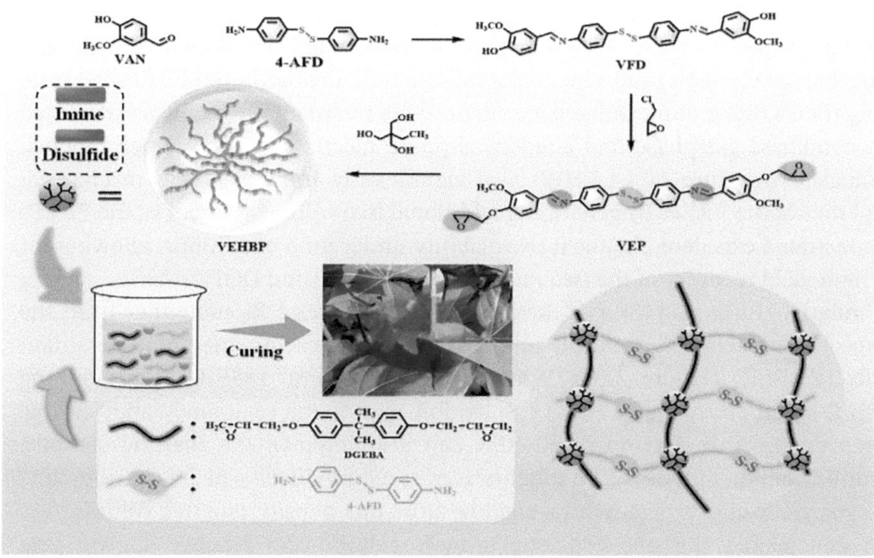

Figure 3.17 Synthetic routes of VEHBP and the relevant hyperbranched epoxy vitrimers. *Source:* [38] / with permission of American Chemical Society.

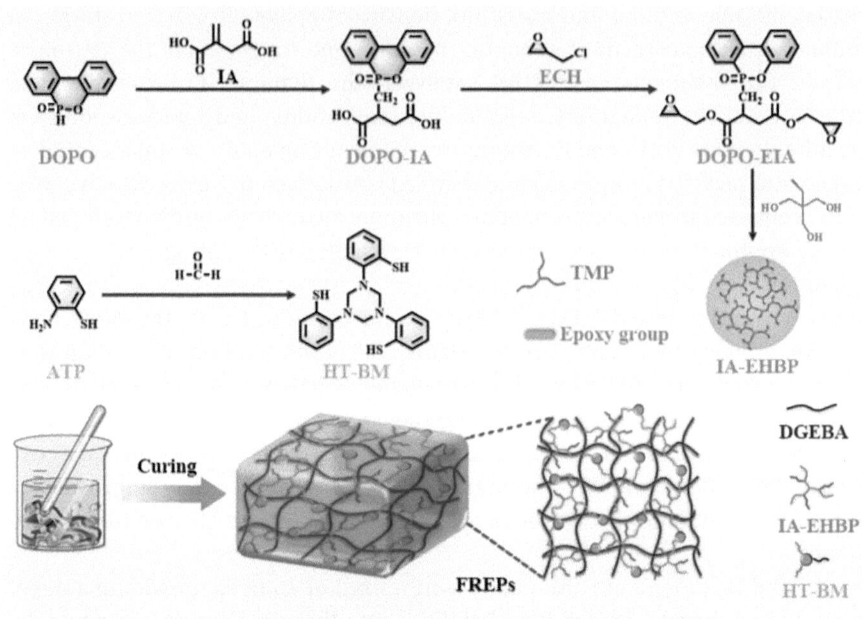

Figure 3.18 Synthetic route of IA-EHBP and the relevant flame-retardant epoxy resins. *Source:* [39] / with permission of American Chemical Society.

9,10-Dihydro-9-oxa-10-phosphaphenanthrene-10-oxide (DOPO) and its derivatives are widely employed in the synthesis of FREPs. Recently, Wei et al. [39] successfully developed the recyclable FREPs using an itaconic acid-derived hyperbranched epoxy resin (IA-EHBP) and (1,3,5-hexahydro-*s*-triazine-1,3,5-triyl) benzyl mercaptan (HT-BM) as curing agents (Figure 3.18). Flame-retardant IA-EHBP was first synthesized via an $A_2 + B_3$ polyaddition between a DOBO-based itaconic acid diepoxide (DOBO-EIA) and TMP. After being cured, the thermoset FREPs incorporating the DOBO group exhibited excellent flame retardancy, which was attributed to a combined gas-phase and condensed-phase mechanism. The hyperbranched topological structure of IA-EHBP also significantly influenced the mechanical properties of the FREPs by generating additional free volume. Moreover, the FREPs demonstrated excellent chemical recyclability under mild conditions, allowing for the high-yield recovery of the two monomer types (ATP and DOPO-IA).

Similarly, Hu et al. [42] also developed a DOPO-based B_3 monomer from the protocatechualdehyde, guaiacol, and DOPO for preparing the flame-retardant HERs (EHBFR) (Figure 3.19). When incorporated into DGEBA, the resulting EHBFR-based epoxy resins (EPs) can exhibit high flame retardancy after curing. Moreover, the introduction of EHBFR can also improve the thermal stability, flexural strength, modulus, and toughness of cured products, which is attributed to the synergistic effects of structural rigidity, cross-link density, and free volume.

In conclusion, the one-step, step-growth polyaddition strategy shows great promise for synthesizing HERs, enabling the precise design of highly branched structures with specific functional groups to meet the requirements of demanding performance and functionality. For example, in aerospace applications, resins

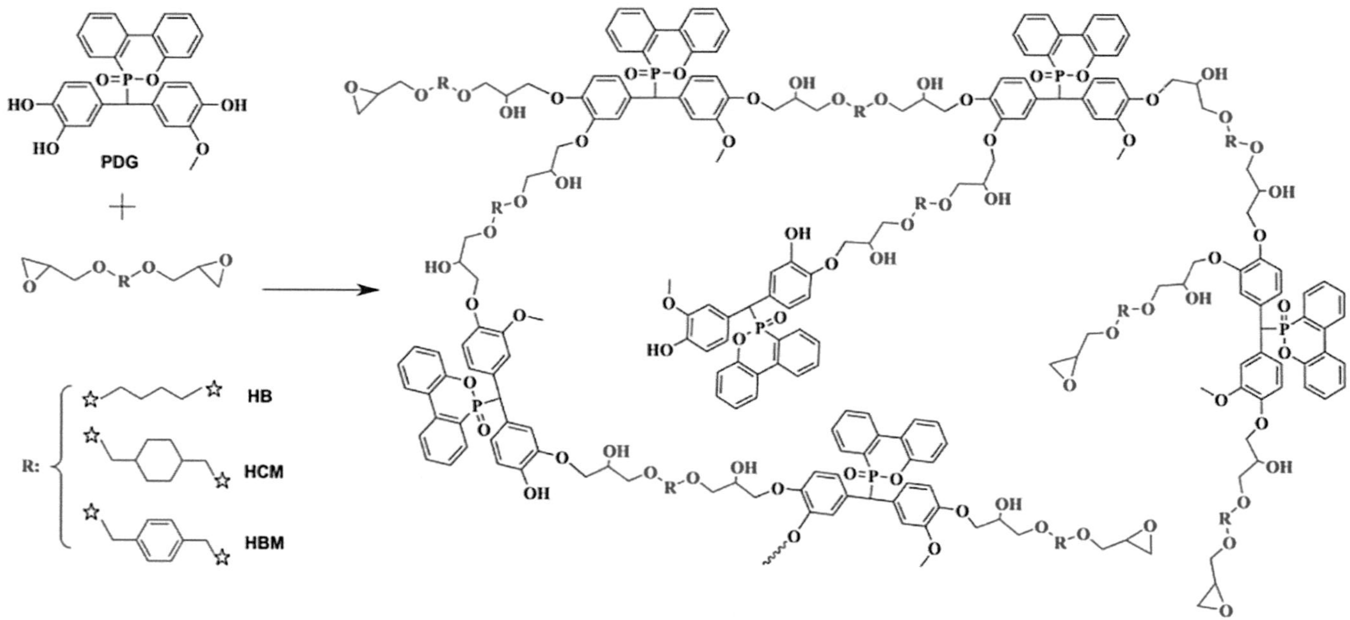

Figure 3.19 Synthetic route of EHBFR and the relevant flame-retardant epoxy resins. *Source:* [42] / MDPI / CC BY 4.0.

produced via this method enhance the mechanical properties and heat resistance of composites when used with high-performance fibers, making them suitable for extreme environments. In electronics encapsulation, it allows for the tailoring of materials with low shrinkage, high insulation, and superior thermal dissipation, supporting the trend toward miniaturized, high-reliability devices.

However, practical challenges remain. The variety of monomers suitable for synthesizing these resins is still limited, and those with specialized structures or functional groups often present significant synthetic difficulties and high costs, restricting the diversity of resin properties. Monomer purity and reactivity also critically influence the reaction—impurities or varying activity levels can lead to side reactions or uneven molecular structures. Furthermore, the step-growth polyaddition typically requires long reaction times to achieve high conversion, limiting production efficiency. Reaction conditions such as temperature, pressure, and catalyst concentration must be tightly controlled, as even minor deviations can considerably impact product quality and performance consistency.

3.3 Post-polymerization of Hyperbranched Polymers

PPM of hyperbranched polymers is a widely adopted strategy for the preparation of HERs. This method, which is a two-step approach, involves first constructing a well-defined hyperbranched core, followed by the subsequent introduction of epoxy functional groups onto its terminal sites. A key advantage of this method is its synthetic flexibility. The hyperbranched core can be independently designed and optimized for its structure and properties (such as DB and MW), while the epoxidation step can be tailored to achieve a specific EEW. Compared to the one-step polymerization method, this strategy significantly broadens the range of suitable monomers for the hyperbranched core, as their selection is no longer constrained by the need to contain epoxy groups themselves. This decoupling of core formation and functionalization makes it a versatile route for developing HERs with tailored functionalities. For example, the preparation of bio-based HERs is facilitated, as bio-derived polyol or polyester cores (e.g., from CA, glycerol, or vegetable oils) can be easily functionalized. A variety of grafting reactions, such as etherification, esterification, Michael addition, and click reactions, have been successfully employed for this epoxy functionalization.

3.3.1 Etherification

The etherification reaction is a fundamental and widely utilized reaction in organic chemistry. This transformation usually involves the formation of an ether linkage (R—O—R′) through the reaction of an alcohol with an alkyl halide, sulfate, or another alcohol under dehydrative conditions [43]. For example, the classical Williamson ether synthesis is celebrated for its efficiency in constructing C—O—C bonds and has been a pivotal structural motif in a vast array of pharmaceuticals, polymers, and functional materials.

For the PPM of the hyperbranched cores, the etherification reaction serves as a powerful tool to introduce epoxy groups [44]. Hyperbranched polyether polyols, polyester polyols, or similar structures can be readily synthesized and subsequently functionalized at their abundant hydroxyl chain ends through etherification. ECH, as a kind of alkyl halide, represents one of the most useful building blocks for etherification reaction. This pivotal step efficiently introduces glycidyl ether groups, thereby converting the core into the HERs.

This section focuses on the synthesis of HERs through a PPM strategy centered on the etherification reaction. It details the preparation of the hyperbranched polyol precursor and investigates the optimization of the subsequent etherification process to achieve a high degree of epoxidation. The structure of the resulting polymers is thoroughly characterized to elucidate the relationship between their branched architecture, epoxy content, and the properties of the cured resins.

Huang et al. [45] successfully synthesized epoxy-functionalized hyperbranched polyphenylene oxide (eHBPPO) via the etherification (Figure 3.20). The hyperbranched polyphenylene oxide (HBPPO), which was synthesized from the copper-catalyzed oxidative coupling of 2,6-dimethylphenol (DMP) and 1,3,5-tris(4-hydroxyphenyl)benzene (THPB), served as the hyperbranched core. Subsequent etherification with ECH in alkaline conditions yielded eHBPPO with controlled branching architecture. The resulting eHBPPO was then introduced to the cyanate ester (CE) resin to enhance its thermal stability, mechanical properties, and dielectric performance.

Remarkably, incorporating 10 wt% eHBPPO markedly lowered the curing peak temperature of CE resin from 267 to 204 °C, driven by multiple synergistic effects: hydrogen bonding between eHBPPO hydroxyl groups and CEs facilitated cyclotrimerization, copolymerization introduced oxazolidinone linkages, and the hyperbranched topology promoted cross-linking density. The resulting CE/eHBPPO composite demonstrated significantly enhanced dielectric performance, with 18.6% and 34.5% improvements relative to neat CE. Dynamic mechanical analysis (DMA) indicated a 28 °C rise in glass transition temperature and a 15% increase in storage modulus, aligning with the restricted molecular mobility predicted by the Fox equation for polymer blends. This work establishes a material design for the hyperbranched CE composites combining low dielectric loss, high thermal stability,

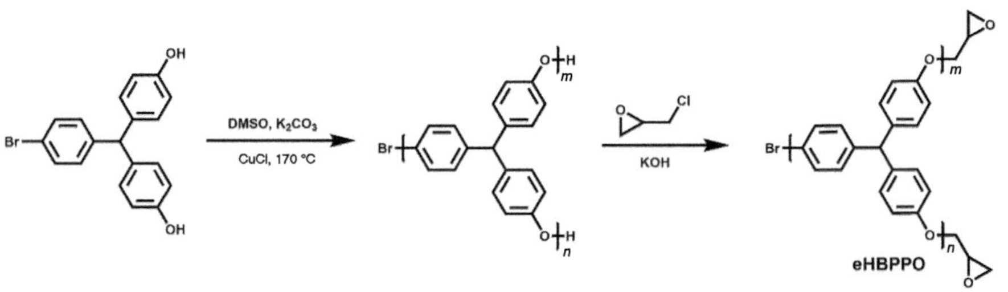

Figure 3.20 Synthetic route of eHBPPO. *Source:* [45] / with permission of John Wiley & Sons.

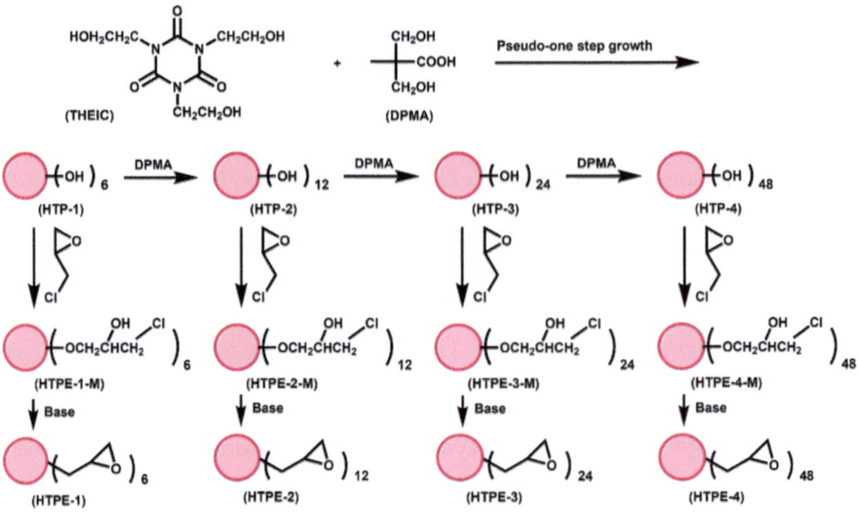

Figure 3.21 Synthetic route of HTPE. *Source:* [46] / with permission of John Wiley & Sons.

and dimensional stability, showing great potential in fabricating high-performance adhesives, advanced composites, etc.

Similarly, Chen et al. [46] reported the synthesis of low-viscosity HER-containing nitrogen heterocycle (HTPE) via an etherification route. Dimethylol propionic acid (DMPA) and tris(2-hydrooxyethyl)-isocyanurate were employed as AB_2 monomer and B_3 monomer, respectively, to construct a hyperbranched core (HTP). This core was subsequently etherified with ECH to yield HTPE, as illustrated in Figure 3.21. To obtain a series of HTPE derivatives (HTPE-2, HTPE-3, and HTPE-4) with distinct hyperbranched architectures and terminal group contents, different core molecules (HTP-2, HTP-3, and HTP-4) were first synthesized via a similar pseudo-one-step growth procedure starting from HTP-1.

The viscosity of HTPE initially decreases and then increases with the increase in their MW, yet remains significantly lower than that of conventional DGEBA systems. These low-viscosity HTPEs could be used to simultaneously toughen and reinforce the common DGEBA, and the mechanical performance of HTPE/DGEBA composites increases first and then decreases with the increase in HTPE content and MW. Compared to those of pure DGEBA, the impact strength, fracture toughness, tensile, and flexural strength of HTPE-2/DGEBA composite can be enhanced by about 192%, 39%, 18%, and 23%, respectively. Mechanistic studies reveal that the improved performance arises from an in situ reinforcing and toughening mechanism, attributed to the intrinsic intramolecular cavities of HTPE and the strong interaction between HTPE and DGEBA.

Tao et al. [47] developed a novel hyperbranched epoxy resin with polyether skeleton (HBPEER), which can be used for the modifier of epoxy/anhydride curing system (Figure 3.22). The hyperbranched polyether skeleton (HBPE) was synthesized through an $A_2 + B'B_2$ synthetic strategy using 1,6-dibromohexane and trifunctional resveratrol as monomers. The 1,6-dibromohexane underwent nucleophilic

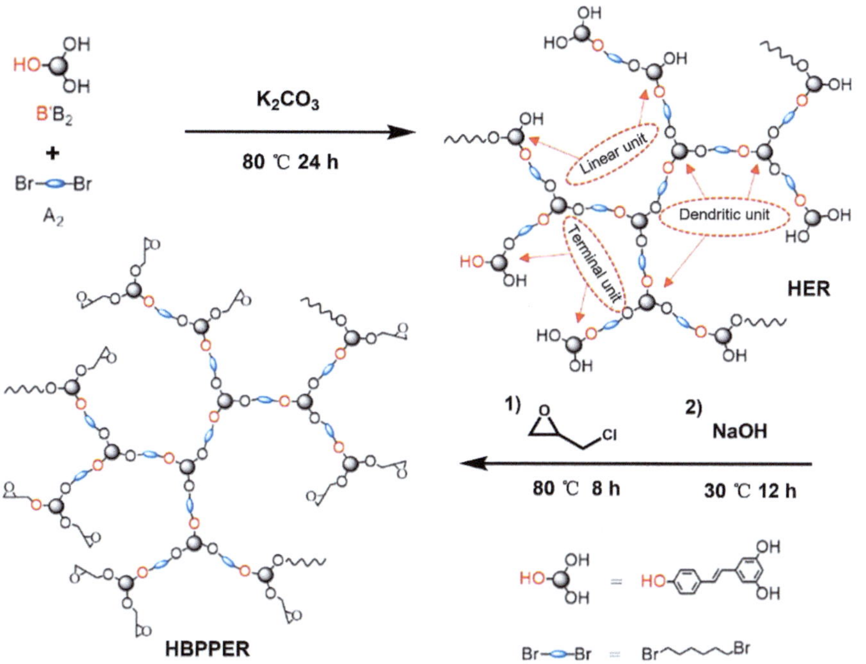

Figure 3.22 Synthetic route of HBPEER. *Source:* [47] / with permission of Elsevier.

substitution with resveratrol to yield HBPE with terminal phenolic hydroxyl groups, which was then subjected to epoxidation through etherification with excess ECH to give the HBPEER. The introduction of HBPEER into DGEBA resulted in the enhancement of viscosity of DGEBA/HBPEER blend, since the HBPEER with a relatively high MW significantly raised the entanglement among the molecular chains. Also, the DGEBA/HBPEER-8 blend showed the obvious shear thinning behavior.

The effect of introduction of HBPEER into epoxy/anhydride system on the properties of the cured thermosets was also investigated. It was found that compared to the neat DGEBA thermoset, the 6 wt% HBPEER-modified epoxy thermoset showed an increase in impact strength, elongation at break, critical stress intensity factor (KIC), critical crack propagation energy release rate (GIC), flexural strength, tensile strength, storage modulus, glass transition temperature (T_g), and initial thermal decomposition temperature ($T_{d5\%}$) by 69.6%, 87.0%, 32.0%, 68.2%, 29.3%, 37.0%, 8.1%, 3.5%, and 1.6%, respectively. This performance improvement was caused by the combined effects of cross-link density improvement, efficient intramolecular cavity, rigid stilbene units, and flexible aliphatic backbone of HBPEER.

Building on these foundational studies, researchers expanded into functionalization and sustainability. Yang et al. [48] developed an epoxy-terminated hyperbranched polyester (HPEEX) through a three-step synthetic strategy. A hydroxyl-terminated hyperbranched polyester (HPE) was first synthesized based on TMP (B_3 core) and AB_2 monomer, followed by the final epoxidation through controlled etherification with ECH (Figure 3.23). HPEEX exhibited high thermal stability, which can be attributed to its compact architecture and enhanced

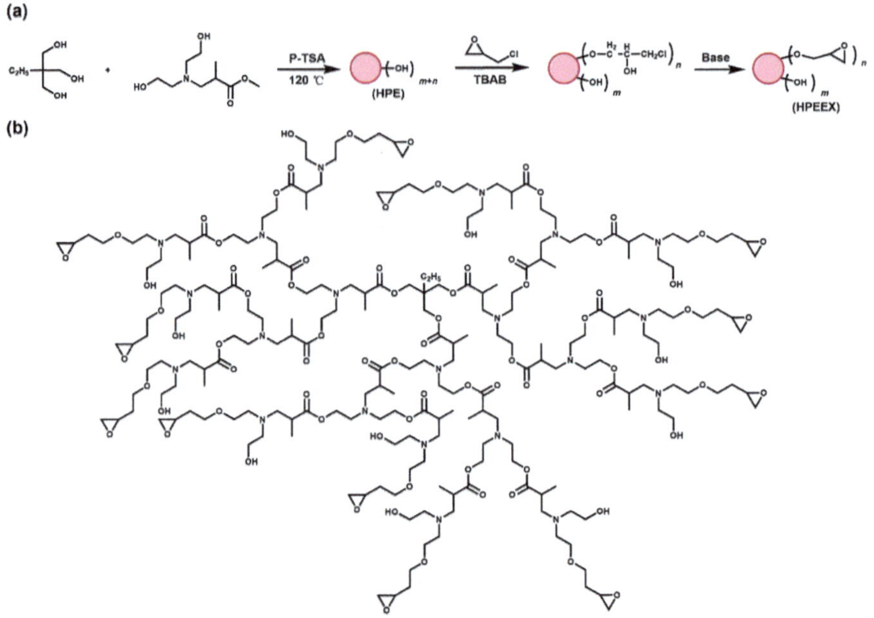

Figure 3.23 Synthetic route of HPEEX (a) and its ideal schematic representation (b). *Source:* [48] / with permission of Springer Nature.

intermolecular interactions resulting from the highly branched and globular structure.

HPEEX can be employed as a cross-linking agent to improve the properties of the waterborne epoxy and their resulting films. The HPEEX-modified waterborne epoxy dispersion (EP-H) demonstrated optimized colloidal characteristics, with a viscosity of 120 ± 5 mPa·s and surface tension of 48.2 mN/m—representing a 58% viscosity reduction and 15% surface energy increase compared to conventional formulations. The cured films demonstrated enhanced performance metrics: pendulum hardness reached 168 seconds, water contact angle 92°, and cross-link density 3.2×10^3 mol/m^3. These enhancements stem from the architectural advantages of HPEEX: (i) a hyperbranched topology that enables efficient space filling; (ii) high epoxy functionality; (iii) formation of a hydrogen-bond network; and (iv) improved cross-linking efficiency.

Bio-based HERs can also be prepared through the etherification strategy. For example, De et al. [49] reported a synthesis route for biodegradable HERs using castor oil as a sustainable feedstock (Figure 3.24). The three-stage synthesis process is initiated with glycerol alcoholysis of castor oil under alkaline conditions, producing monoglyceride intermediates. Subsequent polycondensation with 2,2-bis(hydroxymethyl)propionic acid generated the hyperbranched polyester polyol (HBPP) with controlled branching architecture. Final epoxidation through reaction with BPA and ECH in basic media yielded the target bio-based HER (HBPE).

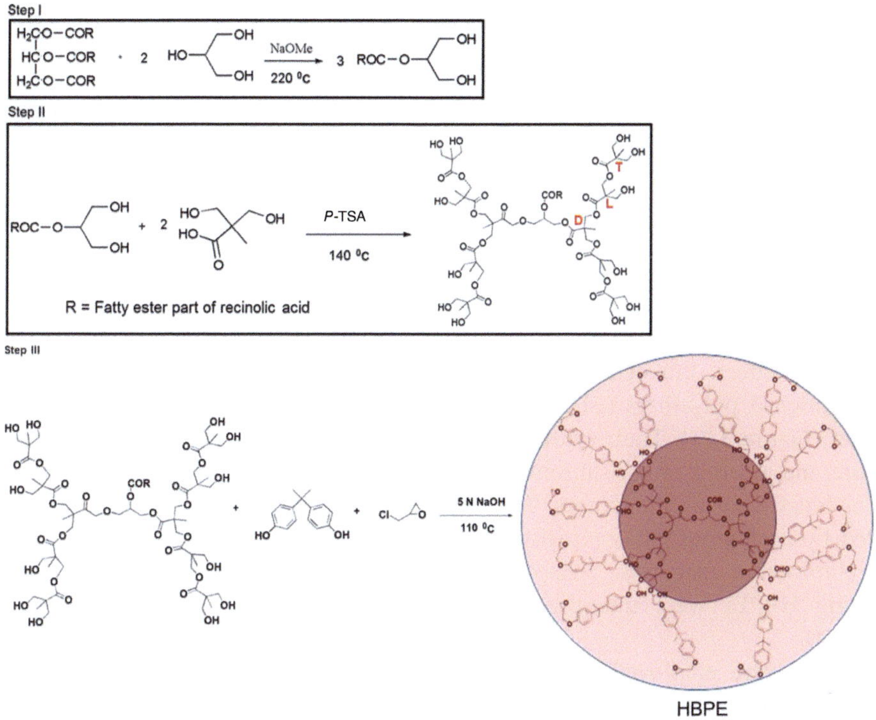

Figure 3.24 Synthetic route of HBPE. *Source:* [49] / with permission of American Chemical Society.

The hyperbranched polymer could be incorporated into the commercial epoxy resins DGEBA to overcome the existing genuine brittleness of epoxy thermoset. When 10 wt% HBPP was mixed with DGEBA, the toughness, scratch hardness, and impact resistance of the cured resins were slightly improved compared to that of neat DGEBA. Notably, the HBPE can also be directly cured by poly(amido-amine) hardener to give the hyperbranched epoxy thermoset. The resulting thermoset exhibited high toughness, arising from the complementary roles of its components. The presence of aromatic rigid moiety of BPA resulted in high tensile strength, whereas the flexible aliphatic polyester units and ether linkage improved elasticity by a plasticization effect. Moreover, the hyperbranched topology of the epoxy increased the free volume within the three-dimensional network, which significantly increased the toughness.

The bio-based HERs also exhibited excellent degradability. Biodegradation studies revealed that the hyperbranched epoxy films could be degraded regularly with exposure time by the hydrocarbon-degraded bacteria, due to the presence of the hydrolyzable ester linkages of HBPP.

Various HPBs have been used as tougheners for epoxy thermosets, with studies indicating that the toughening effects are highly dependent on hyperbranched topological structure. To investigate the influence of the HBP's structure on the

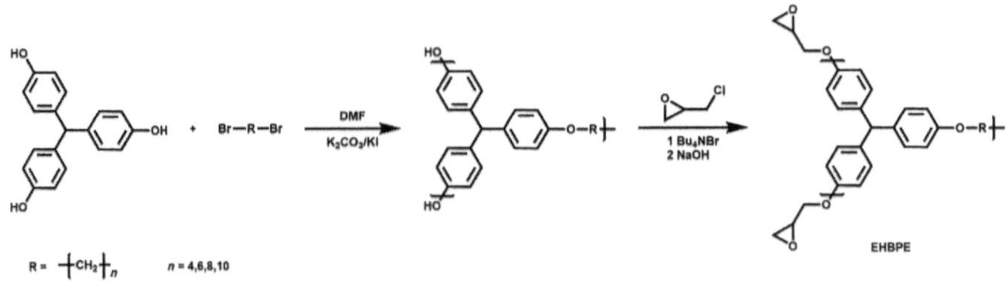

Figure 3.25 Synthetic route of EHBPE with different spacer lengths. *Source:* [50] / with permission of The Royal Society of Chemistry.

toughening effect, Liu et al. [50] synthesized a series of epoxy-functionalized hyperbranched polyethers (EHBPEs) with varying MW and DB through etherification (Figure 3.25). Their synthetic approach utilized triphenylmethane as the B_3 core and a series of α,ω-dibromoalkanes (4C, 6C, 8C, and 10C) with different spacer lengths as A_2 monomers. The branching architecture was precisely controlled by leveraging the competitive kinetics between nucleophilic substitution and elimination reactions. Subsequently, the terminal hydroxyl groups of the resulting HBPEs were quantitatively epoxidized via controlled etherification with ECH.

EHBPEs were employed to toughen the conventional DGEBA/triethylenetetramine (TETA) system, revealing remarkable structure-dependent toughening effects in DGEBA composites. Notably, EHBPE-4C, which has the lowest MW and stiff backbone, and EHBPE-10C, which has the highest MW and flexible backbone, were found to be effective tougheners. They significantly enhanced toughness without compromising tensile strength and T_g. In contrast, EHBPE-6C and EHBPE-8C provided negligible toughening improvement. Hybrids containing proper amounts of EHBPE-4C can be effectively cured, forming a more homogeneous network, which is attributed to its superior toughening performance. For EHBPE-6C and EHBPE-8C with longer spacer length, which possess a high density of terminal groups and significant steric hindrance, the curing process left a considerable fraction of epoxy groups unreacted. This resulted in a less cross-linked structure and larger defects, thereby compromising the mechanical performance. However, as the spacer length further increased in EHBPE-10C, the backbone becomes more flexible, and the terminal groups gain more freedom to react with curing agents and decrease the possibility of defects. These findings confirm that the backbone rigidity and the density of terminal groups are important parameters in shaping the cured network's architecture and the resulting mechanical properties.

Xia et al. [51] synthesized an E-HBP via etherification and systematically investigated the influence of epoxy matrix on the toughening effect of HERs. The synthesis of E-HBP involved the polycondensation of dimethylolpropionic acid (AB_2 monomer) and TMP (B_3 core), followed by terminal epoxidation via etherification with ECH (Figure 3.26). E-HBP was introduced into two different

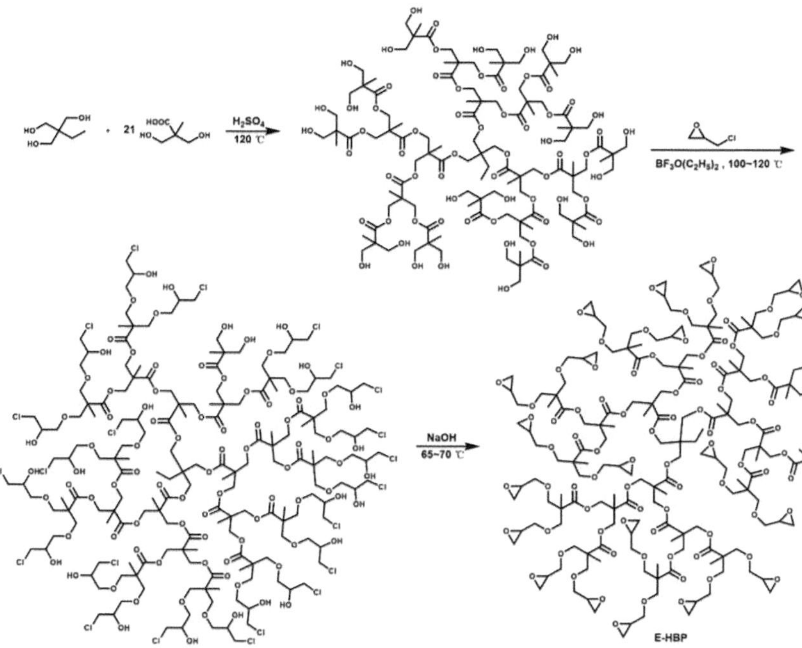

Figure 3.26 Synthetic route of E-HBP. *Source:* [51] / with permission of Springer Nature.

epoxy resin systems: a commercial DGEBA resin with anhydride as curing agent and a tetraglycidyl diaminodiphenyl methane (TGDDM) resin with diamine as curing agent.

In both systems, E-HBP participated in the curing process, then underwent chemical-induced phase separation, and finally formed the gradient transition interface layer (GTIL) structure. This resulting GTIL, with a thickness of approximately 2 μm, serves as a critical interphase that markedly strengthens the interfacial adhesion between the phase-separated E-HBP domains and the epoxy continuous phase. This enhanced interface promotes efficient stress transfer and allows the modifying role of E-HBP to be fully realized. The establishment of this GTIL is the pivotal factor endowing E-HBP-modified epoxies with properties that surpass those of resins modified by other method.

However, it also should be noticed that the properties of the epoxy matrix itself will affect the mechanical behaviors (deformation or cavities) of separated E-HBP particles in the modified epoxy resins during the external loading process, thereby influencing the modification mechanism of E-HBP-modified epoxy resins. SEM results indicated different morphologies of fracture surfaces between the two modified epoxy systems. The cross-linked molecular chains in the TGDDM/DDS system exhibited greater rigidity than those in the DGEBA/MeTHPA system, owing to a higher content of aromatic rings and a greater cross-linking density. This pronounced rigidity difference led to distinct deformation behaviors. In the more flexible DGEBA/MeTHPA matrix, the hysteresis effect during loading was significant, resulting in notable energy dissipation. Here, the separated E-HBP particles themselves underwent deformation, inducing substantial plastic deformation of the surrounding matrix and thereby enhancing toughness. While, for TGDDM/DDS with great rigidity, the separated E-HBP particles could be cavitated quickly and fully before the fracture of the epoxy matrix under the external load. This work demonstrates that E-HBP, synthesized via etherification, serves as a versatile toughener capable of adapting to diverse epoxy formulations through tailored interfacial architectures.

3.3.2 Esterification

In contrast to the nucleophilic substitution characteristic of etherification, esterification offers a distinct and versatile pathway for constructing HERs. This approach involves the reaction between carboxylic acid functionalities and alcohols to form ester linkages, a transformation central to tailoring polymer properties [52]. The method proves valuable for modifying hyperbranched polyol cores, where pendant or terminal carboxylic acid groups serve as ideal anchoring points for epoxy functionalization.

Moreover, ECH is a useful building blocks for esterification of carboxyl-end hyperbranched cores [53]. In this process, the carboxylic acid groups first react with ECH to form ester linkages bearing chlorohydrin intermediates, which are subsequently dehydrochlorinated in a basic medium to yield the final glycidyl ester groups. This two-step sequence-esterification followed by cyclization-efficiently grafts epoxy functionalities onto the hyperbranched scaffold, transforming the core into a multiepoxidized HER. The abundance of carboxylic acid chain ends on hyperbranched polyesters (e.g., Boltorn-type polymers) makes them particularly suitable precursors for the efficient PPM.

This section introduces the esterification route for the post-polymerization strategy to construct HERs in detail. Emphasis is placed on the design and synthesis of hyperbranched precursors with tailored carboxylic acid content and the critical parameters governing the esterification efficiency. The discussion also extends to the role of these HERs as functional additives in commercial epoxy resins, demonstrating their significant contribution to enhancing mechanical, thermal, and toughness properties of the blended systems.

In 2006, Zhang and Jia [54] reported the synthesis of low-viscosity aromatic hyperbranched epoxy resins (HTDE) via esterification between ECH and carboxyl-terminated hyperbranched polymers (Figure 3.27). The carboxyl-ended hyperbranched polymers were synthesized using trimellitic anhydride (TMA) and dihydroxy alcohols as raw materials. By precisely controlling the monomer feed ratios and reaction kinetics, a quasi-one-step synthesis of these carboxyl-terminated hyperbranched polymers was achieved. Subsequent esterification with ECH under alkaline conditions enabled quantitative introduction of epoxy groups at the terminals of the hyperbranched architecture.

Notably, the viscosity of the resulting HERs initially increased and then decreased with an increase in MW, confirming that the DB of the epoxidized aromatic

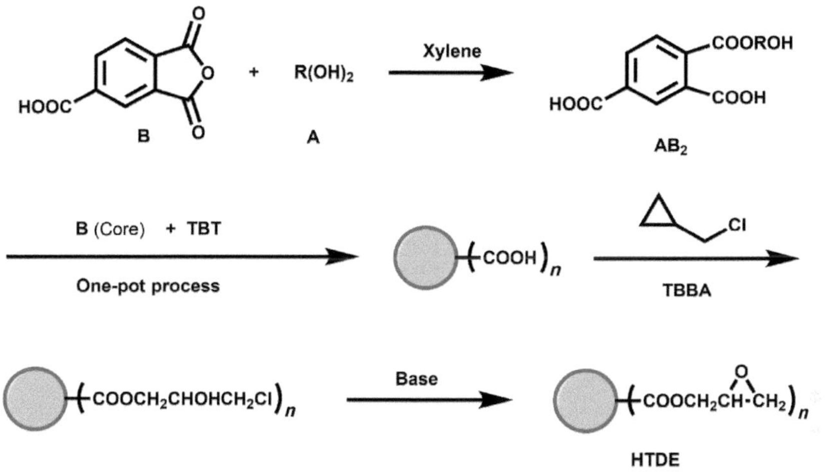

Figure 3.27 Synthetic route of HTDE. *Source:* [54] / with permission of Elsevier.

hyperbranched polymers is high and close to dendrimers. In comparison with conventional epoxy resins, the HERs exhibited exceptionally low viscosity, reaching a remarkable value of 350 cP. As a result, HTDE, which exhibits low-viscosity properties, can serve as a functional thinner of common epoxy resins and also replace organic solvents in coating applications, thereby contributing to environmental protection. Furthermore, these low-viscosity hyperbranched EPs can also be used to enhance the toughness of conventional epoxy thermosets.

Similarly, Zhang et al. [55] reported another example of polyester-based hyperbranched epoxy resins (HTBE) synthesized using the same strategy. The process began with the preparation of carboxyl-terminated hyperbranched polyester (HTB) from inexpensive 1,4-butanediol glycol (BEG) and TMA via a pseudo-one-step approach (Figure 3.28). This was followed by esterification of HTB with ECH to introduce epoxy groups. Using this method, generation-controlled HTBE derivatives with well-defined branching architecture were successfully constructed.

When incorporated into DGEBA, HTBE demonstrated exceptional viscosity reduction capabilities. At a loading of 15–20 wt%, the viscosity of the blended resin decreased to 4000–5000 cP, representing a 66–75% reduction compared to neat DGEBA. This significant viscosity modification establishes HTBE as an effective reactive diluent while maintaining the matrix's intrinsic mechanical properties. Moreover, HTBE also exhibited low EEW, which contributes to an increased curing rate and higher cross-linking density.

As shown in Figure 3.29, Yu et al. [56] further developed a degradable HER (DEHP) synthesized by a reaction between ECH and a carboxyl-ended hyperbranched polyester (DCHP) obtained from polycondensation between CA and maleic anhydride (MA). By adjusting the molar ratios between CA and MA, a series of DEHP-*n* (DEHP-1, DEHP-2, DEHP-3) with different MWs and epoxy

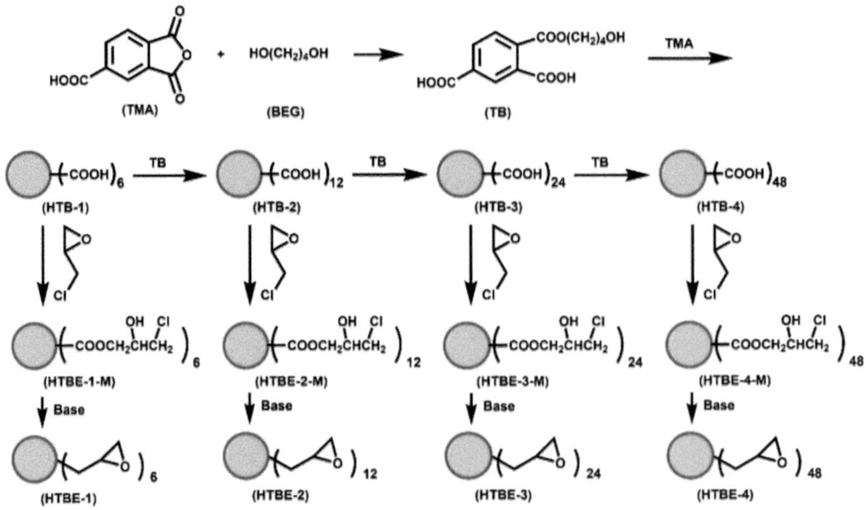

Figure 3.28 Synthetic route of HTBE. *Source:* [55] / with permission of Springer Nature.

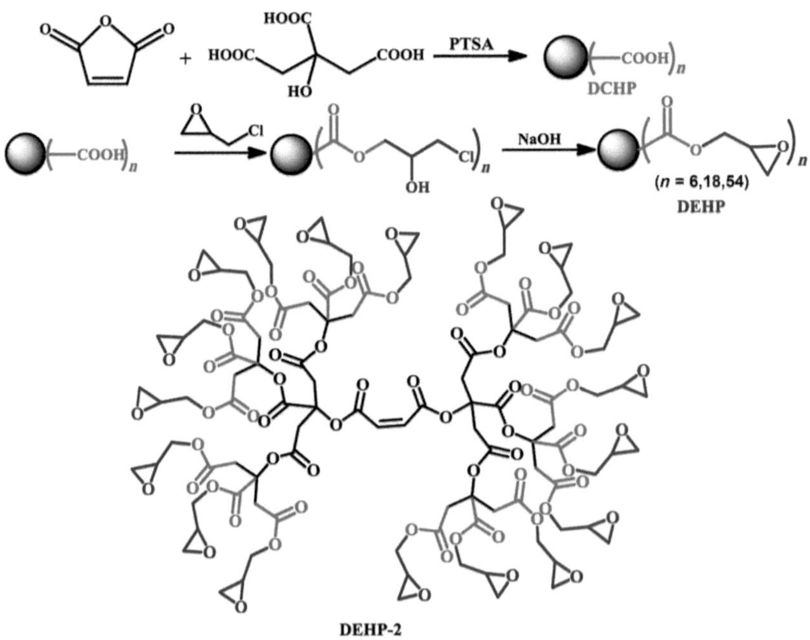

Figure 3.29 Synthetic route of DEHP. *Source:* [56] / with permission of American Chemical Society.

group contents were obtained. All the DEHP-*n* exhibited EEW and low viscosity, suggesting a potential application in solvent-free resins.

DEHP has a remarkable effect to increase the impact, tensile, and flexural strengths of DGEBA, which was influenced by the content and MW of DEHP. With an increase in the DEHP content and the MW of DEHP, the mechanical performance of the DEHP/DGEBA blends increased first and then decreased. The best mechanical performance was achieved by the composite with 12 wt% DEHP-2, which demonstrated 137.8%, 43.4%, and 48.1% increases in the impact, tensile, and flexural strengths of DGEBA, respectively. The intramolecular cavities formed from the non-cross-linkable hyperbranched structure of DEHP would be responsible for the improved toughness, while the increase of the cross-linking density in the blends is attributed to the improved tensile and flexural strengths.

The incorporation of DEHP, with its hydrolyzable ester linkages, provided the blends with environmental merit over conventional DGEBA epoxy, as these hydrolytically labile bonds cleave faster than the ether linkages in DGEBA. The extent of material decomposition was positively correlated with both DEHP content and MW, achieving complete chain scission under mild alkaline conditions. This programmable degradability addresses a key challenge in thermoset recycling, enabling monomer recovery rates exceeding 80% via simple hydrolysis. The combination of enhanced mechanical properties, processability, and programmed degradability positions DEHP as a sustainable alternative for green composite applications.

Xu et al. [57] incorporated the dynamic covalent hexahydro-*s*-triazine structure into a HER to create degradable and recyclable epoxy thermosets. As shown in Figure 3.30, their synthesis involved the polymerization of a hexahydro-*s*-triazine-containing benzyl alcohol (HTA) with phthalic anhydride, which produced a carboxyl-terminated hyperbranched polyester (HCR-HT$_n$). Subsequently, this

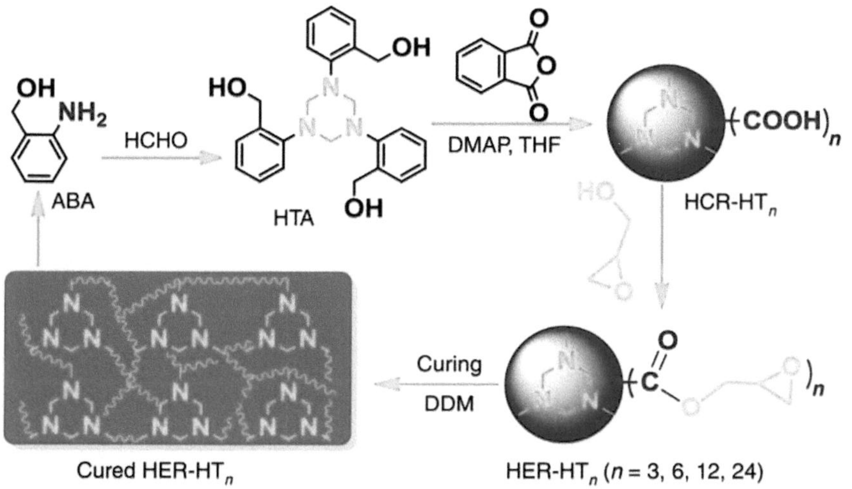

Figure 3.30 Synthetic route and recycling of HER-HT$_n$. *Source:* [57] / with permission of Springer Nature.

polyester was esterified with glycidyl alcohol to produce the final hexahydro-*s*-triazine-containing hyperbranched epoxy resins (HER-HT$_n$). HER-HT$_n$ with different MWs were synthesized by adjusting the molar ratio of HTA to phthalic anhydride.

HER-HT$_n$ was cured by 4,4′-diaminodiphenyl methane (DDM) to prepare the epoxy coating. The T_g, pencil hardness, wear resistance, and adhesion of the cured HER-HT$_n$ coatings all exhibited a gradual increase with the MW of HER-HT$_n$. This enhancement is attributed to the concomitant rise in both the content of rigid benzene ring groups and the cross-linking density. A higher n value provides more reactive epoxy groups per HER-HT$_n$ molecule, thereby promoting a denser cross-linked network.

Due to the presence of dynamic covalent hexahydro-*s*-triazine structure, the cured HER-HT$_n$ coatings exhibited good degradability. Under mild phosphoric acid conditions, the HHT cores underwent selective C—N bond cleavage while preserving aromatic ester linkages. The thermal degradation temperature (~90 °C) used was much lower than the initial degradation temperature for traditional epoxy resins (~350 °C) and thermally sensitive epoxy resins (200–300 °C). More importantly, HTA was recovered by adding a formaldehyde solution to the degradation solution with a yield of 88%, which can be used as a feedstock to synthesize new thermoset hyperbranched polymers. These results demonstrated that HER-HT$_n$ can be degraded, recycled, and reused with high efficiency.

Han et al. [58] developed an energy-efficient microwave-assisted synthesis protocol for the preparation of hyperbranched epoxy resin (MHER). As shown in Figure 3.31, the synthesis started from the polycondensation of 1,2,4-benzenetricarboxylic anhydride (TMA) and ethylene glycol, forming carboxyl-terminated hyperbranched polyesters (MHP). Subsequently, MHP was esterified

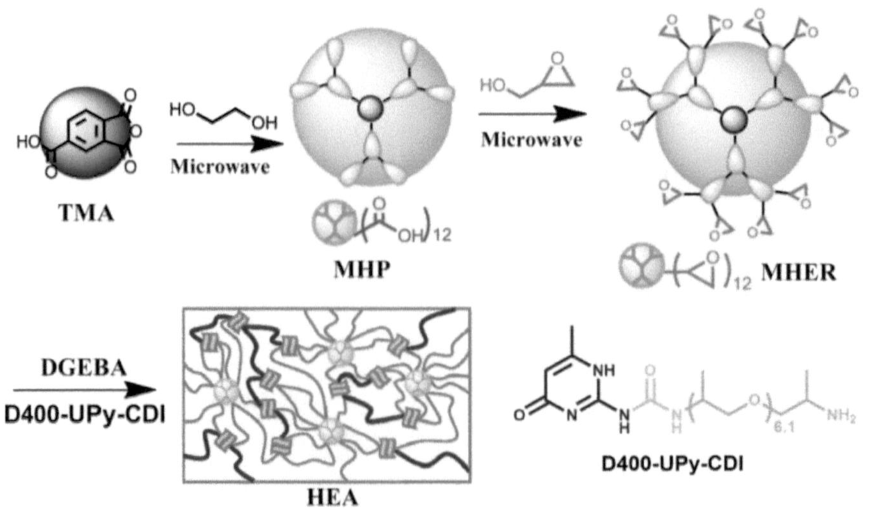

Figure 3.31 Synthetic route of MHER and HEA. *Source:* [58] / with permission of The Royal Society of Chemistry.

with glycidyl alcohol to produce the final MHER. All the reactions were performed under microwave conditions, where the microwave's dielectric heating effect accelerated the ring opening of anhydride groups, as well as the esterification. Compared with the traditional ring-opening/closing reaction for preparing MHER, the reaction time was shortened from 510 to 70 minutes, and the synthesis time of the precursor carboxyl-ended hyperbranched polyester (MHP) also decreased distinctly from 480 to 50 minutes, indicating the high efficiency of the microwave reaction. Moreover, the separation and purification of MHER were greatly simplified, requiring distillation only rather than the tedious process of removing NaCl wastewater.

Microwave irradiation facilitated selective esterification at the hyperbranched polymer periphery, minimizing side reactions such as etherification or cross-linking. This stepwise functionalization strategy yielded MHER with a well-defined core–shell architecture: a rigid aromatic core from TMA derivatives and flexible epoxy-rich outer layers. Gel permeation chromatography analysis revealed a narrow MW distribution and progressive increases in hydrodynamic radius with generational growth, confirming controlled branching propagation.

Various amounts of MHER were incorporated with DGEBA and cured with Upy-contained polyether amine to fabricate a thermoplastic hyperbranched epoxy hot melt adhesive (HEA). The tensile strength, strain, and toughness of HEA-x films increased first and then decreased with the increase in MHER content. Among the formulations, HEA-8, containing 8 wt% MHER, showed the highest performance, with the tensile strength, breaking elongation, and toughness reaching 79.6 MPa, 8.6%, and 4.6 MJ/m^3, respectively. These values represent increases in 78.1%, 168.7%, and 475% compared to neat DGEBA. A similar trend was observed for the lap shear strength of HEA bonded to stainless steel substrates. The mechanical enhancement was attributed to synergistic effect between the hyperbranched architecture of MHER and the quadruple hydrogen bonding provided by UPy motifs. Moreover, the adhesive maintained 82% of its initial bond strength after five thermal reprocessing cycles, demonstrating exceptional recyclability for industrial applications requiring repeated assembly–disassembly cycles. This hierarchical design of hyperbranched epoxy hot melt adhesive successfully decouples the traditional trade-off between adhesive strength and recyclability, achieving an unprecedented combination of lap shear strength and cyclic performance retention.

The esterification reaction emerges as a cornerstone strategy in the synthesis of HERs, offering unparalleled versatility in molecular design and functionality integration. However, esterification's thermodynamic reversibility and inherent susceptibility to hydrolysis necessitate innovative mitigation strategies. Future development in esterification-driven HER synthesis should focus on (i) catalyst innovation to suppress side reactions and enhance regioselectivity and (ii) dynamic ester chemistries combining hydrolytic lability with reprocessability. By synergizing esterification with emerging techniques like dynamic covalent bonding and bio-based monomer utilization, next-generation HERs could achieve unprecedented sustainability-performance equilibria, particularly in applications demanding solvent-free processing or closed-loop recyclability.

3.3.3 Addition Reaction

Synthesizing HERs through the addition reaction represents an important synthetic strategy. This reaction makes use of the terminal reactive groups (such as hydroxyl, amino, thiol, and vinyl groups) of hyperbranched polymer to react with epoxy-containing compounds, thereby introducing epoxy groups to the ends of hyperbranched polymers. The addition reaction features high selectivity and few side reactions, with a fast reaction rate, enabling the introduction of epoxy groups within a relatively short time [59]. Moreover, the addition reaction generally does not require complex reaction steps or cumbersome purification processes, providing convenience for the synthesis of HERs.

Building upon the broader landscape of addition reactions, the hydrosilylation reaction offers a distinct and powerful pathway for epoxy group incorporation [60]. As a fundamental organic transformation, hydrosilylation facilitates the addition of silicon–hydrogen bonds across unsaturated carbon–carbon linkages. This reaction is characterized by its high atom economy, mild conditions, and exceptional selectivity, making it particularly valuable for constructing well-defined polymer architectures.

In the context of hyperbranched polymer modification, hydrosilylation enables precise peripheral epoxidation through the reaction between terminal vinyl/allyl groups and epoxide-functional silanes [16, 61]. The synthetic sequence generally involves first preparing hyperbranched precursors with controlled unsaturated chain ends, followed by their regioselective transformation using reagents such as (glycidoxpropyl)methyldimethoxysilane. This approach allows quantitative grafting of epoxy functionalities onto the branched scaffold while maintaining structural integrity.

Zhang et al. [61] prepared a HER with a silicone skeleton through the hydrosilylation reaction. As shown in Figure 3.32, an allyl-terminated hyperbranched resin with a silicone skeleton (AHRSS) was first synthesized via hydrosilylation of phenyltriallylsilane (PTAS) and 1,1,3,3-tetramethyl-disiloxane (TMDS), catalyzed by halloysite-supported platinum. Subsequently, 1,1,3,3-tetramethyl-disiloxane-propylglycidyl ether (TDPGE), which was obtained from the equimolar addition of TMDS and allyl glycidyl ether (AGE), underwent a similar hydrosilylation process with AHRSS to fabricate the HERSS. Throughout the synthesis, the catalyst, excess reagents, and THF could be recovered and reused. Each step proceeded with a yield exceeding 90%, indicating that this approach is not only suitable for large-scale production but also applicable in solvent-free systems.

The resulting HERSS was employed to modify DGEBA, demonstrating excellent toughening and reinforcing effects. The mechanical properties of the cured HERSS/DGEBA blends initially improved and then declined with increasing HERSS content. At 9 wt% HERSS in the modified DGEBA, the impact, flexural, and tensile strengths were enhanced by approximately 92.5%, 36.0%, and 88.6%, respectively. The fracture surface of the cured DGEBA/HERSS blend exhibited a "sea-island" structure, attributed to the limited compatibility between HERSS and DGEBA. This morphological feature contributed to the improved toughness of the material.

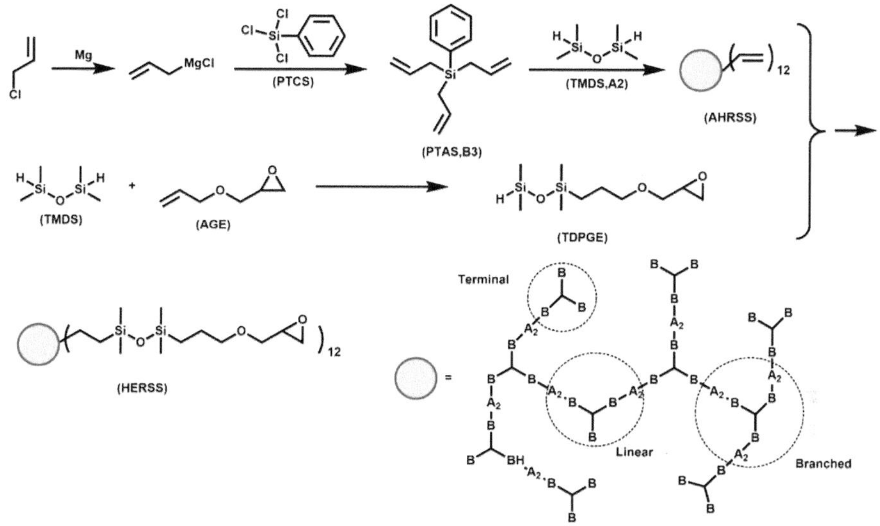

Figure 3.32 Synthetic route of AHRSS and HERSS. *Source:* [61] / with permission of The Royal Society of Chemistry.

On this basis, the same group [16] further investigated the control of the structure and the performance of the HERSS. Three kinds of HERSS (HERSS-1, HERSS-2, HERSS-3, and HERSS-4) with different MWs were synthesized by using various molar ratios of PTAS to TMDS during the hydrosilylation reaction. It was found that the increase in MW of the HERSS results in a decrease in the DB and an appearance of a maximum viscosity, ranging in 0.71–0.84 and 103.5–697.4 mPa s, respectively.

The influence of HERSS structure and MW on the mechanical and micro-phase separation behavior of the cured HERSS-DGEBA blends was systematically investigated. The mechanical properties of the blends exhibited a nonmonotonic dependence on HERSS MW, initially increasing and then decreasing. This trend is attributed to the rising number of epoxy groups in higher MW HERSS, which promotes cross-linking density and thereby enhances tensile and flexural strength. However, this increased cross-linking density concurrently reduces material toughness. Consequently, an optimal balance of strength and toughness was achieved at a medium HERSS MW. Furthermore, SEM analysis revealed that increasing the HERSS MW led to larger and more numerous phase-separated "islands" (composed of cured HERSS) within the "sea-island" morphology. This report demonstrates that by precisely controlling the branching structure and MW of HERs during synthesis, it is possible to effectively regulate the microphase structure and properties of the epoxy thermosets.

Similarly, Pan et al. [62] synthesized an epoxy-terminated hyperbranched PBPS and used it as a toughening additive for DGEBA, aiming to improve their low-temperature toughness. As shown in Figure 3.33, a hyperbranched silicone resin (HHS) was synthesized through the hydrosilylation reaction of 1,3-diethyl-1,1,3,3-tetramethyldisiloxane (TMDVS) and phenyltris(dimethylsiloxy)silane (PTDS), followed by the hydrosilylation addition to AGE. Adding EPTS-12 as a toughening

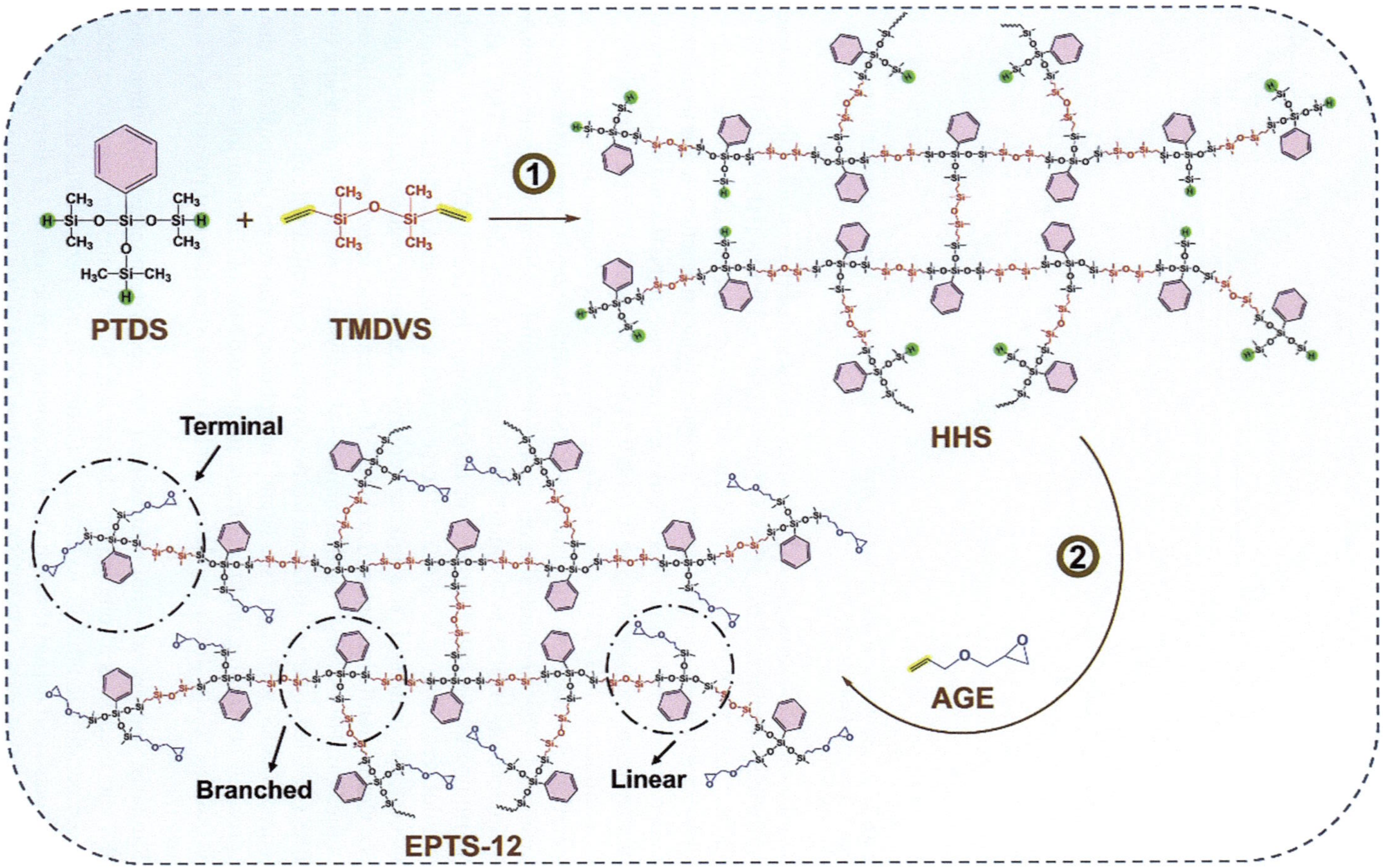

Figure 3.33 Synthetic route of EPTS-12. *Source:* [62] / with permission of Elsevier.

agent to DGEBA can significantly improve the low-temperature toughness of the epoxy resin. With the increased amounts of EPTS-12, the tensile strength and elongation at break of the samples increased first and then decreased to a certain extent, whether at room temperature or low temperature. The modified resin with EPTS-12 content of 5 wt% exhibited the best performance. The max value of the tensile strength and elongation at break at −72 °C were 116.68 MPa and 16.06%, respectively. SEM results showed that the fracture surface of the modified resin had complex crack patterns, which were attributed to the synergistic effect of the hyperbranched structure of EPTS-12 and the flexible Si—O—Si segments, which could effectively absorb fracture energy and improve the toughness of the resin.

The reaction between isocyanates and hydroxyl or amino groups stands as a cornerstone in polymer chemistry, proceeding efficiently even under mild conditions without the necessity of catalysts [63, 64]. Celebrated for its high efficiency and rapid reaction kinetics, this chemistry provides a versatile and reliable route for the precise functionalization and network formation of polymeric materials. In the PPM of hyperbranched architectures, this reaction offers a straightforward pathway to introduce diverse functionalities, including epoxy groups. Isocyanate-terminated hyperbranched polymers can serve as ideal cores, to be quantitatively modified using hydroxyl or amino-bearing epoxides. This pivotal step efficiently installs urethane- or urea-linked epoxy functionalities at the chain ends, thereby converting the core into HERs while simultaneously enhancing potential interfacial adhesion through secondary interactions.

Wang et al. [65] synthesized an epoxy-terminated hyperbranched polyurethane (EHPU) by modifying an isocyanate-terminated hyperbranched polyurethane (HPU) with glycidol. As shown in Figure 3.34, HPU was first synthesized via an $A_2 + B_3$ polyaddition using 4,4′-diphenylmethane diisocyanate (MDI) and TMP as monomers. Subsequently, EHPU was successfully obtained through a grafting reaction between glycidol and HPU, in which the hydroxyl groups of glycidol reacted with the terminal isocyanate groups of HPU. The use of a relatively low temperature (60 °C) for the isocyanate-hydroxyl reaction effectively suppressed

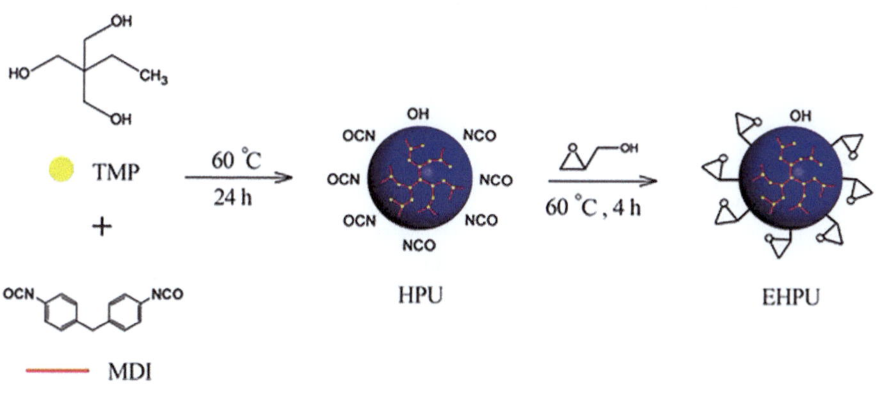

Figure 3.34 Synthetic route of EHPU. *Source:* [65] / with permission of Springer Nature.

potential side reactions between isocyanate and epoxide groups, which typically require higher temperatures (around 120 °C) to occur. By varying the stoichiometric ratio of MDI to TMP, a series of $EHPU_x$ (x represents the MDI/TMP molar ratio) with different epoxy values were obtained.

$EHPU_{1.9}$, which owns both a higher epoxy value and T_g, was used to modify the benzoxazine (BZ) resin, yielding the BE_Y resin system (Y denotes the BZ/EHPU mass ratio). Fully cured BE_Y resin showed only one approximate T_g, indicating the good compatibility between EHPU and BZ resins. Cure kinetic studies performed via the advanced isoconversional method revealed that $EHPU_{1.9}$ catalyzes the cure process of BZ resin, reducing the apparent activation energy by 8.3 kJ/mol. $EHPU_{1.9}$ can also copolymerize with BZ resin through reactions between its terminal epoxide groups and the phenolic hydroxyl groups generated from the ring opening of oxazine rings. The incorporation of epoxy-terminated hyperbranched polyurethane effectively addresses several inherent limitations of BZ resins, such as the requirement for high curing temperatures compared to other thermosetting resins, as well as drawbacks like poor processability, inherent brittleness, and low toughness.

Guo et al. [66] synthesized a series of renewable and self-curable epoxy-ended hyperbranched polymers (DSEHP-*n*, $n = 3, 5, 7$ and 11) via an isocyanate-hydroxyl reaction. As shown in Figure 3.35, the synthesis involved two main steps: first, IHP-*n* intermediates were prepared via a click reaction between toluene-2,4-diisocyanate and 2,2′,2″-(1,3,5-hexahydro-*s*-triazine-1,3,5-triyl)benzyl mercaptan. Subsequently, DSEHP-*n* was obtained by reacting the isocyanate groups of IHP-*n* with the hydroxyl groups of glycidyl alcohol. The presence of urethane groups in DSEHP-*n* enables its self-curing behavior upon heating, as these groups decompose to form reactive species such as olefins and triamines, which further participate in cross-linking reactions. Moreover, DSEHP-*n* can be completely degraded into renewable monomers in a nitric acid/dimethylformamide (HNO_3/DMF) solution, realizing the recycling of materials. Owing to these properties, DSEHP-*n* shows broad application potential in self-curing coatings and comfort-related materials.

Moreover, DSEHP-*n* was employed to modify DGEBA epoxy resin, which was then cured using DETA-AN—a curing agent synthesized from diethylenetriamine (DETA) and acrylonitrile (AN). It was found that the incorporation of DSEHP-*n* led

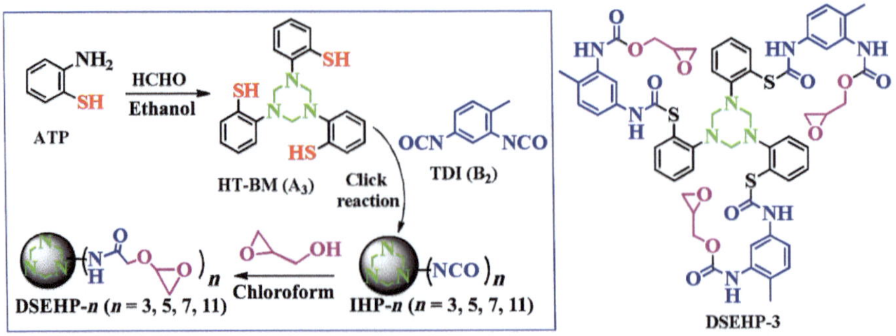

Figure 3.35 Synthetic route of DSEHP-*n*. *Source:* [66] / with permission of Elsevier.

to an initial increase, followed by a decrease in the mechanical properties of the composites. The optimal enhancement was achieved with the addition of 12 wt% DSEHP-5, which resulted in significant improvements in tensile strength, tensile modulus, flexural strength, flexural modulus, impact strength, and elongation at break. The fracture surface of the DSEHP-*n*/DGEBA composites exhibited numerous fibrils or short fibers, indicative of a toughened morphology. The enhancement in mechanical properties is attributed to the combined effects of cross-linking density, intramolecular cavities, hyperbranched topological structure, and good compatibility between DSEHP-*n* and DGEBA. Cross-linking density positively influenced the tensile and flexural strengths, while intramolecular cavities contributed to improved impact resistance. The hyperbranched architecture helped absorb impact energy, transfer load, and increase elongation. In addition, the introduction of DSEHP-*n* significantly improved the low-temperature resistance and degradability of the material. This work provides a new strategy for designing thermosetting materials that combine high strength with degradability, addressing the longstanding challenge of integrating multiple superior properties in traditional thermosets and broadening their potential applications.

The addition reaction strategy has emerged as a robust methodology for synthesizing HERs, wherein terminal functional moieties (e.g., hydroxyl, thiol) of hyperbranched polymers are covalently conjugated with epoxy-bearing reagents. This approach exhibits distinct advantages, including high regioselectivity, rapid reaction kinetics, and simplified purification protocols, while enabling precise modulation of branching density and epoxy group distribution. Notable advancements encompass (i) silicone-incorporated HERs demonstrating exceptional thermal stability for high-temperature applications, (ii) dynamic HERs integrating self-healing and recyclable functionalities via reversible covalent bonds, and (iii) flexible HERs engineered with low-temperature toughness through strategic molecular architecture design.

Key merits of this synthetic paradigm lie in its alignment with green chemistry principles—utilizing solvent-free conditions and recyclable catalytic systems—and the capacity to tailor mechanical/thermal properties via topological optimization. However, challenges persist, particularly the demand for stringent anhydrous/anaerobic reaction environments and the economic barriers posed by high-cost catalysts. Future research should prioritize the development of air-stable catalytic platforms, multifunctional stimuli-responsive systems, and scalable manufacturing processes to broaden the applicability of HERs in aerospace, advanced electronics, and sustainable material technologies.

3.3.4 Click Reaction

Click chemistry has emerged as a useful tool in synthetic polymer chemistry, due to its high efficiency, exceptional selectivity, and robust reaction kinetics under mild conditions [5, 67–71]. Among its versatile reactions, the thiol–ene click reaction is particularly noteworthy, enabling rapid, quantitative, and anti-Markovnikov addition of thiols across carbon–carbon double bonds with remarkable regioselectivity, often mediated by light or thermal initiators. This modular "click" approach provides a powerful and versatile toolkit for the precise control of complex polymer architectures.

In the post-polymerization functionalization of hyperbranched polymers, click chemistry offers an efficient and versatile route for the targeted modification of epoxy groups. Thiol- or alkene-decorated hyperbranched cores can be readily synthesized, followed by facile coupling with complementary epoxide-bearing click partners, such as AGE or other unsaturated epoxy compounds. This pivotal thiol–ene step allows for the quantitative and orthogonal introduction of epoxy functionalities, converting the core into well-defined HERs while incorporating flexible thioether linkages into the polymer backbone.

This section focuses on the construction of HERs via a thiol–ene-based post-polymerization strategy. It introduces the synthesis of thiol/ene-terminal hyperbranched precursors and systematically investigates click reaction parameters to achieve controllable epoxy grafting. The architecture of the resulting HERs is carefully characterized to elucidate the structure–property relationships arising from the high grafting efficiency, with particular emphasis on their influence on the thermal stability, flexibility, and mechanical performance of the cured networks.

In 2016, Zhang et al. [72] reported a fast and environmentally friendly approach to fabricating HEPs with high epoxy content through an ultraviolet (UV)-initiated thiol–ene click reaction. As shown in Figure 3.36, thiol-ended hyperbranched polyester (THBP-12) was prepared by the esterification between 3-mercaptopropionic acid (MPA) and hydroxyl-ended hyperbranched polyester (HyPer H102). Subsequently, a UV-initiated thiol–ene click reaction was performed between THBP-12 and AGE to successfully synthesize the hyperbranched polyester epoxy resin (EHBP-12). All the click reactions took only 5–10 minutes of UV radiation at room temperature and show very high yields between 98.1% and 99.8%. The EEW of the obtained EHBP-12 was approximately 309 g/mol, which was very close to its theoretical value of 301 g/mol.

The UV-initiated click reaction ensures that all the thiol groups can completely react with the double bond of AGE while maintaining the epoxy groups. Based on this, a series of well-defined HERs with controlled functionality, including mono-epoxide, di-epoxides, and tri-epoxides were successfully synthesized based

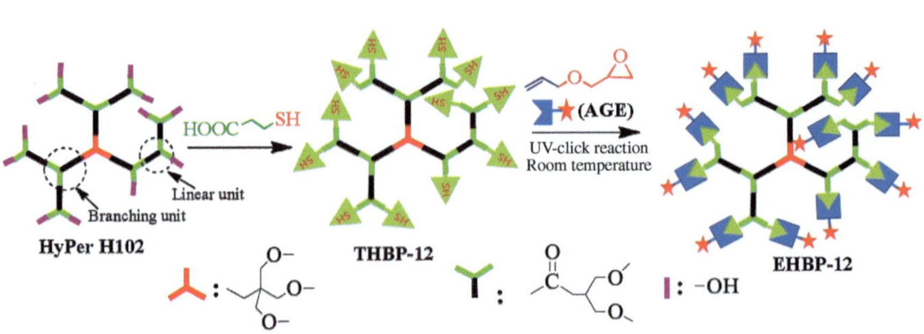

Figure 3.36 Synthetic route of hyperbranched polyester epoxy resin. *Source:* [72] / with permission of Elsevier.

on different thiols precursors and AGE. This new synthetic method is not only rapid and environmentally friendly, with mild reaction conditions, but also avoids the tedious purification process in traditional methods. It opens up a new path for the design and synthesis of HERs and common epoxy resins, showing great potential in industrial applications.

Wang et al. [31] further optimized the synthesis process and successfully prepared HERs with different EEWs, conducting an in-depth investigation into their effects on composite properties. This work provided an important foundation for the performance regulation of HEPs in general. With ongoing research, the synthesis and property studies of HEPs with tailored structures have been increasingly reported. It has been demonstrated that precise control over the molecular structure of HEPs through click reaction can lead to more unique and superior material properties.

In one study, Wang et al. reported a method for synthesizing a series of HERs (EHBP-N) with tunable EEW via a thiol–ene click reaction between thiol-terminated hyperbranched polyesters and AGE. Specifically, thiol-terminated hyperbranched polyesters (THBP-*n*) with different MWs were first synthesized by esterification of hydroxyl-terminated hyperbranched polyester (H102) with 3-MPA. The resulting THBP-*n* was then reacted with AGE under UV irradiation to obtain EHBP-N (Figure 3.37). In this work, by systematically varying the EEW content, the mechanical properties (including tensile, flexural, impact, and adhesive strength) and thermal properties of EHBP-N/DGEBA composites were comprehensively evaluated. The results showed that as the EHBP-N content increased, the mechanical properties of the composites first increased, peaked at around 9 wt% of EHBP-N, and then decreased. In particular, the composite with 9 wt% EHBP-6 in DGEBA exhibited the most outstanding mechanical performance compared to unmodified DGEBA. The tensile strength reached a maximum of 71.1 MPa—an increase of 38.9% over pure cured DGEBA (51.2 MPa). The flexural strength improved by 35.3% to 147.0 MPa, while the impact strength increased by 158.6% to 32.0 kJ/m^2. The adhesive strength also rose significantly by 96.2%, reaching 11.5 MPa. These

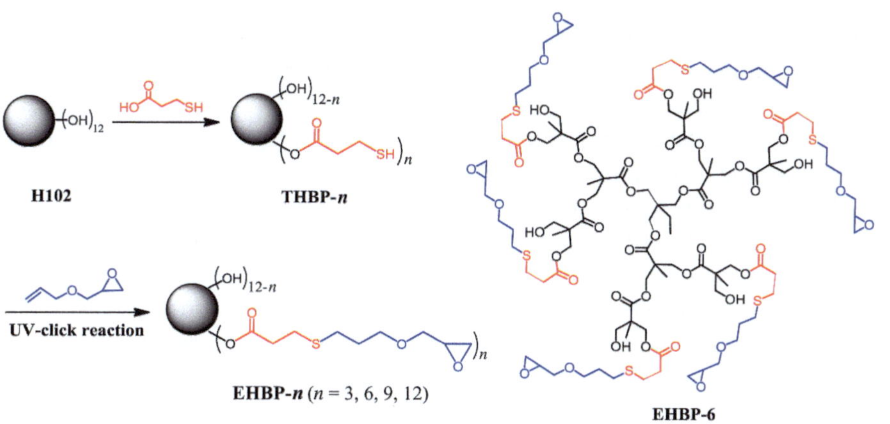

Figure 3.37 Synthetic route of EHBP-*n* (*n* = 3, 6, 9, 12) and chemical structure of EHBP-6. *Source:* [31] / with permission of Elsevier.

improvements can be attributed to the combined effect of cross-linking density and intramolecular cavities. In EHBP-N/DGEBA composites, the intramolecular cavity volume increases with both EHBP-N content and EEW, whereas the cross-linking density increases with EHBP-N content but decreases with higher EEW of EHBP-N. The improved toughness is ascribed to the energy-absorbing capability of the intramolecular cavities within the non-cross-linkable hyperbranched structures. Furthermore, the enhanced adhesion is attributed to strong physical interactions involving the polar hydroxyl groups in the HEPs, as well as interactions between the DETA-AN hardener and the substrate surfaces.

The DB in hyperbranched polymers significantly influences their properties and composites. For instance, the intrinsic viscosity of polymers decreases with increasing DB, and a series of dendritic, hyperbranched, branched, and linear polymers all exhibit behavior consistent with that of pure dendritic polymers. However, precise control over DB has been challenging using conventional synthesis methods for EHPs.

Wang et al. [73] reported a synthetic route to nitrogen–phosphorus skeleton-based epoxy-terminated hyperbranched polymers (NPEHP-*n*) with tunable DB. The process involves three sequential steps (Figure 3.38): First, a thiol–ene click reaction between the methacrylate groups in TMPPM and the thiol groups in an AB_2-type monomer yields nitrogen–phosphorus skeleton hydroxyl-terminated hyperbranched polymers (NPHOH-*n*). These hydroxyl-terminated polymers then undergo esterification with 3-MPA to form nitrogen–phosphorus skeleton thiol-terminated hyperbranched polymers (NPHSH-*n*). Finally, a thiol–ene click reaction between glycidyl methacrylate and NPHSH-*n* produces the target nitrogen–phosphorus skeleton epoxy-terminated hyperbranched polymers (NPEHP-*n*) with systematically varied DB.

This study systematically investigated the relationships among viscosity, time, temperature, and shear rate. It was observed that the viscosity of HBPs decreases with increasing content and DB of NPEHP-*n*. This trend suggests that the highly branched structure of NPEHP-*n* creates free volume that facilitates the movement of epoxy resin chains, thereby reducing viscosity. In addition to rheological behavior, the DB also influences the mechanical and thermal properties of NPEHP-*n*/DGEBA blends. The mechanical performance—including tensile, flexural, and impact strength—of the cured blends improves significantly with higher DB and NPEHP-*n* content, reaching a maximum at approximately 9 wt% of NPEHP-*n*. Specifically, the enhancements in tensile, flexural, and impact strength range from 31.7% to 42.8%, 56.4% to 76.4%, and 172.0% to 198.3%, respectively. In contrast, the glass transition temperature (T_g) of the composites decreases slightly with increasing DB and NPEHP-*n* content, while the thermal decomposition temperatures also decline. The toughening mechanism of the NPEHP-*n*/DGEBA composites can be attributed to in situ reinforcement and enhanced impact resistance. The small molecular size and structural similarity between NPEHP-*n* and DGEBA promote uniform dispersion within the epoxy matrix. As the DB of NPEHP-*n* increases, its molecular conformation becomes more spherical, resulting in a larger intramolecular

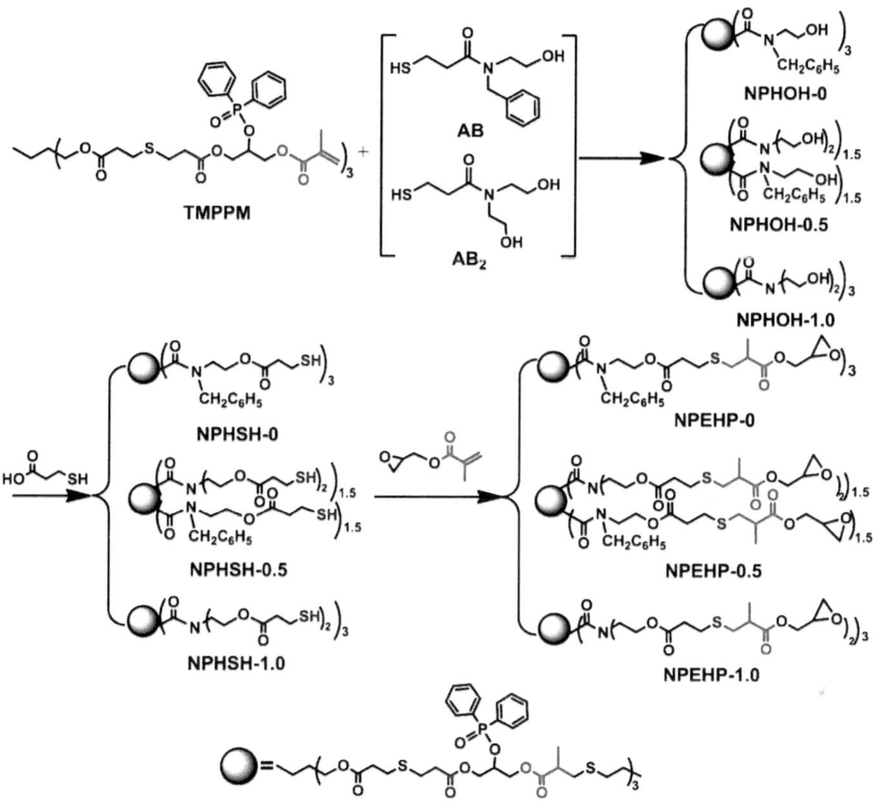

Figure 3.38 Synthetic route of NPEHP-*n* with different degrees of branching. *Source:* [73] / with permission of Elsevier.

cavity volume. These cavities act as energy-absorbing sites during impact, thereby improving the overall toughness of the composite material.

The synthesis of N—P skeleton epoxy hyperbranched polymers (NMHEP-*n*) reported by Wu et al. [17] represents a notable advancement in the structural control of such materials. By employing the highly efficient and selective thiol–ene click reaction, NMHEP-*n* polymers with precisely tunable DB and topological structures were successfully synthesized. This strategy not only facilitates the construction of polymers with tailored architectures but also offers a robust route for achieving targeted molecular design. The DB of epoxy hyperbranched polymers is a key parameter that significantly influences their properties. In this work, DB was systematically regulated by carefully designing the ratio of branched to linear monomers, enabling precise control over the topological structure. For instance, a higher proportion of branched monomers led to more densely branched architectures, whereas an increased fraction of linear monomers yielded more linear polymer configurations.

The synthesis was carried out via a multistep procedure. First, a thiol–ene reaction between thiol-capped THMP and DPPM was conducted. Subsequently, by varying the molar ratio, Michael addition took place between the thiol groups of THMPM and an AB_2-type monomer, producing hydroxyl-terminated hyperbranched polymers (NMHOH-*n*) with different DBs. This step exhibited high sensitivity to the molar ratio, with even minor adjustments resulting in significant variations in the final DB. Next, thiol-terminated hyperbranched polymers (NMHSH-*n*) with distinct DBs were synthesized via esterification of the hydroxyl groups in NMHOH-*n* with the carboxyl groups of 3-MPA. Finally, a thiol–ene click reaction between NMHSH-*n* and AGE yielded N—P skeleton hyperbranched epoxy resins (NMHEP-*n*) with a range of DBs, as illustrated in (Figure 3.39).

The study further investigated how the topological structure and content of NMHEP-*n* influence the properties of NMHEP-*n*/DGEBA composites. Mechanical performance was evaluated using stress–strain analysis. Results showed that as the NMHEP-0.5 content increased, both the stress and strain of the composites rose initially before declining, indicating a nonlinear relationship and the existence of an optimal loading level for mechanical enhancement. When the DB reached approximately 0.5 and the NMHEP-*n* content was around 12 wt%, the stress and strain achieved maximum values—1.33 and 1.44 times those of pure DGEBA, respectively. This suggests that a DB of ~0.5 combined with 12 wt% NMHEP-*n* content can significantly improve the mechanical strength of the composites. A similar trend was observed in impact toughness. The 12 wt% NMHEP-0.5/DGEBA composite exhibited maximum impact duration and impact force of 0.94 ms and 801.58 N, corresponding to increases of 67.86% and 48.58%, compared to those of unmodified DGEBA.

Thermosetting epoxy resins have long faced the challenge of balancing strength with toughness—a fundamental trade-off that has limited their application in fields requiring both properties. Furthermore, amid growing environmental concerns and the global push for sustainable development, achieving both high performance during service and rapid recyclability has become a critical yet difficult target for thermoset epoxy resins. Liu et al. [74] reported a methodology for the precise measurement and control of molecular free volume in epoxy resins. Moreover, they introduced an innovative synthetic route to

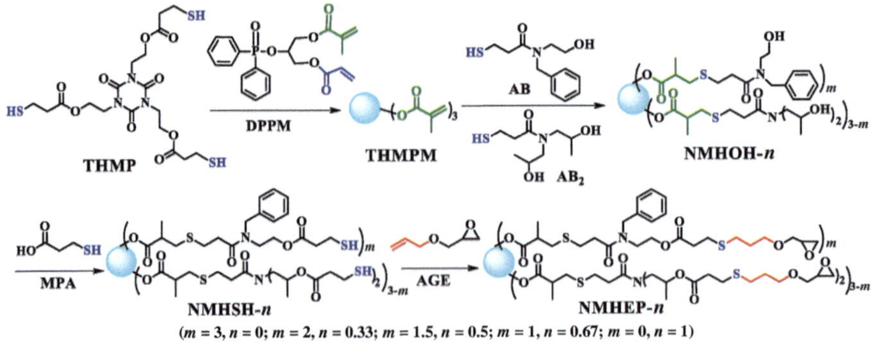

Figure 3.39 Synthetic route of NMHEP-*n*. *Source:* [17] / with permission of Elsevier.

novel HERs (Bn, where n =6, 12, 24) containing both imide and cyanurate structures via an efficient thiol–ene click reaction. The synthesis proceeds as follows: first, 2,2-bis[4-(4-aminophenoxy)phenyl]propane (BAPP) is reacted with TMA to form an imide-based monomer (BTA). This is then combined with 1,3,5-tris(2-hydroxyethyl)cyanuric acid (THEIC) in varying molar ratios to produce hydroxyl-terminated hyperbranched polyimides (BPI-n). Subsequent esterification of BPI-n with 3-MPA yields thiol-terminated hyperbranched polyimides (BSPI-n). Finally, the target HERs (B_n) are obtained via a room temperature thiol–ene click reaction between BSPI-n and AGE, as illustrated in Figure 3.40. This stepwise strategy enables precise structural control over the resulting HERs, offering a pathway to tailor their properties.

Of particular significance in composite applications, these epoxy resins exhibit remarkable properties. They readily integrate into DGEBA, leading to a densified internal molecular architecture and the formation of sub-nanoscale free-volume cavities. The combination of linear DGEBA and hyperbranched Bn creates a homogeneous composite structure characterized by uniformly distributed molecular cavities or free volume. This unique configuration imparts DGEBA with an excellent balance of strength, modulus, toughness, low-temperature resistance, and degradability. A systematic investigation was conducted into the relationship between Bn content, MW, and the mechanical properties of the cured Bn/DGEBA copolymers. Results revealed that as both Bn content and MW increase, the mechanical performance initially improves and then declines. An optimal balance is achieved at a B12 content of 12%, where a pronounced synergistic effect emerges among the topological structure, free-volume characteristics, good compatibility, structural rigidity, and high cross-linking density. For instance, the hyperbranched topology of B12 can deform to absorb energy and effectively transfer load, thereby enhancing both toughness and strength. Optimized free-volume properties—such as reduced cavity size and a narrow size distribution—further contribute to stronger intermolecular interactions and a more compact structure.

These findings not only expand the variety of HERs but also provide valuable insights into their structure–property relationships. Meanwhile, as performance

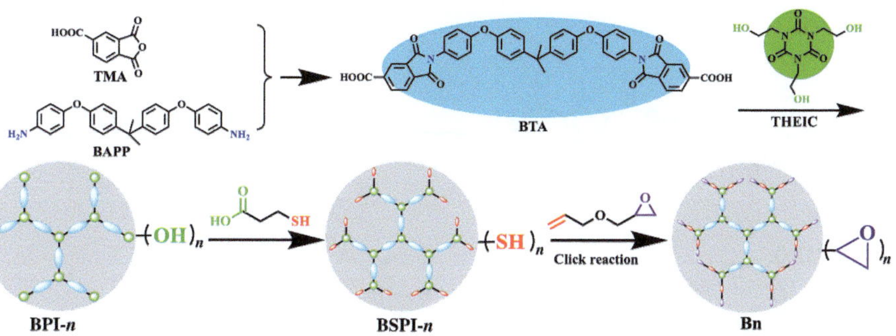

Figure 3.40 Synthetic route of Bn. *Source:* [74] / with permission of John Wiley & Sons.

requirements continue to grow across various sectors, HERs with specialized functionalities have become an emerging research focus. In the context of increasing environmental awareness, degradable and renewable epoxy resins have attracted significant attention. In specific application scenarios such as microelectronic devices and artificial muscles, shape-memory epoxy resins (SMERs) demonstrate unique potential. This evolving demand is driving research on HERs toward the development of materials with tailored functional properties.

Epoxy resins are widely used across various industrial sectors due to their outstanding mechanical strength, chemical stability, and excellent dielectric properties. However, the inherent difficulty in recycling and reusing cured epoxy resins has raised serious environmental concerns. Conventional disposal methods, such as incineration and landfilling, not only lead to resource waste but also contribute to environmental pollution, underscoring the urgent need for sustainable alternatives. In this context, the application of click chemistry—particularly its efficiency, selectivity, and suitability for constructing well-defined polymer architectures—has opened new pathways for designing recyclable epoxy systems.

Inspired by the development of bio-based EHPs and the strategic incorporation of degradable motifs, Chen et al. [75] reported a notable breakthrough: a novel degradable and recyclable high-performance thermosetting HER (EFTH-*n*), synthesized entirely without organic solvents (Figure 3.41). This work represents a significant step toward addressing both environmental sustainability and the performance limitations of conventional epoxy resins. The synthesis of EFTH-*n* was accomplished via a carefully designed three-step route starting from tris(2-hydroxyethyl)isocyanurate (THEIC) and biomass-derived 2,5-furandicarboxylic acid (FDCA). First, a hydroxyl-terminated hyperbranched polymer (HFTH) was prepared, incorporating a dynamic covalent hexahydro-*s*-triazine group—a key structural motif that confers degradability and recyclability. This was followed by esterification with 3-MPA to afford TFTH-*n*. Finally, under UV initiation at room temperature, a thiol–ene click reaction between the thiol groups of TFTH-*n* and the carbon–carbon double bonds of AGE was employed to introduce epoxy functionalities, successfully yielding the degradable HER EFTH-*n*.

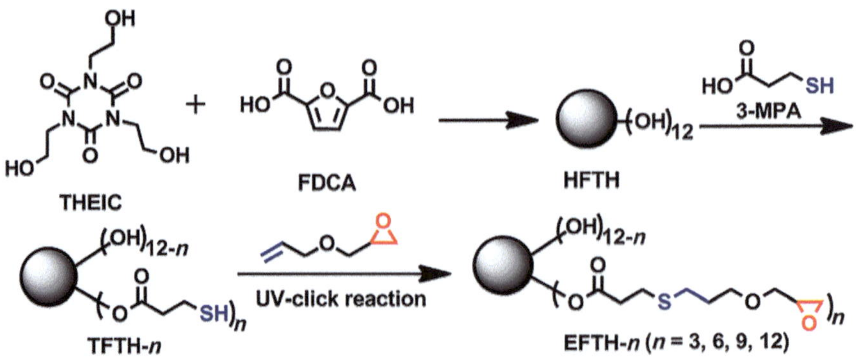

Figure 3.41 Synthetic route of HFTH and EFTH-n ($n = 3, 6, 9, 12$). *Source:* [75] / with permission of The Royal Society of Chemistry.

The cured EFTH-*n*/DGEBA blends exhibited outstanding mechanical properties alongside notable degradability. Mechanically, these blends demonstrated high tensile strength, superior toughness, enhanced modulus, and improved elongation. As both the content and functionality (*n*-value) of EFTH-*n* increased, the mechanical performance initially improved, peaking with the 12 wt% EFTH-6/DGEBA formulation, before eventually plateauing. This optimal performance was attributed to the synergistic effects of multiple factors. The hyperbranched topological structure of EFTH-*n* contributes to efficient energy absorption and effective load transfer, thereby enhancing toughness. Furthermore, EFTH-*n* aids in disentangling the DGEBA molecular chains, leading to an optimized structural arrangement and improved mechanical behavior. Excellent compatibility between EFTH-*n* and DGEBA also promotes homogeneous dispersion and strong interfacial interaction at the molecular level, further elevating the overall performance.

Regarding degradability, the study systematically investigated the influence of EFTH-6 content and functionality (*n*-value) under consistent degradation conditions. Results indicated that increasing the EFTH-6 content significantly enhanced the degradation degree of the cured composites. When the EFTH-6 content reached 12 wt% or higher, the degradation degree reached as high as 99.8%. This exceptional degradability is primarily attributed to the presence of ester bonds within the hyperbranched EFTH-6 structure, which are inherently more labile and susceptible to hydrolysis compared to the robust ether bonds in DGEBA. As the EFTH-6 content increases, the higher density of ester bonds provides more cleavage sites for hydrolysis, thereby facilitating a greater extent of degradation.

In a significant contribution to the field, Lu et al. [76] reported an innovative strategy for synthesizing hyperbranched epoxy resins (SMERs) with outstanding mechanical properties. Utilizing a straightforward and highly efficient thiol–maleimide click reaction, they successfully fabricated a series of HERs (ETMP-*n*, ETHP-*n*) with tunable MWs and tailored architectures. In this study, Lu utilized (*N,N*)-(4,4-methylene phenyl)bismaleimide as the A_2 component, which was reacted with two different B_3-type cross-linkers—trimethylolpropane tris(3-mercaptopropionate) (TMPMP) and 1,3,5-tris(2-hydroxyethyl)isocyanurate-tris(3-mercaptopropionate) (THEIC-MP)—via a thiol-maleimide click reaction. This step yielded hyperbranched precursors (ETHP-*n* and ETMP-*n*) with varied MWs and backbone characteristics. Subsequently, a thiol–ene click reaction with AGE was performed to introduce epoxy functionalities, resulting in the final series of HERs (ETMP-*n*, ETHP-*n*) combining both flexible and rigid structural elements, as illustrated in Figure 3.42.

The mechanical properties of the cured ETMP-*n* films were systematically evaluated. Results showed that the tensile strength initially increased with MW, peaking at 27.63 MPa for ETMP-9. Beyond this point, further increase in MW led to a decline in tensile strength. A similar trend was observed in the cured ETHP-*n* films, which reached a maximum tensile strength of 31.15 MPa for ETHP-9. It is worth noting that the incorporation of a nitrogen-containing heterocyclic rigid structure in ETHP-*n* contributed to its superior mechanical performance, with the tensile strength of cured ETHP-9 being 12.74% higher than that of ETMP-9.

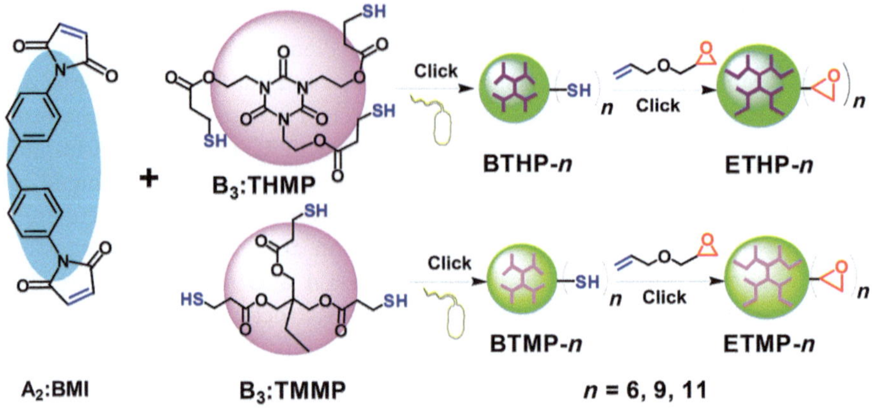

Figure 3.42 Synthetic route of ETHP-*n* and ETMP-*n*. *Source:* [76] / with permission of American Chemical Society.

Nevertheless, the performance of individual HERs often remains inadequate for complex application scenarios that demand multifunctional properties. To overcome these limitations and achieve enhanced material performance, researchers are increasingly focusing on the application of hyperbranched polymers in composite systems. In this context, click chemistry has emerged as a powerful strategy—particularly in the functionalization of carbon fiber surfaces and the strengthening of composite interfaces.

By employing click chemistry, hyperbranched polymers can be precisely grafted onto carbon fibers, significantly improving interfacial adhesion and stress transfer. This approach fully leverages the advantages of each component, enabling synergistic performance complementarity and resulting in composites with well-balanced and superior overall properties.

In this context, Xu et al. [77] developed a novel strategy for modifying carbon fiber fabric and fabricating fully recyclable composites. They synthesized thiol-terminated hyperbranched polyimides (DSPI-*n*) from imide-based monomers and grafted them onto pristine carbon fiber fabric (PCF) via a thiol–ene click reaction, resulting in thiol-functionalized carbon fiber fabric (TCF). This modification significantly enhances the interfacial properties of the resulting composites. Moreover, the composites fabricated in this work can be completely degraded under mild conditions, allowing for the recovery of intact carbon fiber fabric, thereby addressing a crucial challenge in the high-value recycling of composite materials.

The synthesis process began with the preparation of an imide monomer (DTA) from 4,4′-diaminodiphenylmethane (DDM) and TMA. This DTA monomer served as the key building block for two types of imide-skeleton hyperbranched polymers. First, an esterification reaction with THEIC yielded hydroxyl-terminated hyperbranched polyimides (DPI-*n*). These were then converted into thiol-terminated derivatives (DSPI-*n*) via a thiol–ene reaction with MPA. The resulting DSPI-*n* could be grafted onto PCF to produce TCF, or alternatively reacted with AGE to form HERs (DIHE-*n*), as illustrated in Figure 3.43.

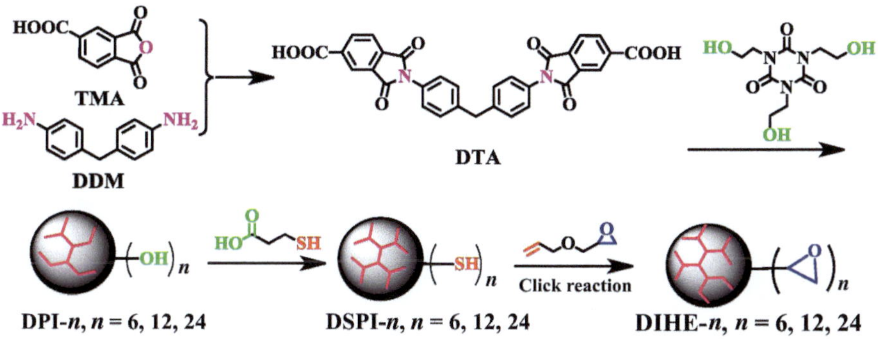

Figure 3.43 Synthetic route of DSPI-*n* and DIHE-*n*. *Source:* [77] / with permission of Elsevier.

Notably, when DSPI-12 was employed to modify the carbon fiber fabric and fabricate the DIHE-12/TCF composite, substantial performance enhancements were achieved. The resulting composite exhibited superior tensile strength compared to the unmodified DIHE-12/PCF counterpart. This reinforcement stems from the multifunctional hyperbranched topological structures of both DSPI-12 and DIHE-12. The grafted DSPI-12 on the carbon fiber surface forms a dense transition layer with DIHE-12 in the matrix, strengthening the interfacial interaction between the modified fabric (TCF12) and the resin. Furthermore, the hyperbranched architecture of these polymers enables effective stress-induced deformation, promoting efficient stress distribution and load transfer throughout the composite. A mechanical interlocking mechanism also develops at the interface, collectively contributing to the enhanced overall performance of the material.

In a pioneering contribution, Liang et al. [78] reported the first preparation of bio-based degradable epoxy resin/carbon fiber composites using biomass-derived FDCA as a building block. These composites exhibit remarkable mechanical properties and strong interfacial interactions. The synthesis commenced with FDCA and TMP as the raw materials, from which a hydroxyl-terminated hyperbranched polymer (OTFP-*n*) was initially prepared. This was followed by an esterification reaction with 3-MPA to yield a thiol-terminated hyperbranched polyester (STFP-*n*). Finally, a thiol–ene click reaction between STFP-*n* and AGE produced the target bio-based degradable HER (ETFP-*n*), as illustrated in Figure 3.44.

STFP-*n* was employed to functionalize PCF, enhancing its wettability and strengthening the interfacial interaction with the polymer matrix. The functionalized fabric (TCF) was then impregnated with ETFP-*n* to fabricate the final composites. Topological analysis via dynamic and static laser scattering revealed that as the MW of ETFP-*n* increased, the shape factor (ρ) and interdiffusion coefficient (D_m) decreased, indicating a structural evolution from loose, linear worm-like chains to more compact ellipsoidal and spherical entangled configurations. Further structural characterization showed that the hydrodynamic radius (R_h) and radius of gyration (R_g) of ETFP-*n* increased with MW. Specifically, ETFP-6, ETFP-12, and ETFP-24 exhibited R_h values of 6.47, 12.32, and 36.55 nm, and R_g

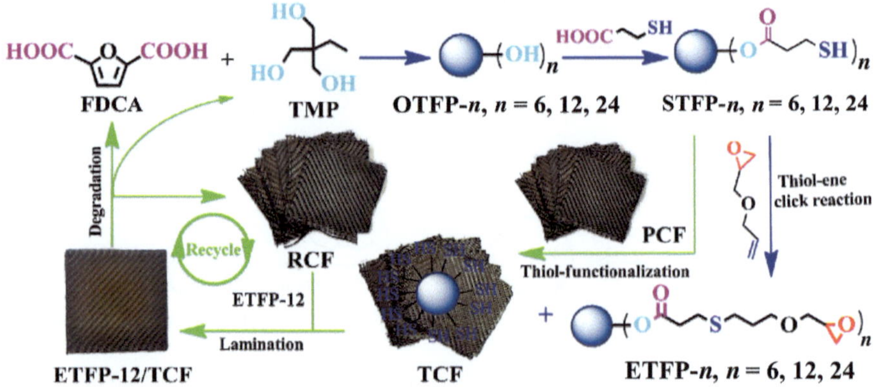

Figure 3.44 Synthetic route and recycling process of hyperbranched epoxy resins/carbon fiber fabric composites. *Source:* [78] / with permission of Elsevier.

values of 10.97, 12.44, and 24.11 nm, respectively. The corresponding shape factor ($\rho = R_g/R_h$) decreased consistently, reflecting a transition from an open, loose architecture in ETFP-6 to a dense core–shell structure in ETFP-24. Rheological measurements at 30 °C showed viscosities of 1.44, 18.55, and 24.54 Pa·s for ETFP-6, ETFP-12, and ETFP-24, respectively. The higher viscosity of ETFP-24 arose from stronger intermolecular friction and entanglement. Notably, viscosity dropped sharply as temperature increased to 80 °C, favoring processability during composite fabrication.

Interfacial performance was critically improved by modifying PCF with STFP-12 to produce TCF. The modified surface showed enhanced wettability and stronger interfacial adhesion with ETFP-12, leading to a 55.98% increase in tensile strength compared to the unmodified composite—surpassing most reported carbon fiber composites.

Moreover, the cured ETFP-12/TCF composites were fully degradable under mild conditions using a phosphoric acid/H_2O_2/ethylene glycol solution at 90 °C within 3 hours. The recovered carbon fibers retained their structural integrity and mechanical properties, enabling closed-loop recycling and high-value reuse in new composites with excellent retention of performance.

3.4 Conclusions and Perspectives

HERs have evolved over decades from laboratory curiosities into a class of high-performance polymers with significant potential in toughening modifications, composite materials, electronic packaging, and sustainable materials. Their unique hyperbranched topology, abundant terminal functional groups, low viscosity, and tunable molecular parameters offer revolutionary solutions to the inherent brittleness, high viscosity, and nonrecyclability of conventional epoxy resins. However, as this review has revealed, despite remarkable progress in synthetic methodologies, a significant gap remains between fundamental research and the large-scale industrial application of HERs. This section aims to systematically outline the core challenges in current HER research and development and, based on this, provide a perspective on future research directions and development prospects.

3.4.1 Challenges

3.4.1.1 Challenges in Synthesis and Precise Structural Control

Despite the development of various synthetic strategies, achieving precise and reproducible control over the structure of HERs remains a challenge in the field.

1) Complexity of Molecular Parameter Control: Key structural parameters of HERs, such as the DB, MW and its distribution (polydispersity index [PDI]), and the type and number of end groups, are decisive for their final properties. However, these parameters are often coupled and difficult to control independently. For instance, in $A_2 + B_3$ polymerizations, increasing the extent of reaction not only typically increases both MW and DB but also significantly elevates the risk of gelation. Most current synthetic methods still only allow "coarse-tuning" of DB and MW, unable to achieve the monodispersity and perfect symmetry of dendrimers, which limits the ultimate performance and predictability of HERs.
2) Limitations and Cost of the Monomer Library: There is still a limited variety of structurally diverse and cost-effective monomers suitable for the synthesis of HER. One-pot polymerization method (especially proton transfer polymerization and step-growth polymerization of AB_n monomers) requires extremely high monomer purity and reactivity. The synthesis of many monomers with special functionalities (e.g., intrinsic flame retardancy, high heat resistance) is often cumbersome and expensive, hindering the development of functional HERs. While the PPM strategy broadens the choice of core precursors, the types of capping reagents available for efficient grafting of epoxy groups are relatively limited, relying primarily on ECH and its derivatives, which restricts the diversity of end-group chemistry.
3) Side Reactions and Structural Defects: Side reactions are difficult to completely avoid during synthesis, especially in reactions involving epoxy groups. For example, epoxy groups can undergo hydrolysis under alkaline conditions; excess ECH may self-polymerize during esterification or etherification; and etherification between hydroxyl and epoxy groups at elevated temperatures can lead to altered DB or pregelation. These side reactions introduce uncontrollable structural defects, affecting the EEW, functionality, and ultimately the homogeneity of the cured network structure.
4) Barriers to Scale-up Production: Many efficient synthetic methods (e.g., click chemistry) perform excellently on a laboratory scale but face challenges when scaled up for industrial production. For instance, UV-initiated thiol–ene click reactions become less efficient when dealing with highly viscous or opaque systems; some catalysts (e.g., platinum, copper) are expensive, may remain as residues affecting performance, and are difficult to recover; reactions requiring strict anhydrous/anaerobic conditions increase production complexity and cost.

Furthermore, drastic viscosity changes during the reaction, exothermic control, and postprocessing (e.g., treatment of large amounts of saline wastewater) are engineering problems that must be solved for industrialization.

3.4.1.2 Challenges in Functionalization and Sustainability

For future applications, the functionalization and enhanced sustainability of HERs are inevitable trends, but they come with new challenges.

1) Functional Integration and Stability: Integrating multiple functionalities (e.g., flame retardancy, self-healing, conductivity, shape memory) into a single HER molecule requires ingenious molecular design. However, different functional groups may have chemical reactivity or compatibility issues. For example, the introduction of flame-retardant elements (e.g., phosphorus, nitrogen) may affect curing kinetics or reduce the hydrolytic stability of the material. Dynamic covalent bonds (e.g., ester, imine, hexahydro-*s*-triazine), while enabling recyclability, may also cleave under service conditions, affecting long-term durability.
2) Economic Viability and Performance Balance of Sustainable Pathways: Although breakthroughs have been made in bio-based and recyclable HERs, their large-scale application still faces hurdles. Bio-based monomers (e.g., vanillin, itaconic acid, FDCA) are often more expensive than their petroleum-based counterparts, and their chemical structures often impart high rigidity and insufficient toughness to the resulting resins. The degradation conditions for recyclable HERs (e.g., acidic or alkaline environments) can be harsh, and the purification and repolymerization processes for the degradation products are complex, raising questions about economic feasibility.
3) Lack of Life Cycle Assessment (LCA): Current environmental impact assessments for HERs, particularly novel sustainable HERs, often focus on single attributes like "degradable" or "recyclable." There is a lack of systematic and comprehensive LCA data. Whether these materials truly have lower energy consumption, carbon footprint, and overall environmental impact than conventional epoxies throughout their life cycle—from raw material acquisition and synthesis to processing, use, and final disposal—still requires substantial data support.

3.4.2 Perspectives

Addressing the aforementioned challenges, future HER research will move toward greater precision, functionality, intelligence, and sustainability. The following directions deserve focused attention:

3.4.2.1 Innovation of Synthesis Methods

1) Developing Novel Controlled Polymerizations and Click Chemistries: Explore new click reactions that are insensitive to moisture and oxygen and offer higher efficiency, such as copper-free strain-promoted azide-alkyne cycloaddition (SPAAC) and Diels–Alder reactions, to simplify synthetic conditions. Develop new mechanisms for the ring-opening polymerization of epoxy groups

themselves to achieve living/controlled polymerization, enabling precise control over MW and chain architecture. Introduce green synthetic methods like enzyme catalysis and electrochemical synthesis into HER preparation to improve reaction selectivity and atom economy.

2) Leveraging Artificial Intelligence (AI) and Big Data for Molecular Design: Utilize machine learning and AI to integrate existing HER synthesis data (monomer types, reaction conditions, catalysts) with performance data (DB, MW, mechanical properties, etc.), establishing predictive structure–property models. Use these models to inversely guide the design of synthetic routes, rapidly screening for the optimal molecular structure and synthesis protocol to achieve target properties, enabling "rational design" and drastically reducing experimental trial-and-error costs and time.
3) Advancing Characterization and Monitoring Techniques: Employ advanced characterization techniques such as high-field nuclear magnetic resonance (NMR), high-resolution mass spectrometry, and small-angle X-ray/neutron scattering to more accurately resolve the complex structure of HERs, particularly end-group distribution and chain topology. Introduce online monitoring techniques like infrared (IR) and Raman spectroscopy during synthesis to track reaction progress and functional group conversion in real time, providing immediate data for process control and optimization, ensuring product quality and reproducibility.

3.4.2.2 Toward Sustainable Development and a Circular Economy

1) Expanding Bio-based Monomer Resources and Enhancing Performance: Develop and utilize a wider range of abundant, low-cost bio-based feedstocks, such as lignin degradation products, cellulose/hemicellulose derivatives, and modified vegetable oils. Overcome the performance shortcomings (e.g., brittleness) associated with bio-based monomers through chemical modification and molecular design, developing high-performance bio-based HERs whose properties can match or even surpass those of petroleum-based products.
2) Optimizing Degradation and Recycling Strategies: Develop degradation methods with milder conditions and lower energy consumption. Beyond utilizing dynamic covalent bonds like esters, hexahydro-*s*-triazines, and acetals, explore supramolecular chemistry based on physical interactions (e.g., hydrogen bonding, host-guest recognition) to construct reversible cross-linked networks. Investigate closed-loop recycling processes, focusing not only on the efficient recovery of monomers but also on the purification and repolymerization of the recovered monomers to ensure the performance of the regenerated material.
3) Promoting Industrial Application and Standard Establishment: Strengthen industry–academia collaboration. Conduct pilot-scale and industrial demonstration studies for promising HER systems to address engineering and equipment issues in scale-up production. Simultaneously, promote the establishment of performance testing and evaluation standards for HERs and their composites, providing a basis for quality control and market access to accelerate their commercialization.

3.4.3 Conclusion

As a rising star in the epoxy resin family, HERs, with their unique structural and performance advantages, offer a promising pathway to addressing the bottleneck problems of traditional thermosets. We are at a critical turning point, moving from "being able to synthesize" to "synthesizing with precision," from "improving a single property" to "synergistically enhancing comprehensive performance," and from "linear consumption" to "circular utilization." Future research requires close collaboration among chemists, materials scientists, chemical engineers, and computational theorists. Through multidisciplinary integration, we can collectively overcome the core challenges in synthesis, structure–property relationships, and sustainability. It is foreseeable that as these challenges are gradually addressed, HERs will play an increasingly important role in fields such as aerospace, advanced electronics, green transportation, and sustainable packaging, opening a new chapter in the development of high-performance polymeric materials.

Acknowledgments

The authors gratefully acknowledge financial support from the National Natural Science Foundation of China (U23A20691, 52373017), Natural Science Foundation of Hubei Province (2025AFB798), and Innovation Group of National Ethnic Affairs Commission of China (MZR20006).

References

1 Shundo, A., Yamamoto, S., and Tanaka, K. (2022). Network formation and physical properties of epoxy resins for future practical applications. *JACS Au* 2 (7): 1522–1542. https://doi.org/10.1021/jacsau.2c00120.
2 Weiss, H. (1957). Anhydride curing agents for epoxy resins. *Ind. Eng. Chem.* 49 (7): 1089–1090. https://doi.org/10.1021/ie50571a024.
3 Montarnal, D., Capelot, M., Tournilhac, F., and Leibler, L. (2011). Silica-like malleable materials from permanent organic networks. *Science* 334 (6058): 965–968. https://doi.org/10.1126/science.1212648.
4 Charlesworth, J. (1980). An analysis of the substitution effects involved in diepoxide-diamine copolymerization reactions. *J. Polym. Sci., Polym. Chem. Ed.* 18 (2): 621–628. https://doi.org/10.1002/pol.1980.170180221.
5 Hoyle, C.E., Lowe, A.B., and Bowman, C.N. (2010). Thiol-click chemistry: a multifaceted toolbox for small molecule and polymer synthesis. *Chem. Soc. Rev.* 39 (4): 1355–1387. https://doi.org/10.1039/B901979K.
6 Dušek, K., Bleha, M., and Luňák, S. (1977). Curing of epoxide resins: model reactions of curing with amines. *J. Polym. Sci., Polym. Chem. Ed.* 15 (10): 2393–2400. https://doi.org/10.1002/pol.1977.170151009.

7 Jyotish, J., Mahapatra, B.K., Moharana, S., and Mahaling, R.N. (2025). Chemistry and types of epoxy resins. In: *Recent Advances on Waterborne Epoxy Coatings: Synthesis, Properties and Applications* (ed. S. Pradhan, S. Moharana, and S. Mohanty), 9–35. Singapore: Springer Nature Singapore.
8 Aziz, T., Haq, F., Farid, A. et al. (2024). The epoxy resin system: function and role of curing agents. *Carbon Lett.* 34 (1): 477–494. https://doi.org/10.1007/s42823-023-00547-7.
9 Li, J., Zhang, B., Yuan, X. et al. (2025). Benzene ring structural design strategy toward well-balanced thermal and electrical properties in epoxy dielectric polymers. *Compos. Part B* 297: 112308. https://doi.org/10.1016/j.compositesb.2025.112308.
10 Domun, N., Hadavinia, H., Zhang, T. et al. (2015). Improving the fracture toughness and the strength of epoxy using nanomaterials – a review of the current status. *Nanoscale* 7 (23): 10294–10329. https://doi.org/10.1039/C5NR01354B.
11 Zhang, J., Zhang, Z., Huang, R., and Tan, L. (2025). Advances in toughening modification methods for epoxy resins: a comprehensive review. *Polymers* 17 (9): 1288.
12 Jiang, Y., Li, J., Li, D. et al. (2024). Bio-based hyperbranched epoxy resins: synthesis and recycling. *Chem. Soc. Rev.* 53 (2): 624–655. https://doi.org/10.1039/D3CS00713H.
13 Kim, Y.H. and Webster, O.W. (1992). Hyperbranched polyphenylenes. *Macromolecules* 25 (21): 5561–5572. https://doi.org/10.1021/ma00047a001.
14 Ghosh, A., Banerjee, S., and Voit, B. (2015). Aromatic hyperbranched polymers: synthesis and application. In: *Porous Carbons – Hyperbranched Polymers – Polymer Solvation* (ed. T.E. Long, B. Voit, and O. Okay), 27–124. Springer International Publishing.
15 Mezzenga, R., Boogh, L., Månson, J.-A.E., and Pettersson, B. (2000). Effects of the branching architecture on the reactivity of epoxy–amine groups. *Macromolecules* 33 (12): 4373–4379. https://doi.org/10.1021/ma991906w.
16 Zhang, D., Liang, E., Li, T. et al. (2013). The effect of molecular weight of hyperbranched epoxy resins with a silicone skeleton on performance. *RSC Adv.* 3 (24): 9522–9529. https://doi.org/10.1039/C3RA00023K.
17 Wu, J., Liang, N., Liu, X. et al. (2022). Controllability on topological structures and properties of hyperbranched epoxy resins. *Prog. Org. Coat.* 165: 106735. https://doi.org/10.1016/j.porgcoat.2022.106735.
18 Niu, S., Yan, H., Li, S. et al. (2016). A multifunctional silicon-containing hyperbranched epoxy: controlled synthesis, toughening bismaleimide and fluorescent properties. *J. Mater. Chem. C* 4 (28): 6881–6893. https://doi.org/10.1039/C6TC02546C.
19 Zhen, X., Li, H., Xu, Z. et al. (2021). Demethylation, phenolation, and depolymerization of lignin for the synthesis of lignin-based epoxy resin via a one-pot strategy. *Ind. Crop. Prod.* 173: 114135. https://doi.org/10.1016/j.indcrop.2021.114135.

20 Chang, H.-T. and Fréchet, J.M.J. (1999). Proton-transfer polymerization: a new approach to hyperbranched polymers. *J. Am. Chem. Soc.* 121 (10): 2313–2314. https://doi.org/10.1021/ja983797r.

21 Gadwal, I., Binder, S., Stuparu, M.C., and Khan, A. (2014). Dual-reactive hyperbranched polymer synthesis through proton transfer polymerization of thiol and epoxide groups. *Macromolecules* 47 (15): 5070–5080. https://doi.org/10.1021/ma500920z.

22 Emrick, T., Chang, H.-T., and Fréchet, J.M.J. (1999). An $A_2 + B_3$ approach to hyperbranched aliphatic polyethers containing chain end epoxy substituents. *Macromolecules* 32 (19): 6380–6382. https://doi.org/10.1021/ma990687b.

23 Emrick, T., Chang, H.-T., and Fréchet, J.M.J. (2000). The preparation of hyperbranched aromatic and aliphatic polyether epoxies by chloride-catalyzed proton transfer polymerization from AB_n and $A_2 + B_3$ monomers. *J. Polym. Sci. A Polym. Chem.* 38 (S1): 4850–4869. https://doi.org/10.1002/1099-0518(200012)38:1+<4850::aid-pola230>3.0.co;2-g.

24 Ma, L.-j., Wang, H.-q., He, L.-f., and Li, X.-y. (2011). Hyperbranched epoxy resins prepared by proton transfer polymerization from an $A_2 + B_3$ system. *Chin. J. Polym. Sci.* 29 (3): 300–307. https://doi.org/10.1007/s10118-011-1034-x.

25 Lv, J.-Y., Meng, Y., He, L.-F. et al. (2012). Synthesis of a hyperbranched polyether epoxy through one-step proton transfer polymerization and its application as a toughener for epoxy resin DGEBA. *Chin. J. Polym. Sci.* 30 (4): 493–502. https://doi.org/10.1007/s10118-012-1166-7.

26 Luo, L., Meng, Y., Qiu, T., and Li, X. (2013). An epoxy-ended hyperbranched polymer as a new modifier for toughening and reinforcing in epoxy resin. *J. Appl. Polym. Sci.* 130 (2): 1064–1073. https://doi.org/10.1002/app.39257.

27 Cook, A.B., Barbey, R., Burns, J.A., and Perrier, S. (2016). Hyperbranched polymers with high degrees of branching and low dispersity values: pushing the limits of thiol–yne chemistry. *Macromolecules* 49 (4): 1296–1304. https://doi.org/10.1021/acs.macromol.6b00132.

28 Khalyavina, A., Schallausky, F., Komber, H. et al. (2010). Aromatic–aliphatic polyesters with tailored degree of branching based on AB/AB_2 and ABB^*/AB_2 monomers. *Macromolecules* 43 (7): 3268–3276. https://doi.org/10.1021/ma100037n.

29 Fei, X., Tang, Y., Wei, W. et al. (2018). One-pot synthesis of tetramethyl biphenyl backboned hyperbranched epoxy resin as an efficient toughening modifier for two epoxy curing systems. *Polym. Bull.* 75 (10): 4571–4586. https://doi.org/10.1007/s00289-018-2269-2.

30 Boonlert-uthai, T., Samthong, C., and Somwangthanaroj, A. (2019). Synthesis, thermal properties and curing kinetics of hyperbranched BPA/PEG epoxy resin. *Polymers* 11 (10) 1545.

31 Wang, Y., Chen, S., Chen, X. et al. (2019). Controllability of epoxy equivalent weight and performance of hyperbranched epoxy resins. *Compos. Part B* 160: 615–625. https://doi.org/10.1016/j.compositesb.2018.12.103.

32 Hu, C., Jiang, J., Hu, X. et al. (2025). Hyperbranched epoxy resin with flexible siloxane skeleton for improving the comprehensive performance of epoxy thermosets. *Polym. Adv. Technol.* 36 (8): e70306. https://doi.org/10.1002/pat.70306.

33 Patil, D.M., Phalak, G.A., and Mhaske, S.T. (2017). Synthesis of bio-based epoxy resin from gallic acid with various epoxy equivalent weights and its effects on coating properties. *J. Coat. Technol. Res.* 14 (2): 355–365. https://doi.org/10.1007/s11998-016-9853-x.

34 Qiao, H., Chen, M., Chen, B. et al. (2023). Understanding the curing kinetics of boron-based hyperbranched polysiloxane reinforced and toughened epoxy resin by rheology. *Chem. Eng. J.* 467: 143542. https://doi.org/10.1016/j.cej.2023.143542.

35 De, B. and Karak, N. (2013). Novel high performance tough hyperbranched epoxy by an $A_2 + B_3$ polycondensation reaction. *J. Mater. Chem. A* 1 (2): 348–353. https://doi.org/10.1039/C2TA00011C.

36 Boro, U. and Karak, N. (2017). Tannic acid based hyperbranched epoxy/reduced graphene oxide nanocomposites as surface coating materials. *Prog. Org. Coat.* 104: 180–187. https://doi.org/10.1016/j.porgcoat.2016.10.039.

37 Zhong, L., Hao, Y., Zhang, J. et al. (2022). Closed-loop recyclable fully bio-based epoxy vitrimers from ferulic acid-derived hyperbranched epoxy resin. *Macromolecules* 55 (2): 595–607. https://doi.org/10.1021/acs.macromol.1c02247.

38 Hao, Y., Zhong, L., Li, T. et al. (2023). High-performance recyclable and malleable epoxy resin with vanillin-based hyperbranched epoxy resin containing dual dynamic bonds. *ACS Sustain. Chem. Eng.* 11 (30): 11077–11087. https://doi.org/10.1021/acssuschemeng.3c01337.

39 Wei, F., Zhang, J., Wu, C. et al. (2023). Closed-loop recycling of tough and flame-retardant epoxy resins. *Macromolecules* 56 (14): 5290–5305. https://doi.org/10.1021/acs.macromol.3c00650.

40 Srinivasa Rao, V. and Samui, A.B. (2008). Effect of molecular architecture and size of mesogen on phase behavior and photoactive properties of photoactive liquid crystalline hyperbranched polyester epoxies containing benzylidene moiety. *J. Polym. Sci. A Polym. Chem.* 46 (2): 552–563. https://doi.org/10.1002/pola.22405.

41 Theis, J., Ritter, H., and Klee, J.E. (2009). Microwave-assisted one-pot synthesis of hyperbranched epoxide-amine adducts. *Macromol. Rapid Commun.* 30 (16): 1424–1427. https://doi.org/10.1002/marc.200900157.

42 Hu, J., Zhang, L., Chen, M. et al. (2023). Synthesis of hyperbranched flame retardants with varied branched chains' rigidity and performance of modified epoxy resins. *Polymers* 5 (2): 449.

43 Sörensen, K., Pettersson, B., Boogh, L., and Månsson, J.-A. E. (1996). A thermosetting material (WO 1996/19537 A1). World Intellectual Property Organization.

44 Lv, J., Meng, Y., He, L. et al. (2012). Novel epoxidized hyperbranched poly(phenylene oxide): synthesis and application as a modifier for diglycidyl ether of bisphenol A. *J. Appl. Polym. Sci.* 128 (1): 907–914. https://doi.org/10.1002/app.38270.

45 Huang, P., Gu, A., Liang, G., and Yuan, L. (2011). Synthesis of epoxy-functionalized hyperbranched poly(phenylene oxide) and its modification of cyanate ester resin. *J. Appl. Polym. Sci.* 123 (4): 2351–2359. https://doi.org/10.1002/app.34791.

46 Chen, S., Zhang, D., Jiang, S., and Jia, D. (2011). Preparation of hyperbranched epoxy resin containing nitrogen heterocycle and its toughened and reinforced composites. *J. Appl. Polym. Sci.* 123 (6): 3261–3269. https://doi.org/10.1002/app.35012.

47 Tao, X., Wei, M., Hu, X. et al. (2024). Novel resveratrol-based hyperbranched epoxy resin suitable for comprehensive modification of epoxy resins. *Polymer* 302: 127093. https://doi.org/10.1016/j.polymer.2024.127093.

48 Yang, D., Chen, Z., Rong, X. et al. (2013). Formulation and characterization of epoxidized hydroxyl-terminated hyperbranched polyester and its application in waterborne epoxy resin. *J. Polym. Res.* 21: https://doi.org/10.1007/s10965-013-0331-1.

49 De, B., Gupta, K., Mandal, M., and Karak, N. (2014). Biodegradable hyperbranched epoxy from castor oil-based hyperbranched polyester polyol. *ACS Sustain. Chem. Eng.* 2 (3): 445–453. https://doi.org/10.1021/sc400358b.

50 Liu, T., Nie, Y., Zhang, L. et al. (2015). Dependence of epoxy toughness on the backbone structure of hyperbranched polyether modifiers. *RSC Adv.* 5 (5): 3408–3416. https://doi.org/10.1039/C4RA10974K.

51 Xia, M., Yang, H., Ling, J. et al. (2018). The mechanical behaviors of epoxy-terminated hyperbranched polyester (E-HBP) as toughener in different epoxy resins. *Adv. Compos. Hybrid Mater.* 1 (2): 310–319. https://doi.org/10.1007/s42114-018-0027-4.

52 Fu, J.F., Shi, L.Y., Yuan, S. et al. (2008). Morphology, toughness mechanism, and thermal properties of hyperbranched epoxy modified diglycidyl ether of bisphenol A (DGEBA) interpenetrating polymer networks. *Polym. Adv. Technol.* 19 (11): 1597–1607. https://doi.org/10.1002/pat.1175.

53 Vora, A., Nasrullah, M.J., and Webster, D.C. (2007). Synthesis and characterization of novel epoxy- and oxetane-functional reversible addition–fragmentation chain transfer agents. *Macromolecules* 40 (24): 8586–8592. https://doi.org/10.1021/ma071065q.

54 Zhang, D. and Jia, D. (2006). Synthesis of novel low-viscosity liquid epoxidized aromatic hyperbranched polymers. *Eur. Polym. J.* 42 (3): 711–714. https://doi.org/10.1016/j.eurpolymj.2005.09.013.

55 Zhang, D., Jia, D., and Zhou, Z. (2009). Synthesis and characterization of low viscosity aromatic hyperbranched polyester epoxy resin. *Macromol. Res.* 17 (5): 289–295. https://doi.org/10.1007/BF03218864.

56 Yu, Q., Liang, Y., Cheng, J. et al. (2017). Synthesis of a degradable high-performance epoxy-ended hyperbranched polyester. *ACS Omega* 2 (4): 1350–1359. https://doi.org/10.1021/acsomega.7b00132.

57 Xu, Z., Liang, Y., Ma, X. et al. (2020). Recyclable thermoset hyperbranched polymers containing reversible hexahydro-*s*-triazine. *Nat. Sustainability* 3 (1): 29–34. https://doi.org/10.1038/s41893-019-0444-6.

58 Han, X., Wu, Y., Han, H. et al. (2023). Super-strong reusable hot melt adhesives prepared from hyperbranched epoxy resin. *Mater. Chem. Front.* 7 (13): 2676–2682. https://doi.org/10.1039/D3QM00090G.

59 Gong, C. and Fréchet, J.M.J. (2000). End functionalization of hyperbranched poly(siloxysilane): novel crosslinking agents and hyperbranched–linear star block copolymers. *J. Polym. Sci. A Polym. Chem.* 38 (16): 2970–2978. https://doi.org/10.1002/1099-0518(20000815)38:16%3C2970::AID-POLA150%3E3.0.CO;2-%23.

60 Kang, N., Du, Z., Li, H., and Zhang, C. (2011). Synthesis, characterization of an epoxy-terminated carbosiloxane and its application as a modifier in epoxy systems. *Polym. Eng. Sci.* 52 (4): 868–874. https://doi.org/10.1002/pen.22153.

61 Zhang, D., Liang, E., Li, T. et al. (2013). Environment-friendly synthesis and performance of a novel hyperbranched epoxy resin with a silicone skeleton. *RSC Adv.* 3 (9): 3095–3102. https://doi.org/10.1039/C2RA22969B.

62 Pan, Z., Lin, L., and Cao, Z. (2024). The synthesis of an epoxy-terminated hyperbranched polysiloxane toughener and its application at low temperature. *React. Funct. Polym.* 195: 105828. https://doi.org/10.1016/j.reactfunctpolym.2024.105828.

63 Liu, Y., Wang, Y., Jiang, X., and Zhang, X. (2015). Cure behavior and enhanced thermal properties of benzoxazine thermosets with isocyanate-terminated hyperbranched polyurethane of different degree of branching. *Polym. Bull.* 72 (8): 2105–2125. https://doi.org/10.1007/s00289-015-1392-6.

64 Liu, Y., Zhang, X., and Gao, N. (2015). Synthesis and characterization of isocyanate-functionalized hyperbranched polyurethane and its cocuring with benzoxazine. *Polym. Eng. Sci.* 55 (3): 604–613. https://doi.org/10.1002/pen.23925.

65 Wang, Y., Jiang, X., Zhang, C. et al. (2017). Synthesis of epoxide functionalized hyperbranched polyurethane and its blending with benzoxazine: cure kinetics and thermal properties. *Polym. Bull.* 74 (10): 4209–4222. https://doi.org/10.1007/s00289-017-1949-7.

66 Guo, W., Chen, S., Cheng, J. et al. (2019). Synthesis of renewable and self-curable thermosetting hyperbranched polymers by a click reaction. *Prog. Org. Coat.* 134: 189–196. https://doi.org/10.1016/j.porgcoat.2019.05.009.

67 Park, E.J., Lee, W.-H., and Bae, C. (2016). Versatile functionalization of aromatic polysulfones via thiol-ene click chemistry. *J. Polym. Sci. A Polym. Chem.* 54 (19): 3237–3243. https://doi.org/10.1002/pola.28210.

68 Jin, K., Leitsch, E.K., Chen, X. et al. (2018). Segmented thermoplastic polymers synthesized by thiol–ene click chemistry: examples of thiol–norbornene and thiol–maleimide click reactions. *Macromolecules* 51 (10): 3620–3631. https://doi.org/10.1021/acs.macromol.8b00573.

69 Lowe, A.B. (2014). Thiol–ene "click" reactions and recent applications in polymer and materials synthesis: a first update. *Polym. Chem.* 5 (17): 4820–4870. https://doi.org/10.1039/C4PY00339J.

70 Li, Y., Zhang, Y., Goswami, M. et al. (2020). Liquid crystalline networks based on photo-initiated thiol–ene click chemistry. *Soft Matter* 16 (7): 1760–1770. https://doi.org/10.1039/C9SM01818B.

71 Hoyle, C.E. and Bowman, C.N. (2010). Thiol–ene click chemistry. *Angew. Chem. Int. Ed.* 49 (9): 1540–1573. https://doi.org/10.1002/anie.200903924.

72 Zhang, D., Liu, C., Chen, S. et al. (2016). Highly efficient preparation of hyperbranched epoxy resins by UV-initiated thiol-ene click reaction. *Prog. Org. Coat.* 101: 178–185. https://doi.org/10.1016/j.porgcoat.2016.08.010.

73 Wang, Y., Chen, S., Guo, W. et al. (2019). The precise effect of degree of branching of epoxy-ended hyperbranched polymers on intrinsic property and performance. *Prog. Org. Coat.* 127: 157–167. https://doi.org/10.1016/j.porgcoat.2018.11.014.

74 Liu, X., Wu, H., Xu, W. et al. (2024). Ultrastrong and high-tough thermoset epoxy resins from hyperbranched topological structure and subnanoscaled free volume. *Adv. Mater.* 36 (9): e2308434. https://doi.org/10.1002/adma.202308434.

75 Chen, X., Chen, S., Xu, Z. et al. (2020). Degradable and recyclable bio-based thermoset epoxy resins. *Green Chem.* 22 (13): 4187–4198. https://doi.org/10.1039/D0GC01250E.

76 Lu, Y., Xu, H., Liang, N. et al. (2022). High mechanical strength of shape-memory hyperbranched epoxy resins. *ACS Appl. Polym. Mater.* 4 (8): 5574–5582. https://doi.org/10.1021/acsapm.2c00625.

77 Xu, H., Hu, J., Liu, X. et al. (2023). Interface strengthening and high-value recycling of epoxy resin/carbon fiber fabric composites. *Chem. Eng. J.* 465: 142998. https://doi.org/10.1016/j.cej.2023.142998.

78 Liang, N., Liu, X., Hu, J. et al. (2023). Influence of topological structure on mechanical property of recyclable bio-based hyperbranched epoxy/carbon fiber fabric composites. *Chem. Eng. J.* 471: 144329. https://doi.org/10.1016/j.cej.2023.144329.

4

Applications of Hyperbranched Epoxy Resins

Zejun Xu[1], Zi Wang[1], and Xingping Zhou[2]

[1] *Key Laboratory of Catalysis and Energy Materials Chemistry of Ministry of Education & Hubei Key Laboratory of Catalysis and Materials Science, Hubei R&D Center of Hyperbranched Polymers Synthesis and Applications, School of Chemistry and Materials Science, South-Central Minzu University, 182 Minyuan Road, Hongshan District, Wuhan 430074, China*
[2] *School of Chemistry and Chemical Engineering, Huazhong University of Science and Technology, 1037 Luoyu Road, Wuhan 430074, Hubei Province, China*

4.1 Introduction

Epoxy resins (EPs) are recognized as strategically important thermosets with the broadest range of applications. Their intrinsic properties—including mechanical strength, chemical inertness, thermal and corrosion resistance, effective electrical insulation, and strong adhesive properties—make them suitable for diverse uses such as high-performance composites, anti-corrosion coatings, adhesives, semiconductor encapsulation, and electrical insulation products [1–6]. The unique chemical architecture of EPs is key to their critical function across multiple sectors. This makes them indispensable in fields ranging from construction and transportation to electronics, insulation, and aerospace [7–10]. Global production capacity for EPs reached 4.69 million tons in 2024. Demand is projected to rise steadily, driven by expanding end-use industries like transportation, aerospace, marine coatings, and composites. As shown in Figure 4.1, coatings represent the largest application sector, accounting for approximately 50% of total EPs consumption.

In 2024, China accounted for approximately 51.0% of global EPs consumption, reaching 2.155 million tons. This dominant share contrasts with the considerably smaller portions held by Western Europe and the United States (each around 10%) and South Korea (approximately 4%). Growth in this regional market is expected to be further propelled by rising demand from the automotive sector. Epoxy–matrix composites have thus established themselves as critical materials in modern industry. However, their performance is intrinsically limited by the nature of conventional thermoset networks. Upon curing, epoxies form densely crosslinked structures that exhibit negligible plastic deformation. Concurrently, polymerization-induced volumetric shrinkage generates internal stresses, collectively rendering the materials susceptible to crack initiation and propagation. This inherent brittleness poses a

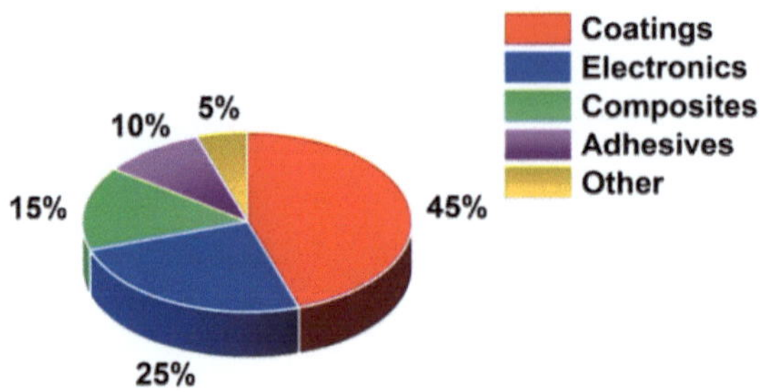

Figure 4.1 The global consumption composition of EPs by sector demand.

significant constraint on their application in next-generation, high-performance composites [11, 12]. Improving the fracture resistance of EPs has centered on both homogeneous and heterogeneous toughening strategies. A common and effective method involves incorporating a secondary phase into the epoxy matrix to enhance the toughness of the cured network [13]. This results in a heterogeneous microstructure within the cured epoxy, comprising the original matrix and a discrete toughening phase. By definition, this second phase consists of a component chemically distinct from the base resin [14, 15]. Researchers have developed diverse strategies to toughen EPs, including rubber (elastomers, REs), thermoplastics (TPs), core–shell polymers (CSPs), liquid crystal polymers (LCPs), block copolymers (BCPs), nanomaterials (NMs), hyperbranched polymers (HBPs), bio-based materials (BBMs), and topological structures (TSs) [16–22].

Heterogeneous toughening approaches, for instance, encompass the formation of interpenetrating or semi-interpenetrating polymer networks (IPNs/SIPNs), in situ polymerization strategies, and the incorporation of toughening agents such as REs, TPs, CSPs, LCPs, BCPs, and NMs. In contrast, homogeneous toughening is typically achieved by blending the epoxy matrix with BBMs, flexible chain segments, or HBPs [23–30].

To analyze historical trends in EP toughening, the authors surveyed published studies from multiple databases and quantified the publication output for each method over time. Research prior to 1974 was limited, focusing primarily on the incorporation of REs and TPs. A significant expansion in methodological approaches occurred from the mid-1970s through the 1980s, with the introduction of techniques involving BCPs, NMs, and TSs [31, 32]. The following decade witnessed the adoption of core–shell latex particles (CSLPs), thermotropic LCPs , and BBMs for epoxy toughening. Subsequently, between 1998 and 2009, HBPs emerged as a novel toughening agent. As summarized in Figure 4.2, research efforts in recent years have predominantly focused on second-phase (heterogeneous) toughening strategies, while homogeneous approaches have received comparatively less attention. This disparity is exemplified by publication counts from 2010 to 2021, which show 1995 articles on NM-based toughening versus only 121 on HBP-mediated homogeneous toughening.

Despite the fact that EP boasts numerous outstanding properties and a broad scope of applications, the high crosslinking degree of its cured products results in inferior toughness. Consequently, research on epoxy toughening ranks among the most critical topics in the practical application of epoxy. While researchers

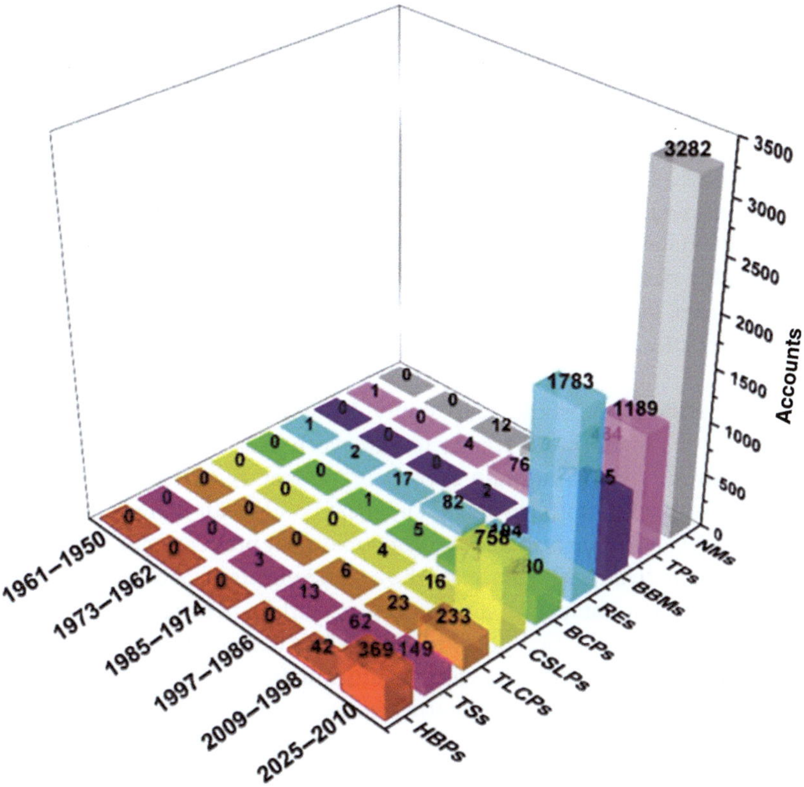

Figure 4.2 The number of published papers with different toughening methods.

have developed a series of toughening approaches—such as rubber toughening, engineering plastic toughening, and nanoparticle (NP) toughening—through years of efforts, and these methods have effectively enhanced the toughness of cured epoxy [33–39], traditional toughening techniques mentioned above often struggle to balance other essential properties of the material. For instance, rubber toughening tends to reduce the material's strength and stiffness; engineering plastic toughening faces issues like high viscosity, poor compatibility with EP, and strict requirements for the curing process to achieve phase separation-induced toughening; NP toughening also leads to increased viscosity, and it is challenging to realize uniform dispersion of NPs. In contrast, HBP modifiers can largely avoid the aforementioned problems. Since HBPs were first used for epoxy toughening in 1996, a variety of HBP-based toughening agents have been developed [2, 24, 26, 40, 41].

As a type of thermosetting polymer featuring a unique three-dimensional network structure and highly branched molecular configuration, hyperbranched EP (HER) has exhibited extensive application potential in multiple fields in recent years, thanks to its excellent physical performance, processability, and structural designability [42–49]. The following content will provide a review of the application fields of hyperbranched EP.

4.2 Adhesive Materials

HERs have emerged as a promising class of materials in the adhesive industry, primarily attributed to their unique architectural features—including low viscosity, high functionality, and excellent compatibility—that distinguish them from traditional linear EPs. These inherent advantages enable HERs to address the escalating demands of modern bonding scenarios involving diverse substrates such as metals, composite materials, and ceramics. However, to fully exploit their potential and ensure reliable performance in practical applications, HER-based adhesives must meet a series of stringent requirements, which are closely intertwined with substrate characteristics, performance metrics, processing compatibility, and safety compliance.

From the perspective of substrate adaptability, HERs need to be tailored to the surface properties of different adherends to achieve robust interfacial bonding. For metallic substrates (e.g., aluminum alloys, carbon steels), HERs must exhibit adequate surface wettability to form stable chemical bonds with surface hydroxyl groups and oxide layers, while also being compatible with pretreatment processes (e.g., phosphating, anodization) to eliminate the adverse effects of oil contamination and oxide impurities. For composite substrates (e.g., carbon fiber-reinforced composites, glass fiber [GF]-reinforced plastics [GFRPs]), a delicate balance between compatibility with the resin matrix and permeability into fiber interfaces is essential to prevent interfacial debonding caused by stress concentration. For ceramic or glass substrates, adjusting the molecular chain structure to enhance polarity matching with inorganic substrates is necessary to ensure effective bonding under ambient or moderate–low-temperature conditions.

In summary, the successful application of HERs in adhesives hinges on the synergistic fulfillment of these multidimensional requirements. As the demand for high-performance, environmentally friendly adhesives continues to grow, further optimization of HER formulations and processing technologies to meet these requirements will remain a key research focus in the field.

Zhang et al. fabricated hyperbranched polyamine adhesives based on sucrose, employing air oxidation and hyperbranched crosslinking strategies [50] (Figure 4.3). Three sucrose-PA adhesives were prepared with sucrose/PA molar ratios of 3:1, 5:1, and 10:1, maintaining a consistent solid content of approximately 50%. The synthesis involved a 1-hour sucrose hydrolysis step, followed by a 1-hour reaction at 90 °C. The plywood was hot-pressed at 200 °C for bonding, and its tensile shear strength was evaluated according to the Chinese National Standard (GB/T 9846-2015). As shown in Figure 4.4 the dry shear strength of plywood bonded with the three adhesives showed minimal variation across the different molar ratios, with all values exceeding 0.7 MPa. However, since dry shear strength primarily reflects the inherent strength of the wood substrate rather than the adhesive's bonding performance, these data provide limited insight into the adhesive efficacy itself.

Liu et al. presented a versatile strategy for fabricating a HBP elastomer pressure-sensitive adhesive (PSA) via one-pot polymerization of thioctic acid with multi-vinyl monomers within a thermoreversible agarose network [51] (Figure 4.5).

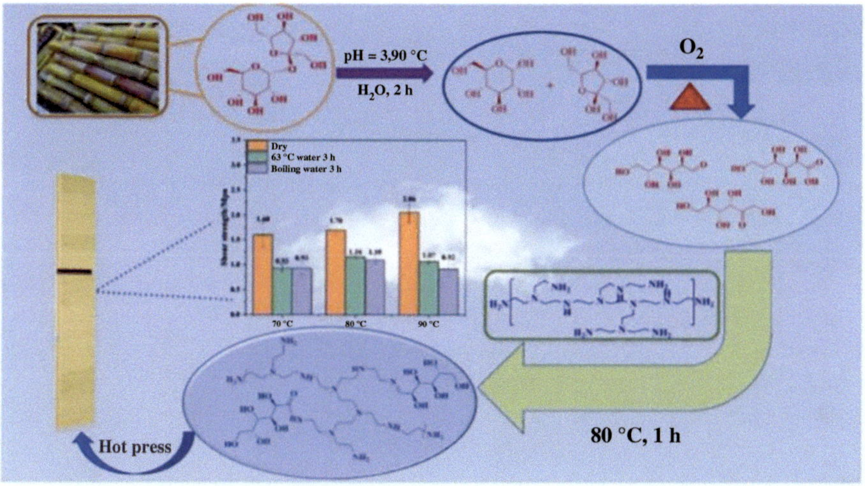

Figure 4.3 Diagram of preparation of sucrose-based hyperbranched polyamine adhesive. *Source:* Schreib-Engel / Pixabay / CC-0.

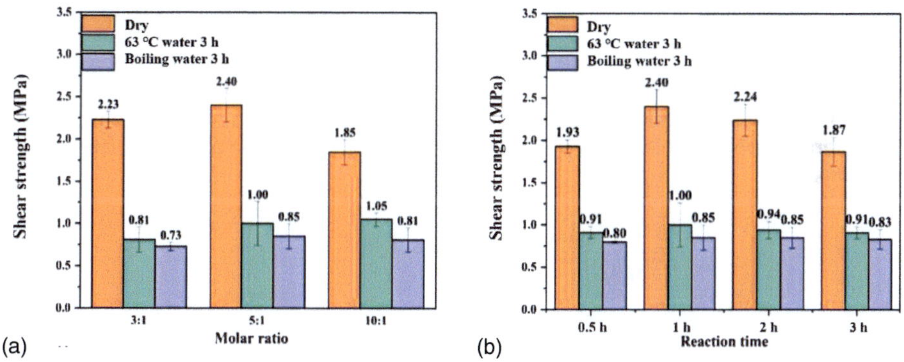

Figure 4.4 (a) Shear strength as a function of sucrose/PA molar ratio. (b) Shear strength as a function of reaction time.

The poly(thioctic acid,TA) (polyTA)-pentaerythritol tetraacrylate (PETEA)-polyethylene glycol diacrylate (PEGDA)-agarose (Agar) elastomer PSA exhibits rapid adhesion to surfaces such as pig skin in air. This capability is attributed to a high surface density of carboxyl groups, which enables instant and robust adhesion to diverse substrates through the formation of multiple noncovalent interactions, including hydrogen bonds, electrostatic attraction, hydrophobic interactions, and ionic complexation.

Achieving strong instant adhesion under wet conditions—highly desirable in biomedical, industrial, and underwater applications—remains challenging, largely due to the interfering hydration layer on substrate surfaces. The present PSA addresses this issue through two key design features: an inherently hydrophobic backbone and a surface rich in adhesive carboxyl groups (Figure 4.6). Upon contact with a wet substrate, the hydrophobic backbone first displaces the interfacial water

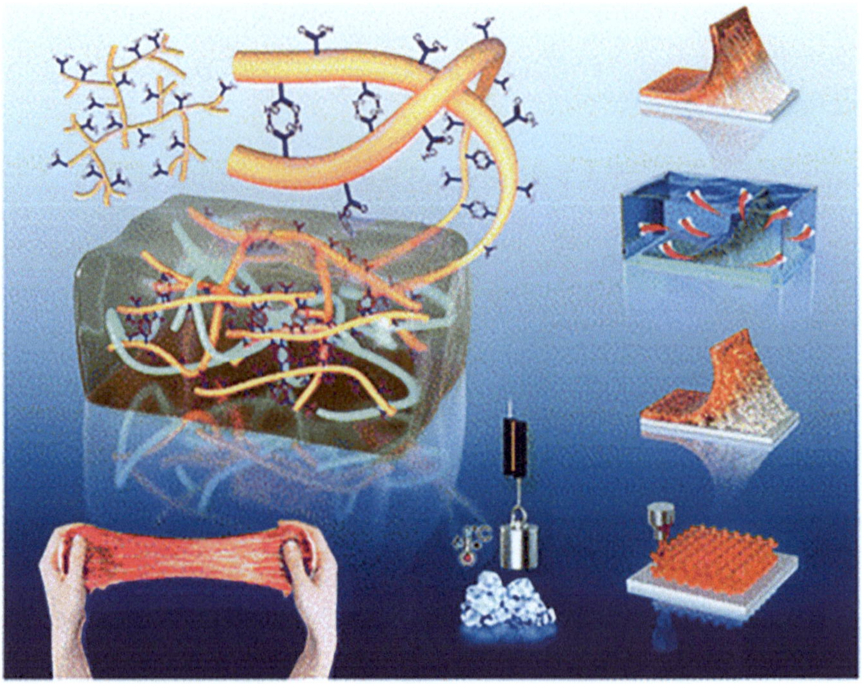

Figure 4.5 Schematic illustration of the structure and versatile applications of the hyperbranched polymer elastomer PSA. *Source:* [51] / with permission of Royal Society of Chemistry.

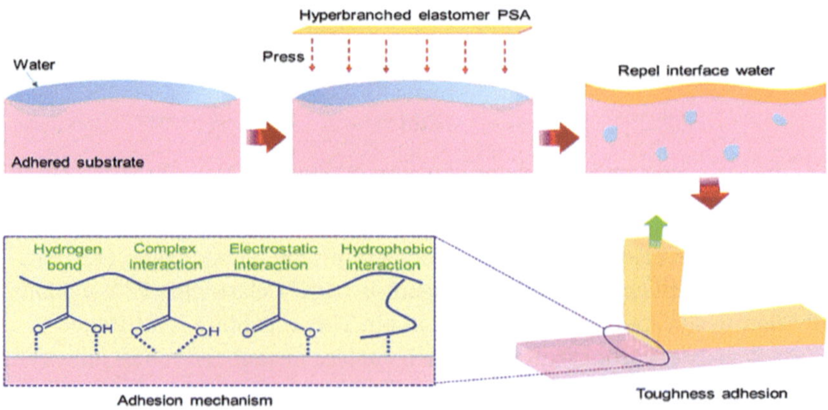

Figure 4.6 Underwater adhesion mechanism.

film, while the surface carboxyl groups rapidly form abundant noncovalent bonds (e.g., hydrogen bonding and electrostatic interactions) with the substrate, creating strong interfacial adhesion.

The hydrophobic nature of the PSA backbone is confirmed by water contact angle (CA) measurements. To further visualize its water-displacing ability, a macroscopic drainage test was performed: a dyed water droplet placed on the PSA surface rolled off completely when tilted, leaving no residue. This demonstrates the excellent drainage capacity of the hyperbranched elastomer, which is essential for achieving rapid and strong underwater adhesion.

The development of high-performance, multifunctional bio-based adhesives to replace petroleum-derived systems remains a significant challenge. Inspired by biomineralization, a strong and versatile soybean protein adhesive was engineered using a soybean protein matrix crosslinked with 4-carboxyphenylboronic acid (CPBA), a self-synthesized hyperbranched catechol polymer (DBA@HBPA) as a catechol donor, and $AgNO_3$ as a silver source. The incorporation of DBA@HBPA markedly improved processing, boosting prepressing strength by 156% to 0.89 MPa. Two synergistic mechanisms conferred outstanding multifunctionality: a dynamic quinone–catechol equilibrium mediated by in situ-formed silver NPs (Ag NPs) and the inherent mildew resistance and antibacterial activity of Ag NPs and boron species. Consequently, the adhesive exhibited a prolonged pot life (12 days), exceptional mildew resistance over 30 days, and potent antibacterial action (zero colony count). The wet shear strength of plywood bonded with this adhesive reached 1.28 MPa, a 106% improvement over pure soy protein isolate (SPI). The formulation also demonstrated flame retardancy (limiting oxygen index [LOI] = 37.4%) and low volatile organic compound (VOC) emissions (4.26×10^6 mg/m^2/h), illustrating a novel biomimetic route to high-performance biomass adhesives.

The individual contributions of DBA@HBPA, CPBA, and Ag to interfacial adhesion were systematically studied (Figure 4.7a). The adhesion strength of pure SPI adhesive to wood chips was 0.34 MPa. Adding DBA@HBPA increased the strength to 1.04 MPa (a 206% gain), primarily due to robust hydrogen bonding between catechol groups and the wood surface. Subsequent crosslinking with CPBA, forming borate ester bonds, slightly reduced the strength to 0.91 MPa. The final addition of Ag, which promotes partial catechol oxidation to quinone, led to a further minor decrease to 0.89 MPa for the full SPI-DBA@HBPA-CPBA-Ag system.

To decouple the role of catechol oxidation in durability, adhesion strength was monitored over three days at 4 °C to suppress microbial effects. The Ag-containing adhesive maintained its initial strength, whereas the strength of the SPI-DBA@HBPA-CPBA control (lacking Ag) declined from 0.91 to 0.78 MPa. This stability is attributed to the dynamic, reversible quinone–catechol redox system, which mitigates aging and ensures long-term adhesive efficacy.

Rheological properties are crucial for porous substrate bonding. The zero-shear viscosity of pure SPI was 160,900 mPa·s, with no observable flow (Figure 4.7b,c). Adding spherical DBA@HBPA particles, which act as a lubricant, reduced viscosity by 60.3% to 63,930 mPa·s and restored fluidity. Crosslinking with CPBA increased viscosity to 83,080 mPa·s by restricting molecular mobility. The incorporation of Ag further raised the viscosity to 91,410 mPa·s due to additional crosslinking sites.

Coating uniformity is critical for application. While pure SPI adhesive distributed unevenly on veneers and tended to stick to rollers, forming a discontinuous layer, all formulations containing DBA@HBPA showed excellent, uniform dispersion (Figure 4.7d,e). This is ascribed to an optimal viscosity and the surface-wetting

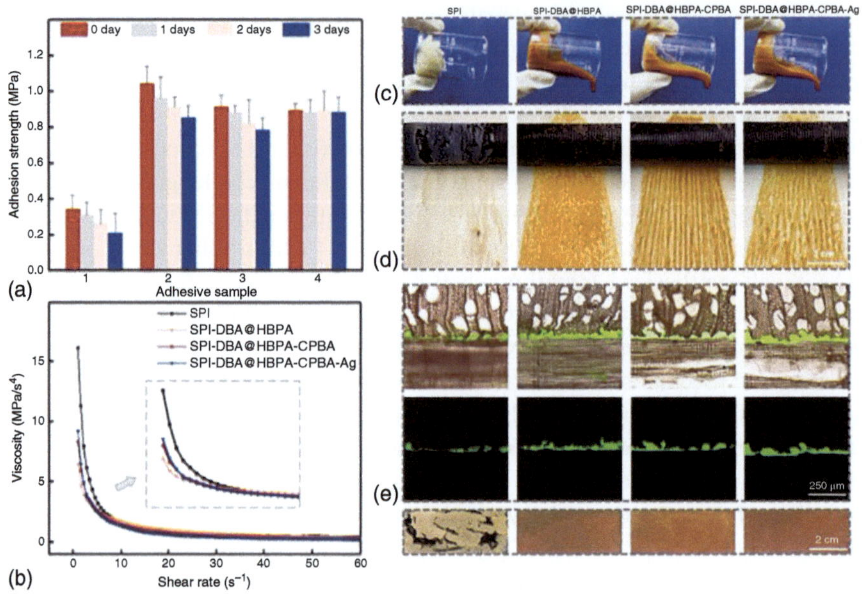

Figure 4.7 Physical properties of the prepared adhesives. (a) Adhesion strength measured initially and after storage at 4 °C for 1–3 days (1, SPI; 2, SPI-DBA@HBPA; 3, SPI-DBA@HBPA-CPBA; 4, SPI-DBA@HBPA-CPBA-Ag). (b) Apparent viscosity. (c) Visual assessment of fluidity. (d) Coating uniformity on veneer substrates. (e) Cross-sectional SEM image of the bonded plywood. *Source:* (c–f): Liu, Z. et al. / with permission of Elsevier.

capability of catechol groups, facilitating the formation of a continuous, high-quality adhesive layer.

From an environmental sustainability standpoint, soybean meal (SM) adhesives present a promising alternative to petroleum-derived counterparts. However, a fundamental trade-off often exists between the strength and toughness of adhesive materials. Overcoming this compromise to develop protein-based adhesives with high water-resistant bonding strength, excellent toughness, and added functionalities represents a major challenge. To address this, Li et al. devised a crosslinking strategy employing dual hyperbranched siloxane nanoclusters to fabricate high-performance SM-based adhesives (Figure 4.8) [52].

A design strategy was devised to integrate the key attributes of PSA and hot-melt adhesives (HMAs) into a single material. This led to the development of an ion-terminated HBP, synthesized via Michael addition in a water/methanol mixture, which combines PSA-like tack, HMA-like low viscosity and cohesive strength, and inherent stress-sensing capability [53].

Figure 4.9a,b presents the lap-shear stress–displacement profiles and corresponding strength values of the adhesive on various substrates. When employed as a PSA, it exhibits high bond strengths to stainless steel (2.85 MPa), iron (2.67 MPa), aluminum (2.26 MPa), polyamide (1.04 MPa), and polyimide (0.85 MPa). These values substantially exceed those reported for other PSA systems—including ionogels, hydrogels, elastomers, and composites—as benchmarked in Figure 4.9c.

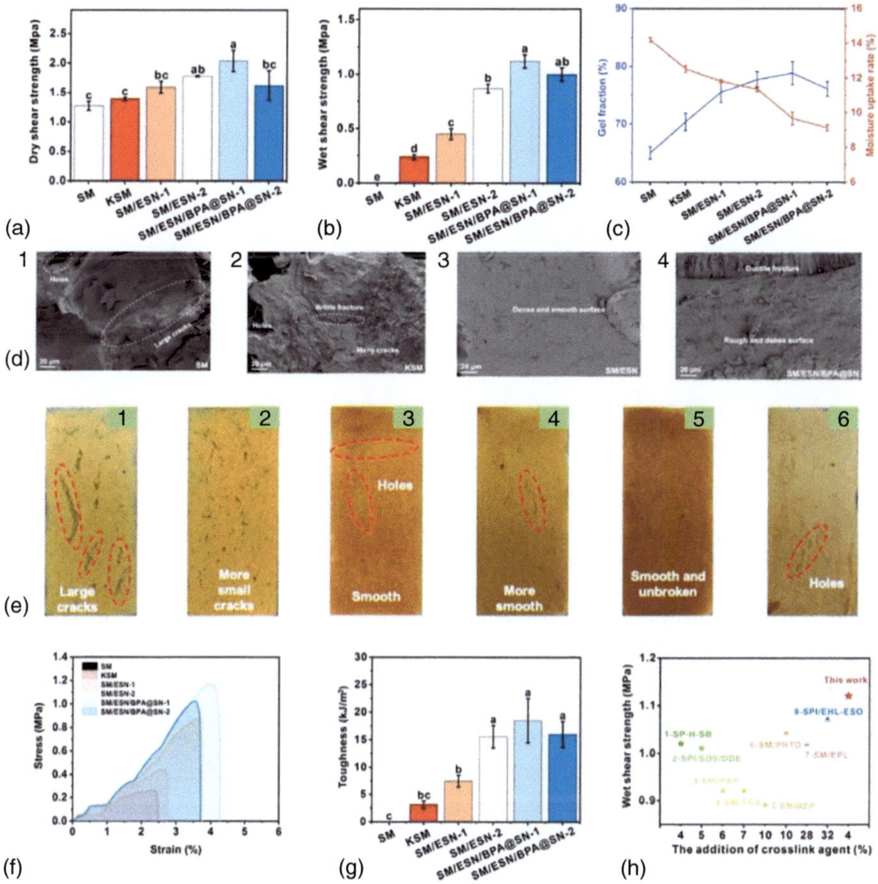

Figure 4.8 Assessment of adhesive properties. (a) Dry shear strength. (b) Wet shear strength. (c) Gel fraction and moisture absorption. (d) Fracture surface morphology (SEM). (e) Cured layer flatness. (f) Wet shear stress–strain curves. (g) Toughness derived from wet shear tests. (h) Comparative wet shear strength of various modified soy protein adhesives. *Source:* [52] / with permission of American Chemical Society.

To our knowledge, the strength achieved on stainless steel represents the highest value documented for a PSA.

This exceptional performance arises from a synergistic combination of factors: (i) strong interfacial ionic interactions, which surpass typical van der Waals forces; (ii) a low glass-transition temperature (T_g) and rapid polymer relaxation, enabling effective surface wetting; and (iii) a high density of hydrogen bonds within the matrix, which elevates cohesive strength and shifts the failure mode to predominantly cohesive rather than interfacial. Notably, when tested at 0 °C, the lap-shear strength on stainless steel increases markedly to 6.96 MPa—an unusually high value for PSAs. Fracture behavior transitions from ductile at room temperature (above T_g) to brittle at 0 °C (below T_g), indicating that the enhanced bonding strength at low temperature is primarily due to improved cohesive strength within the adhesive.

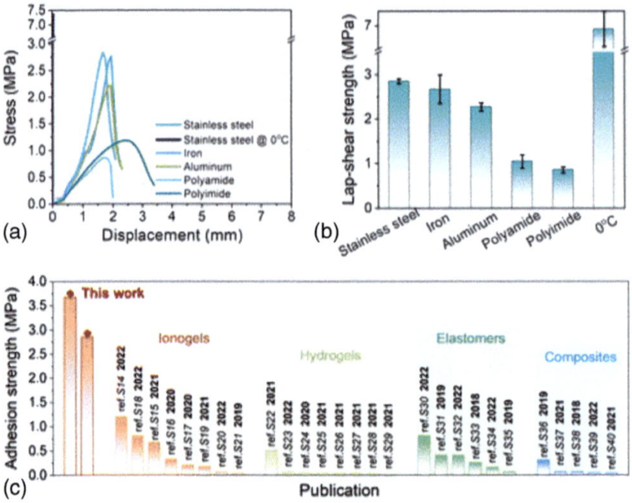

Figure 4.9 Adhesive performance of the hyperbranched polymer. (a) Lap-shear stress–displacement curves and (b) corresponding lap-shear strength of the material (denoted D230-0.9) as a pressure-sensitive adhesive (PSA). The 0 °C data point represents bonding to stainless steel tested at that temperature. (c) Lap-shear strength benchmarked against other reported PSAs. *Source:* Liu, Z. et al. / with permission of American Chemical Society.

Recent advances in bone adhesives for median sternotomy have yet to adequately address the challenge of bone defect regeneration. Healing is often compromised by poor adhesive strength, insufficient bioactivity, and a lack of antimicrobial properties. Bioinspired strategies, particularly those emulating marine organisms, offer a promising alternative. To this end, Mohammad Dinari and co-workers designed a polyhedral biomimetic adhesive via a dual crosslinking strategy, incorporating a catecholamine-functionalized HBP (polydopamine-co-acrylate, PDA), bredigite (BR) NPs, and Fe^{3+} ions [54]. These hybrid adhesives exhibit robust, substrate-dependent adhesion to poly(methyl methacrylate), glass, bone, and skin tissues, achieved through a combination of irreversible covalent and reversible non-covalent crosslinking. Adhesion strength correlates with BR NP content; notably, formulations containing 2 wt% BR demonstrate a bone tissue adhesion strength of 2.3 ± 0.8 MPa, approximately threefold greater than that of pure PDA adhesives. Beyond mechanical performance, the adhesives are bioactive—accelerating apatite formation in vitro—and display concentration-dependent antibacterial activity against *Staphylococcus aureus*. Furthermore, they promote favorable cellular responses: enhanced spreading and osteogenic differentiation of MG63 human osteosarcoma cells, attributable to controlled ion release and the flexible double-network structure. In summary, multifunctional PDA/BR hybrid adhesives, which combine osteoconductivity, mechanical strength, and antimicrobial properties, show promise as bone adhesives for sternotomy closure.

Lap-shear testing on Poly(methyl methacrylate) (PMMA) and glass substrates (following modified ASTM F2255-05) confirms that incorporating BR NPs reinforces the adhesive matrix, enhancing both strength and toughness (Figure 4.10).

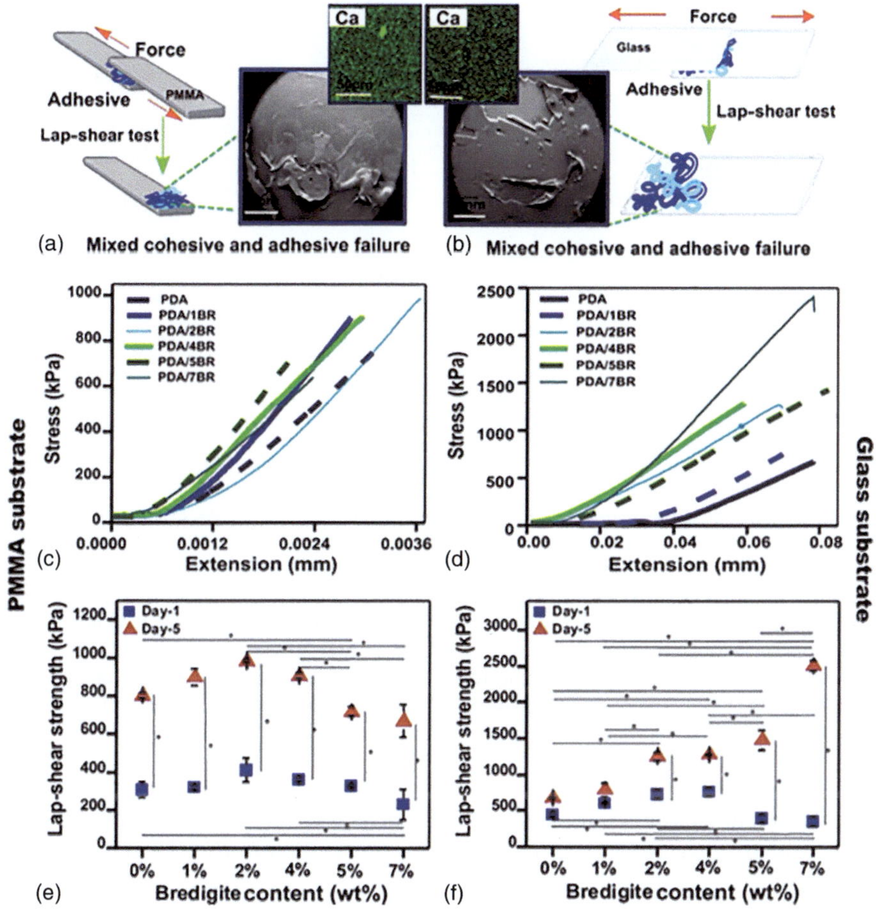

Figure 4.10 Lap-shear performance of PDA adhesives reinforced with bredigite nanoparticles. Schematic of the lap-shear test and cross-sectional SEM images of PDA/BR adhesives on (a) glass and (b) PMMA substrates. Representative stress–extension curves after 5 days of curing on (c) glass and (d) PMMA. Adhesive shear strength after 1 and 5 days of curing on (e) glass and (f) PMMA substrates. Data are mean ± s.d. ($n = 5$). *$P < 0.05$. *Source:* [54] / with permission of American Chemical Society.

The adhesion mechanism derives from the versatile bonding capacity of catechol groups, which form hydrogen bonds, electrostatic and π-interactions, and coordination bonds with substrates. The flexibility and spatial arrangement of cationic and catechol moieties are key to optimizing the adhesion–cohesion balance. Additionally, as established in prior work, metal-coordination bonds serve as reversible, dynamic cross-links that contribute to cohesive strength in the polymer network.

Previous work described a dual-component, solvent-free bioadhesive that utilizes thiol–isocyanate cross-linking [55] (Figure 4.11). A catalyst-free, solventless bioadhesive system was developed using a thiolated HBP and isocyanate-modified PEG (PEGNCO). These low-viscosity components gel within 1.8–3 minutes via a thiol–isocyanate click reaction. The material achieves strong tissue adhesion

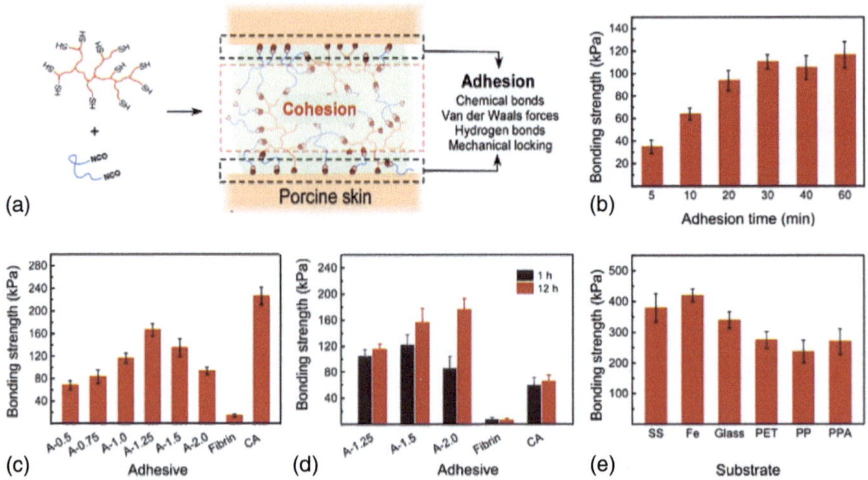

Figure 4.11 Adhesion performance of the HBPTE–PEGNCO bioadhesive. (a) Schematic of the bonding mechanism on porcine skin, highlighting contributions from interfacial adhesion (black circles) and matrix cohesion (red circles). (b) Adhesion strength of formulation A-1.0 on porcine skin as a function of time. (c) Comparison of adhesion strength after 1 hour among HBPTE–PEGNCO adhesives, fibrin glue, and cyanoacrylate (CA) on porcine skin. (d) Adhesion strength of formulations A-1.25, A-1.5, A-2.0, and CA on wet porcine skin after 1 and 12 hours. (e) Adhesion strength of A-1.0 on various substrates: stainless steel (SS), iron, glass, polypropylene (PP), polyethylene terephthalate (PET), and polyphthalamide (PPA).

through covalent isocyanate linkages and demonstrates excellent biocompatibility in vitro and in vivo, offering a novel design for high-performance, fast-curing bioadhesives.

Han et al. synthesized a low-viscosity HER (MHER) via an efficient microwave-assisted method, which significantly shortened the reaction time and enhanced the product yield [49]. A polyether amine functionalized with ureidopyrimidinone (UPy) and imidazole groups (D400-UPy-CDI) was synthesized. This compound was then blended with the microwave-synthesized HER (MHER) and diglycidyl ether of bisphenol A (DGEBA) to produce a series of TP hyperbranched epoxy hot-melt adhesives (HEAs).

The tensile strength, strain, and toughness of the HEA films displayed a non-monotonic dependence on MHER content, initially rising and then declining (Figure 4.12). The HEA-8 formulation exhibited optimal mechanical performance, with a tensile strength of 79.6 MPa, breaking elongation of 8.6%, and toughness of 4.6 MJ/m^3—corresponding to increases of 78.1%, 168.7%, and 475% relative to the MHER-free control (HEA-0). These mechanical trends align with the measured adhesion strengths, demonstrating the tunability of the adhesive's overall performance.

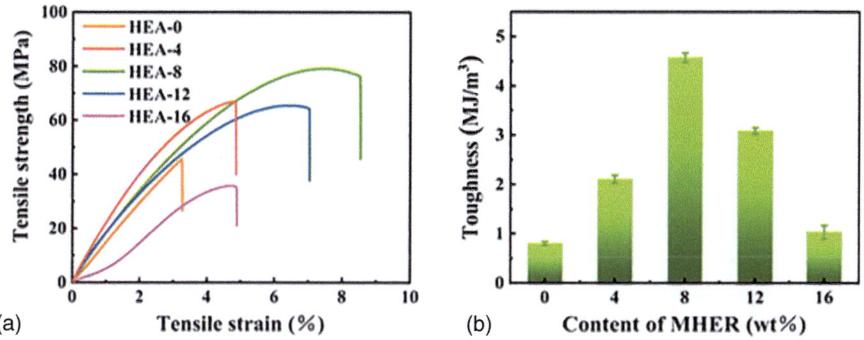

Figure 4.12 Mechanical and viscoelastic properties of the adhesives. (a) Tensile stress–strain curves. (b) Tensile toughness.

4.3 Coating

HERs have garnered significant attention in the coating industry as a revolutionary alternative to conventional linear EPs, owing to their distinctive three-dimensional architectural structure. This structural feature endows HERs with a suite of superior properties, including ultra-low viscosity (eliminating the need for excessive VOC-based diluents), high density of reactive epoxy groups (facilitating rapid and complete curing), and excellent compatibility with various pigments, fillers, and co-resins—attributes that directly address the long-standing challenges of traditional epoxy coatings, such as high VOC emissions, prolonged curing cycles, and limited adaptability to complex substrates. As modern coating applications (ranging from automotive and aerospace to marine and industrial maintenance) increasingly demand high performance, environmental friendliness, and multifunctional integration, HERs have emerged as a pivotal material to meet these evolving requirements, though their practical application still relies on optimizing key performance metrics and addressing substrate-specific challenges.

According to Figure 4.13, with the increase of the AHPU content, the gel fraction of the AHPU/EB600 film tends to increase first and then decrease [56]. Correspondingly, the swelling ratio and water absorption of the AHPU/EB600 film decline initially before rising. Generally, when the AHPU content reaches 12%, the film exhibits the highest gel fraction, alongside the lowest swelling ratio and water absorption. This phenomenon is likely due to the high crosslinking density of the film—when 12% AHPU is added, this dense crosslinking restricts the film's swelling behavior. However, as the AHPU content further increases to 15%, the amount of flexible HBP chains rises significantly. Even though the crosslinking density remains high, this increase in flexible chains leads to a higher swelling ratio and water absorption.

A novel hard coating based on hyperbranched polyurethane acrylate (HBPUA) was developed for protecting 5G mobile phone covers. This coating is characterized

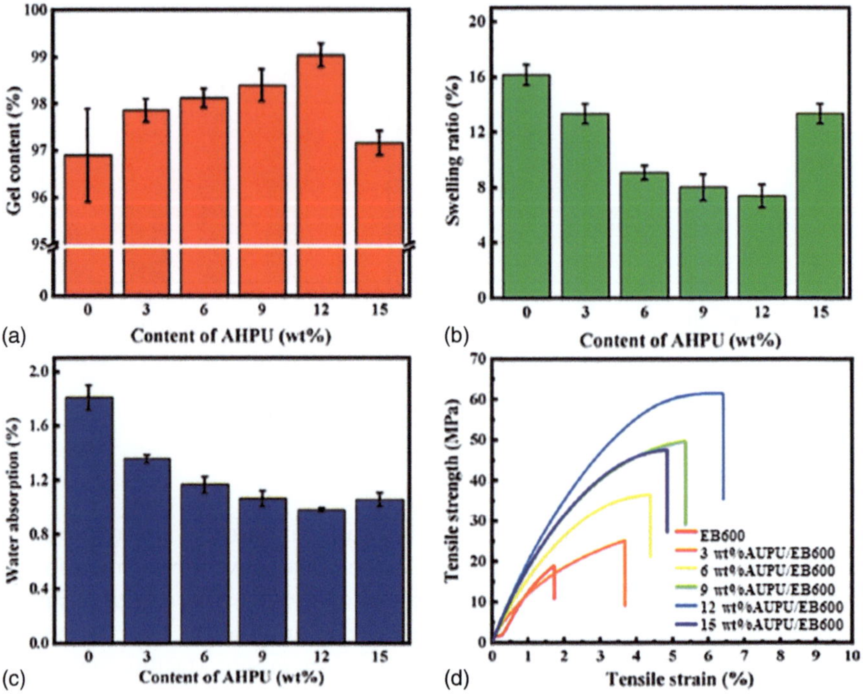

Figure 4.13 (a) Gel fraction. (b) Swelling ratio. (c) Water absorption. (d) Tensile stress–strain curves of the cured AHPU/EB600 films.

by a highly rigid hyperbranched core and was applied as a protective film [57]. As shown in Figure 4.14, coatings based on PETMP achieve a gel content of up to 95 wt%, exceeding that of neat HBPUA. The gel content increases across the series S1 < S2 < S3, a trend consistent with the double-bond conversion rate and indicative of a corresponding rise in chemical crosslinking. Notably, the S1 coating attains a high gel content of 92.3 ± 0.4 wt% despite a conversion rate of only 38.5%. This is attributed to the high functionality of the oligomers: even attachment of a single terminal C=C group to the polymer network renders the modified oligomer insoluble in acetone during Soxhlet extraction, thereby explaining the high gel content at low double-bond consumption.

Zhang et al. reported a non-silicone elastomeric coating comprising hyperbranched poly(ε-caprolactone) (HPCL) crosslinked by an amphiphilic poly(ethylene glycol)-*block*-poly(propylene glycol)-*block*-poly(ethylene glycol) tri-BCP [58]. The mechanical response of the coatings, characterized by AFM force–indentation curves and surface modulus mapping (Figure 4.15), reveals a clear contrast. While polydimethylsiloxane (PDMS) displays the largest indentation at 800 nN—reflecting its soft, elastic nature—HPCL-T0 exhibits minimal deformation and the highest modulus. This significant stiffening in HPCL-T0 is due to partial crystallization of the polymer network.

Liu et al. designed a transparent, flexible, dynamic omniphobic liquid-like coating with a pencil hardness of 8H [59]. Hyperbranched EP (HBEP) and

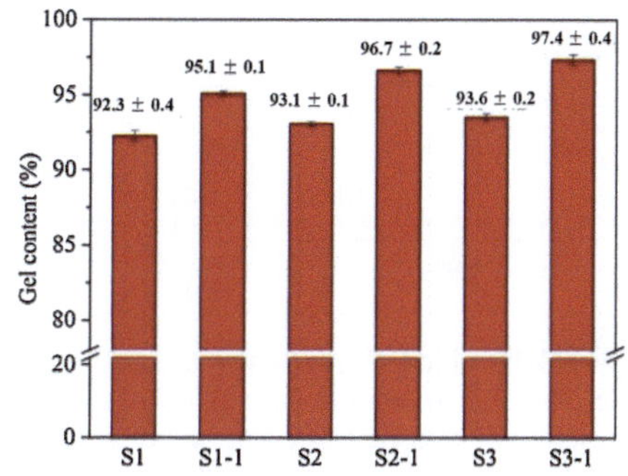

Figure 4.14 Gel content of the HBPUA coatings.

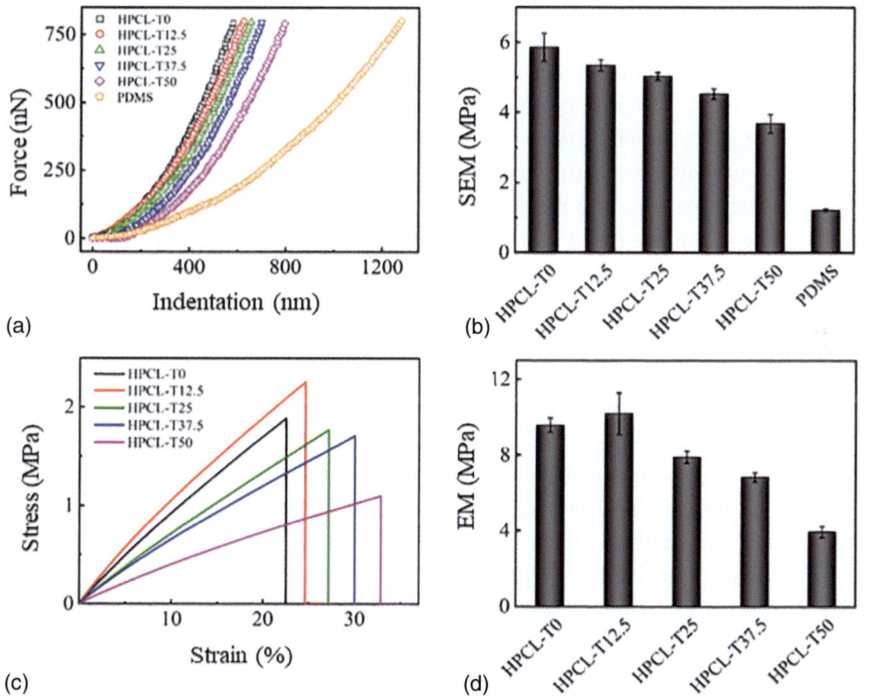

Figure 4.15 Mechanical characterization of HPCL-Tx coatings. (a) AFM force–indentation curves at 25 °C. (b) Corresponding surface elastic modulus (obtained by AFM). (c) Tensile stress–strain curves at 25 °C. (d) Elastic modulus derived from tensile tests.

hexamethylene diisocyanate trimer (HDIT) undergo a reaction to form a rigid oxazolidinone structure (OX) with a high degree of crosslinking. This specific structural component imparts ultrahigh hardness to the resulting coating.

The liquid-repellent capability of omniphobic liquid-like coatings serves as the fundamental basis for their antifouling performance, typically evaluated by measuring the CAs and sliding angles (SA) of droplets with different surface tensions. This repellency primarily arises from the surface enrichment of HO-PDMS. By varying

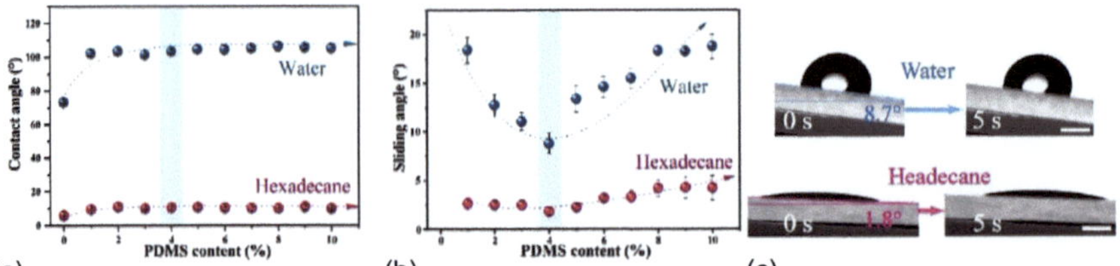

(a) (b) (c)

Figure 4.16 Liquid repellency and antifouling performance of the HPOX/PDMS coating. (a, b) Variation in contact angles (CAs) and sliding angles (SAs) for water and hexadecane as a function of coating composition. (c) Time sequence images of water (top) and hexadecane (bottom) droplets sliding on a tilted surface (scale bar: 1 mm). *Source:* [59] / with permission of Elsevier.

the PDMS content, the CA and SA for both high-surface-tension (10 μL water) and low-surface-tension (5 μL hexadecane) liquids were systematically assessed.

As shown in Figure 4.16a,b, increasing the PDMS concentration gradually raises the CA while steadily lowering the SA. This trend results from the grafted liquid-like PDMS chains, which convert the droplet–substrate contact from a solid–solid to a solid–liquid interface, thereby markedly reducing sliding resistance. The optimal performance was achieved at 4 wt% PDMS. At this loading, the coating exhibits a water CA of 103.4° and SA of 8.7°, and a hexadecane CA of 10.5° with an SA of 1.8° (Figure 4.16c).

Metallic compound cathodes were synthesized by coating Metal-organic frameworks (MOFs) with a HBP, followed by calcination [60]. The multiple functional sites in Zr_2N_2O facilitate the conversion of soluble lithium polysulfides (LiPSs) to insoluble Li_2S_2/Li_2S, while simultaneously enhancing ion and electron transport. This synergy improves sulfur utilization and rate capability. Furthermore, nitrogen and oxygen atoms in Zr_2N_2O chemically adsorb LiPSs, effectively suppressing the shuttle effect. Additionally, the small particle size and porous structure of the composite shorten the Li^+ diffusion path. Consequently, the S/Zr_2N_2O/NC-6 cathode exhibits a Li^+ diffusion coefficient seven to eight times higher than that of simply N-doped carbon. Benefiting from these combined effects—efficient chemical adsorption, facilitated ion transport, and multifunctional sites—a cell with the S/Zr_2N_2O/NC-6 cathode demonstrates a capacity retention of 81.8% after 300 cycles at 1C. Even under demanding conditions (sulfur loading: 8.16 mg/cm^2; lean electrolyte: 7.79 μL/mg), it maintains good performance, delivering an areal capacity competitive with commercial lithium-ion batteries.

The electrochemical performance of MOF-derived cathodes was systematically evaluated (Figure 4.17a). Cells with S/Zr_2N_2O/NC cathodes outperformed the S/NC control, with the S/Zr_2N_2O/NC-6 variant delivering the highest capacity. Initial capacities at 1C were 891.9, 857.0, 735.6, and 741.9 mAh/g for S/Zr_2N_2O/NC-3, -6, -9, and S/NC, respectively, stabilizing at 529.2, 701.8, 586.4, and 491.2 mAh/g after 300 cycles. The superior 81.8% retention of the NC-6 cathode may involve an initial activation process. As shown in Figure 4.17b, this cathode also exhibits excellent rate performance, with capacities of 960.8, 930.9, 817.2, 730.6, 654.6, 605.6, and

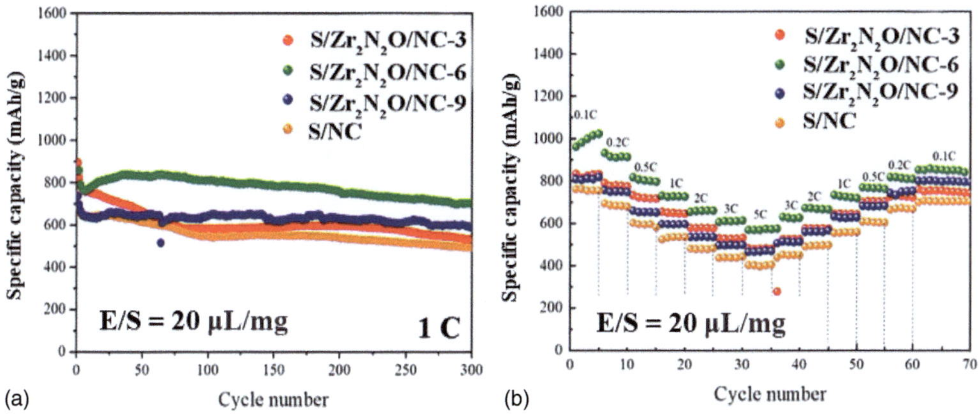

Figure 4.17 Electrochemical performance evaluation. (a) Cycle capacities of various cathodes at 1C. (b) Rate capability tested between 0.1C and 5C.

566.8 mAh/g at current rates from 0.1C to 5C. When the rate was returned to 0.1C, the capacity recovered with minimal decay, underscoring the markedly superior electrochemical performance of the $S/Zr_2N_2O/NC$-6 cathode.

4.4 Matrix for Composites

In the context of rapid advancements in aerospace, automotive, and electronic packaging industries, high-performance composite materials have become pivotal for addressing the demands of lightweight, energy efficiency, and structural reliability. EPs are widely employed as matrix materials in fiber-reinforced composites due to their excellent adhesion, chemical resistance, and dimensional stability [9, 31, 61–63]. However, traditional linear EPs suffer from inherent limitations, such as high viscosity during processing, poor toughness of the cured network, and insufficient thermal stability under extreme service conditions, which severely restrict their application in high-end composite systems.

HERs, characterized by their three-dimensional globular architecture and abundant terminal epoxy groups, have emerged as promising alternatives to overcome these drawbacks. The unique TS of HERs endows them with low melt viscosity (facilitating fiber wetting and composite fabrication), high reactivity (enabling rapid curing), and tunable mechanical/thermal properties (via molecular design). These advantages make HERs effective modifiers or co-matrixes for composites, as they can simultaneously enhance the toughness, thermal stability, and processability of the composite matrix without compromising modulus or adhesion.

Despite significant progress in laboratory-scale research, challenges including high synthesis cost, inconsistent batch performance, and interface compatibility with specific reinforcing fibers (e.g., carbon fibers) remain to be addressed for the large-scale industrial application of HER-based composites [44, 64, 65].

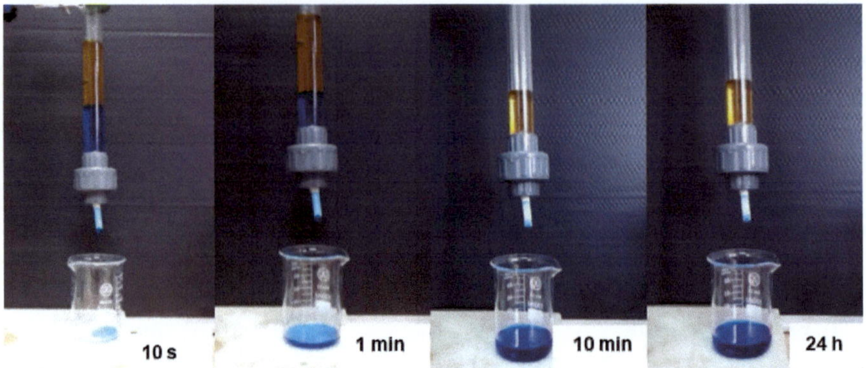

Figure 4.18 Schematic of the oil/water separation process using the modified membrane. An oil-in-water dispersion was introduced into the upper glass tube. Water selectively permeates through the membrane, while the oil phase (orange) is retained above. (For interpretation of the references to color in this figure legend, the reader is referred to the web version of this article.) *Source:* [66] / with permission of Elsevier.

4.4.1 Glass Fiber Composites

GFRPs are widely used in aerospace, automotive, and renewableenergy sectors due to their high specific strength and low cost [65–68]. However, traditional linear EPs, the dominant matrix for GFRPs, face critical limitations: high viscosity (often >3000 mPa·s at 25 °C) hinders wetting and impregnation of GF bundles, leading to voids and reduced interfacial bonding; low reactive group density causes prolonged curing (8–12 hours at 80 °C) and insufficient crosslinking, limiting GFRPs' resistance to heat (typically <120 °C) and mechanical fatigue. Emerging demands exacerbate these issues: aerospace GFRPs require $T_g > 150\,°C$ and interlaminar shear strength (ILSS) ≥70 MPa; automotive lightweighting needs fast-curing matrices (<1 hour at 60 °C) for high-efficiency production; wind turbine blades demand GFRPs with excellent moisture resistance (water absorption <0.8% after 1000 hours immersion) to withstand harsh environments. HERs address these gaps via their low viscosity (<800 mPa·s at 25 °C) for improved fiber wetting, high terminal epoxy density for rapid curing (<2 hours at 80 °C) and high T_g, and tunable terminal groups for enhanced fiber–matrix adhesion. Yet, industrial adoption requires optimizing HER–fiber compatibility, reducing synthesis costs, and ensuring GFRP performance consistency—critical for meeting sector-specific requirements.

Meng et al. [66] reported inexpensive, superhydrophilic glass fiber membranes grafted with hyperbranched polyethyleneimine (HPEI) for oil/water separation. The pure-water flux under a 10-cm gravity head was first evaluated. While the pristine membrane showed a flux of ~2600 L/m^2/h, the modified membranes (GF-g-HPEI60000, GF-g-HPEI10000, and GF-g-HPEI600) exhibited slightly higher values of 2700, 2800, and 2800 L/m^2/h, respectively—consistent with the enhanced hydrophilicity observed in dye-staining tests (Figure 4.18).

Based on its optimal permeation, hydrophilicity, and low flux variance, the GF-g-HPEI600 membrane was selected for separating surfactant-free oil-in-water

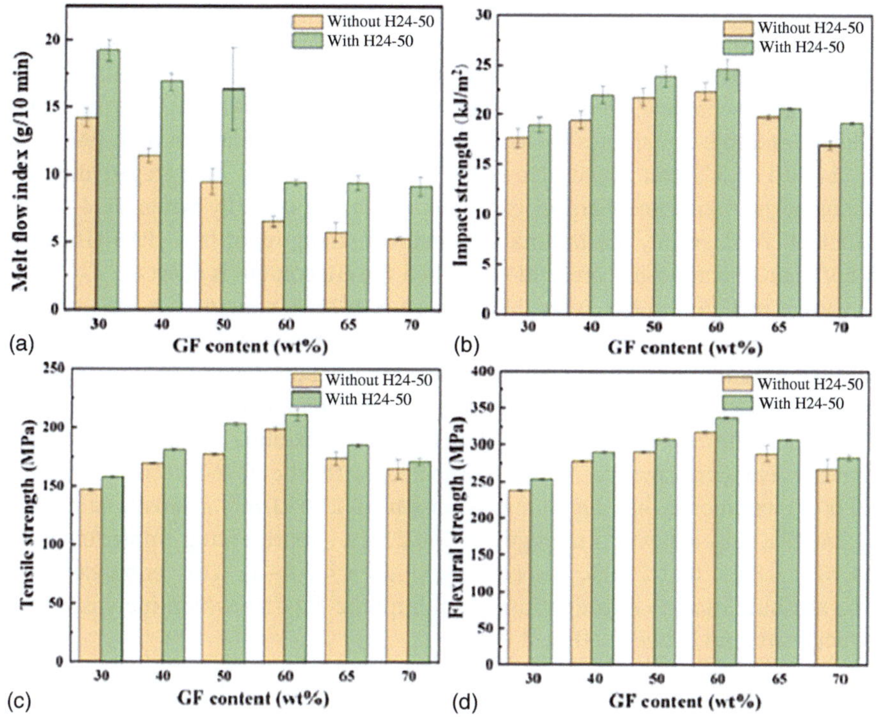

Figure 4.19 Properties of PA6/glass fiber (GF) and H24-50/PA6/GF composites: (a) melt flow index, (b) impact strength, (c) tensile strength, (d) flexural strength.

emulsions. As illustrated in Figure 4.18, the membrane was placed in a dead-end filtration cell. Using an *n*-dodecane/water dispersion (water dyed with methylene blue, oil with Sudan I) and gravity as the sole driving force, complete water permeation was achieved within 24 hours, with the oil retained atop the membrane due to its underwater oleophobicity. Rapid separation was accomplished in seconds, demonstrating a simple and cost-effective operation.

Liang et al. [67] synthesized amphiphilic hyperbranched polyester (H24-m) containing long fatty chain stearate and hydroxyl groups from hydroxyl-terminated hyperbranched polyester (H24) and stearic acid chloride (SC), which was used to prepare high-performance GF/polyamide 6 (PA6) composites. It is widely recognized that GF can enhance PA6. Figure 4.19 presents the influence of GF content on the flowability and mechanical properties of PA6/GF composites and H24-50/PA6/GF composites (with 0.90 wt% H24-50). According to Figure 4.19a, when the GF addition increases from 30 to 70 wt%, the melt flow index (MFI) of both PA6/GF and H24-50/PA6/GF composites declines significantly. Meanwhile, their mechanical strengths—including impact strength, tensile strength, and flexural strength—first increase and then decrease (Figure 4.19b–d).

It is quite evident that the properties of all H24-50/PA6/GF composites are far superior to those of the corresponding PA6/GF composites with the same GF content. This further highlights the significance of H24-50's amphiphilicity in improving the interfacial interaction between PA6 and GF. A common observation is that the MFI of the composites deteriorates as GF content increases, which is due to the fact that GF cannot be melted under the processing temperature. Fibrous GF exhibits high mechanical strength because it is capable of distributing and transferring loads, and this contributes to the high strength and toughness of both PA6/GF and H24-50/PA6/GF composites. However, excessive GF tends to aggregate during the short-term thermomechanical extrusion process, especially when H24-50 is insufficient. As a result, the optimal mechanical strengths of both PA6/GF and H24-50/PA6/GF composites are achieved when the GF content is 60 wt%.

As shown in Figure 4.20a, with the increase of H24-50 content, the MFI of the composites rises sharply from 6.54 to 10.86 g/10 min. In contrast, their mechanical strengths (including impact strength, tensile strength, and flexural strength) show a slight increase, as presented in Figure 4.20b–d. Among these composites, the GP60-50-11 composite (containing 1.10 wt% H24-50) exhibits relatively excellent comprehensive performance. When compared with the GP60 composite, the GP60-50-11 composite has an impact strength of 25.0 kJ/m^2, a tensile strength of 212.4 MPa, and a flexural strength of 240.2 MPa—representing improvements of 11.78%, 6.80%, and 7.18%, respectively (Figure 4.20b–d). This enhancement can be attributed to the introduction of amphiphilic H24-50, which increases the interfacial strength between GF and PA6.

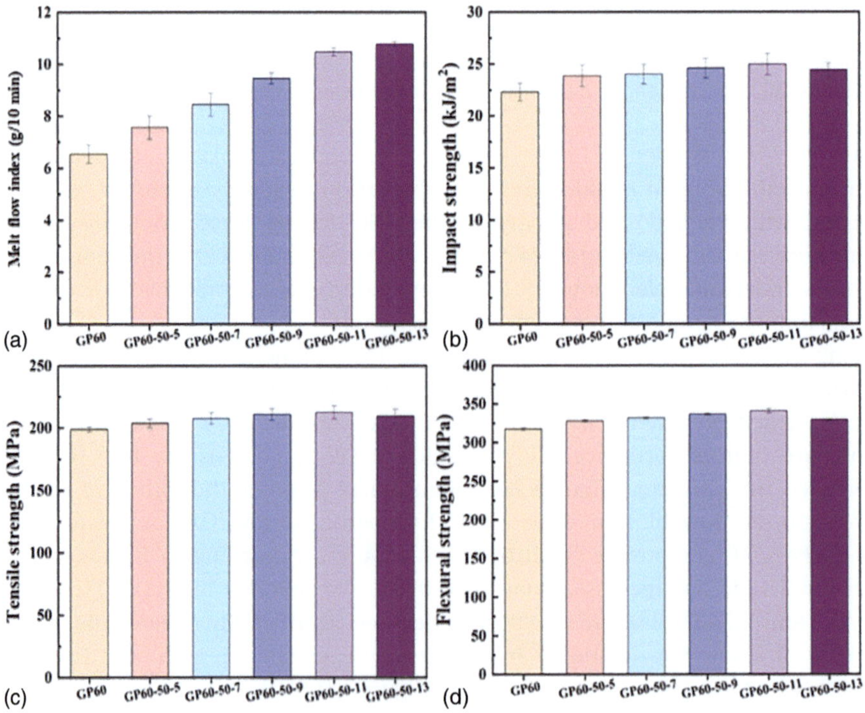

Figure 4.20 Effect of H24-50 content on the properties of composites: (a) melt flow index, (b) impact strength, (c) tensile strength, (d) flexural strength.

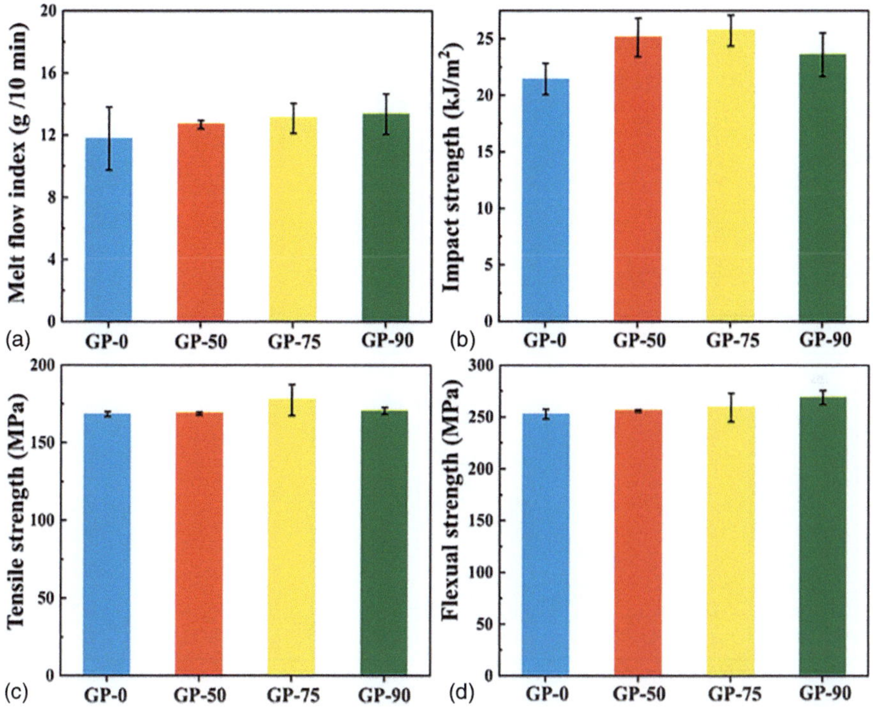

Figure 4.21 Mechanical and flow properties of the composites with different additives. (a) Melt flow index. (b) Impact strength. (c) Tensile strength. (d) Flexural strength.

Yu et al. [65] synthesized a series of HBP additives (HBP-12-n, where n = 50, 75, 90 denotes the esterification degree) via esterification of hydroxyl-terminated HBPs with stearic acid and examined their effects on the mechanical and melt flow properties of GF/PA6 composites. The esterification degree modulates the polarity of HBP-12-n and, consequently, the interfacial interaction between PA6 and GF, thereby influencing both processability and mechanical performance (Figure 4.21).

Adding HBP-12-n increased the MFI by approximately 12.7–13.3%, with higher esterification degrees yielding gradually greater improvements (Figure 4.21a). Mechanical properties—impact, tensile, and flexural strength—showed a non-monotonic dependence on esterification degree, first increasing and then either declining or stabilizing. Optimal overall performance was achieved at a moderate esterification degree of 75%. Compared with the unmodified composite (GP-0), the formulation containing 0.70 wt% HBP-12-75, 49.65 wt% GF, and 49.65 wt% PA6 (designated GP-75) exhibited increases of 20.1% in impact strength, 5.4% in tensile strength, and 2.5% in flexural strength. The concurrent enhancement of MFI and impact strength is notable. This improvement is attributed to the amphiphilic nature of HBP-12-n. Its weakly polar internal polyester structure ensures compatibility with the PA6 matrix, while the polar hydroxyl groups form hydrogen bonds with the GF surface. This dual affinity enables HBP-12-n to act as an interfacial bridge,

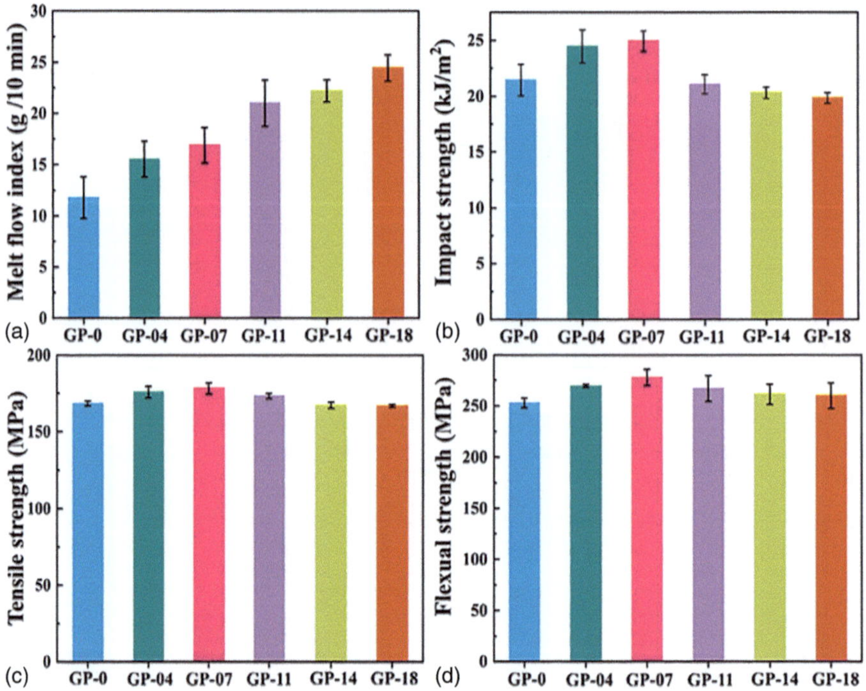

Figure 4.22 Properties of GP-*n* composites with various mass ratios of HyPer C181 and HBP-12-75. (a) MFI, (b) impact strength, (c) tensile strength, and (d) flexural strength.

mitigating the discontinuity and poor stress transfer that typically occur at the PA6-GF interface in uncompatibilized systems.

Processing large, thin-walled automotive components demands polymer composites with far greater flowability than that needed for smaller parts. However, achieving ultrahigh flowability in GF/PA6 composites remains challenging. To address this, a carboxyl-terminated HBP (HyPer C181) was introduced into the previously optimized HBP-12-75/GF/PA6 system. Figure 4.22 illustrates the effect of the HyPer C181-to-HBP-12-75 mass ratio on the MFI, impact strength, tensile strength, and flexural strength of the resulting hybrid composites, where the total additive content (HBP-12-75 + HyPer C181) was fixed at 1.40 wt%.

4.4.2 Carbon Fiber Composites

Carbon fiber-reinforced polymers (CFRPs) are pivotal in aerospace, high-end automotive, and precision engineering due to their ultrahigh specific strength and modulus. However, conventional linear epoxy matrices, the mainstay of CFRPs, present critical bottlenecks: high viscosity (>2500 mPa·s at 25 °C) impairs uniform impregnation of carbon fiber tows, inducing voids that degrade interlaminar bonding; low crosslink density limits CFRPs' heat resistance (T_g < 130 °C) and anti-fatigue performance, failing to meet harsh service conditions. Emerging demands intensify these challenges: aerospace CFRPs require T_g > 180 °C, ILSS

≥85 MPa, and fast curing (<1.5 hours at 100 °C) for manufacturing efficiency; high-performance automotive CFRPs need improved impact resistance (≥60 kJ/m^2) and moisture resistance (water absorption <0.5% after 1000 hours). HERs address these gaps via low viscosity (<600 mPa·s at 25 °C) for superior fiber wetting, high terminal epoxy density for rapid curing and high T_g, and tunable structures to enhance fiber–matrix interfacial adhesion. Yet, industrial application requires optimizing HER-carbon fiber compatibility, reducing synthesis costs, and ensuring CFRP performance stability, which is key to meeting sector-specific demands.

Hyperbranched structure makes the composite material have better toughness and impact resistance, and improves the safety and service life of the product [2, 40, 41, 67, 69]. A low-viscosity liquid HER (Hyper E102) was introduced as an additive to modify short-carbon-fiber-reinforced composites (SCFRCs). Its influence on thermal and mechanical performance was systematically evaluated. The overall properties of the SCFRCs exhibited a non-monotonic dependence on Hyper E102 content, initially improving and then declining. At an optimal loading of 12 phr, the composite showed increases of 16.7% in tensile strength, 74.9% in fracture toughness, 95.3% in notched impact strength, and 194.5% in unnotched impact strength. The addition of Hyper E102 significantly enhanced the impregnation of the epoxy matrix into the fiber bundles, promoting stronger interfacial adhesion. It also reduced internal stress and mitigated stress concentration within the composite. Furthermore, the critical crack length was markedly extended, contributing to improved fracture toughness and damage tolerance.

Zhang et al. [69] designed and synthesized sulfhydryl hyperbranched polyimide (DSPI-*n*) and degradable imide skeleton HER (DIHE-*n*) using sulfhydryl click reaction (Figure 4.23).

Compared with pristine carbon fiber (PCF) in Figure 4.24, both the tensile strength and modulus of produced acid-treated PCF (ACF) were slightly lower than those of PCF due to the partial structure deterioration of carbon fibers. But the thiol group functionalization on PCF without acid treatment effectively improved their performance. By thiol–ene click reaction between thiol group of DSPI-*n* and double bonds of PCF, the formed crosslinked thin interlayer effectively fills into the grooves on the carbon fiber surface to alleviate stress concentration. The thiol–ene click reaction converted some sp^2-C structures into sp^3-C structures, producing a crosslinked structure among PCF. Molecular chain winding, intermolecular hydrogen bonds on the surface of TCF12, and the rigid ring structure of DSPI-*n* were beneficial to improve the strength and modulus. Additionally, the deformable hyperbranched TS of DSPI-*n* can transfer the load to strengthen the TCF. As shown in Figure 4.24a, the TCF12 from DSPI-12 shows a higher mechanical property than that of TCF6 and TCF24 from DSPI-6 and DSPI-24. Mercaptan and PCF formed a crosslinked structure by click reaction, improving the mechanical properties of the modified carbon. The thiol content per gram of DSPI-6 is higher than that of DSPI-12 and DSPI-24, which can provide higher crosslinked density. However, with the increase of molecular weight, DSPI-12 and DSPI-24 have bigger size and stronger topological deformability, which is beneficial to stress and load transfer, also improving mechanical strength. Therefore, carbon fiber (TCF12) modified by DSPI-12 with

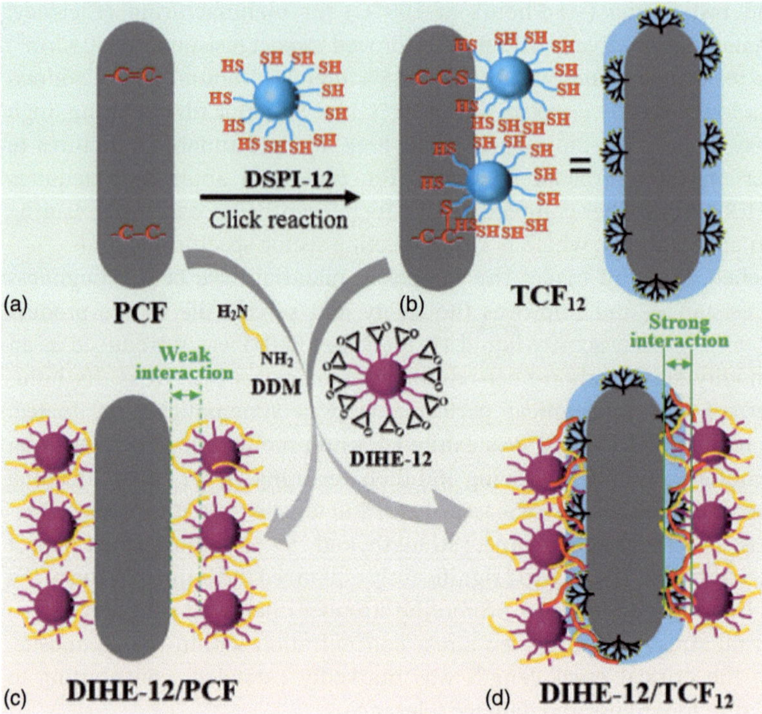

Figure 4.23 Interface strengthening scheme of the CF-reinforced composites: (a) PCF, (b) TCF12, (c) DIHE-12/PCF, and (d) DIHE-12/TCF12.

the relatively high crosslinked density and strong deformability shows the best mechanical properties. The TCF12 showed a 59.51% and 25.46% increase in tensile strength (5.20 ± 0.61 GPa) and modulus (327.83 ± 33.48 GPa), compared with PCF of 3.26 ± 0.37 GPa and 261.31 ± 18.52 GPa, respectively.

Relative to the PCF control (Figure 4.24), ACF exhibited slightly reduced tensile strength and modulus due to partial degradation of the fiber structure. In contrast, direct thiol functionalization of PCF without acid pretreatment significantly enhanced performance. Through a thiol–ene click reaction between DSPI-*n* derivatives and the double bonds on PCF, a crosslinked thin interphase was formed, filling surface grooves and alleviating stress concentration. This reaction also converted some sp^2-C to sp^3-C, creating a crosslinked network on the fiber. Noncovalent interactions (e.g., chain entanglement and hydrogen bonding) and the rigid ring structure of DSPI-*n* further contributed to improved strength and modulus. Moreover, the deformable hyperbranched topology of DSPI-*n* facilitated efficient load transfer to the fiber.

As shown in Figure 4.24a, TCF12 modified with DSPI-12 displayed the highest mechanical properties, outperforming TCF6 and TCF24 (modified with DSPI-6 and DSPI-24, respectively). Although DSPI-6 possesses a higher thiol content per gram—potentially enabling greater crosslink density—DSPI-12 and DSPI-24 offer larger molecular dimensions and stronger topological deformability, which

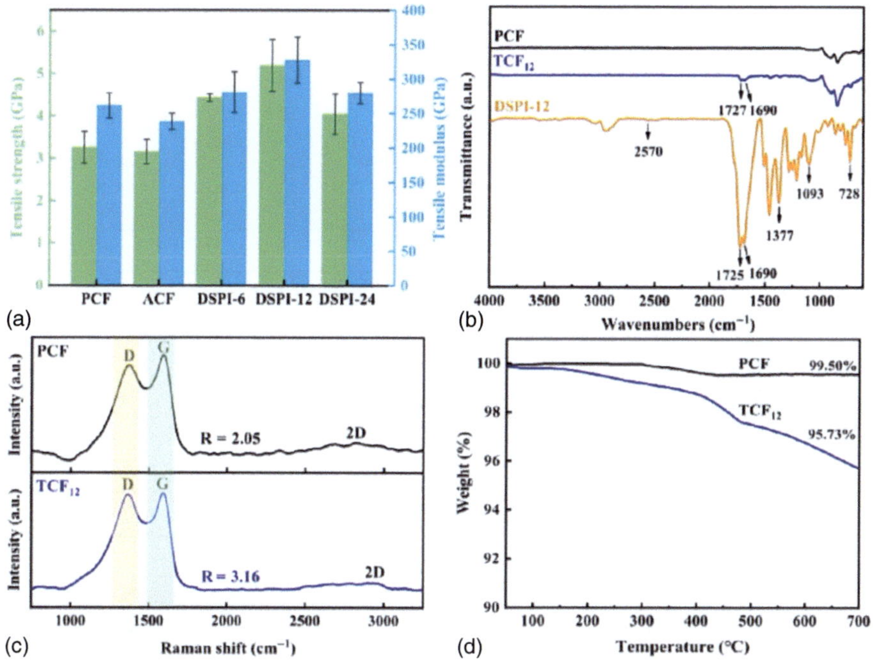

Figure 4.24 Interfacial strengthening strategy for carbon fiber-reinforced composites. Representative schemes for (a) pristine CF (PCF), (b) thiol-functionalized CF (TCF12), (c) DIHE-12/PCF, and (d) DIHE-12/TCF12.

enhances stress dissipation and load transfer. Consequently, TCF12 balances relatively high crosslink density with pronounced deformability, yielding optimal performance. Compared with PCF (tensile strength: 3.26 ± 0.37 GPa; modulus: 261.31 ± 18.52 GPa), TCF12 achieved increases of 59.51% in strength (5.20 ± 0.61 GPa) and 25.46% in modulus (327.83 ± 33.48 GPa). The bio-based degradable hyperbranched epoxy resins (ETFP-n)/TCF composites were prepared by impregnating the TCF fabric in a mixture of ETFP-n and diaminodiphenylmethane (DDM) followed by hot press molding. The tensile strength, tensile modulus, flexural strength, flexural modulus, interlaminar shear strength (ILSS), and interfacial shear strength (IFSS) of ETFP-12/TCF composites are improved by 55.98%, 44.53%, 77.77%, 76.79%, 65.52%, and 78.09%, compared with the corresponding performance of ETFP-12/PCF composites in Figure 25(a)–(b), respectively. The influence of the topological structure of ETFP-n on composites mechanical properties was then investigated. The tensile strength, modulus, and stress-strain property of both R1CF and R2CF change a little with the corresponding performance of PCF in Figure 25(c)–(d), indicating intact recycling with a perfect microstructure of carbon fiber fabric of the resultant composites. Moreover, compared with the performance of ETFP-12/PCF composites, the tensile strength, tensile modulus, ILSS, flexural strength, and flexural modulus of ETFP-12/R2CF composites in Figure 25(e)–(f) show high retention of about 97%, 99%, 95%, 98%, and 88%, respectively, suggesting closed loop recycling and high-value reusing of carbon fiber fabric.

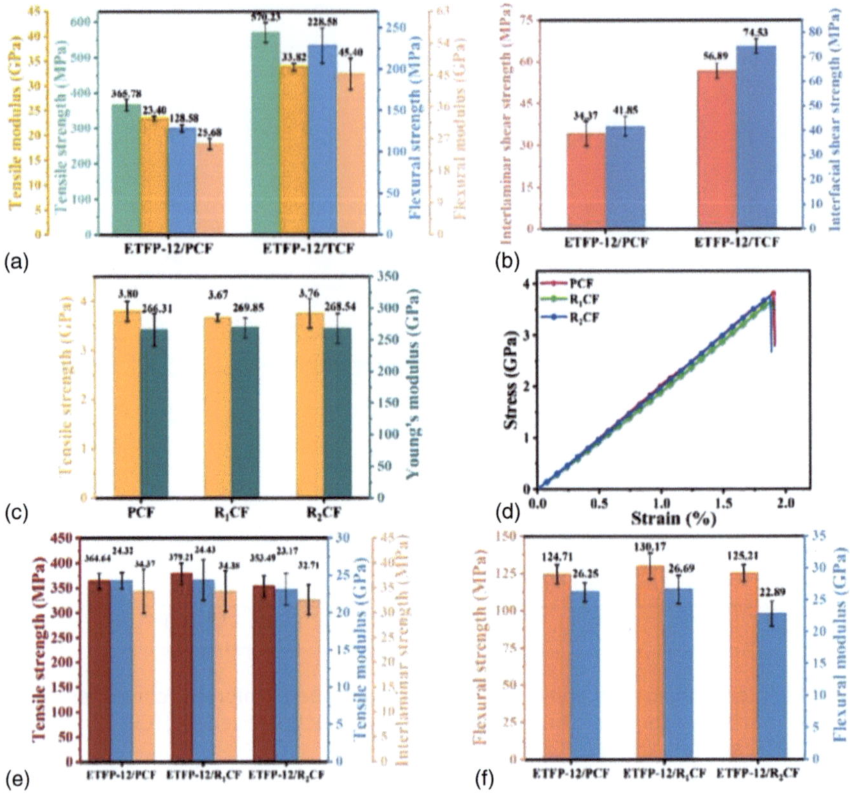

Figure 4.25 Mechanical performance of carbon fibers and derived composites. (a, b) ETFP-12/PCF and ETFP-12/TCF composites. (c, d) Pristine (PCF) and recycled carbon fibers (R1CF, R2CF). (e, f) Corresponding ETFP-12-based composites.

A closed-loop recyclable vitrimer matrix was prepared from hydroxy-terminated HBPs, rosin-derived fumaropimaric acid (FPA), and glycerol triglycidyl ether (GTE) [70].

Increasing the HBP content in FPA/GTE vitrimers from 0% to 15% raised the storage modulus at 25 °C from 1798 to 2453 MPa. A further increase to 20% HBP, however, reduced the modulus to 2286 MPa (Figure 4.26a). This non-monotonic trend was also reflected in the crosslink density—calculated from the rubbery-plateau modulus using Flory's theory—which first increased from 0.86×10^{-3} to 1.31×10^{-3} mol/cm^3 and then declined to 1.16×10^{-3} mol/cm^3. Gel content measurements followed a similar pattern. The initial enhancement is attributed to hydrogen bonding interactions between hydroxyl-terminated HBP and polar groups in the vitrimer network, which effectively increase the crosslink density. At excessive HBP loading (20%), the flexible aliphatic chains of HBP promote chain mobility, thereby reducing the overall crosslink density.

Consistent with the crosslink-density trend, the glass-transition temperature (T_g, obtained from tan δ peaks) initially rose from 75.6 to 90.5 °C and then decreased to 86.1 °C as the HBP content increased from 0% to 20% (Figure 4.26b).

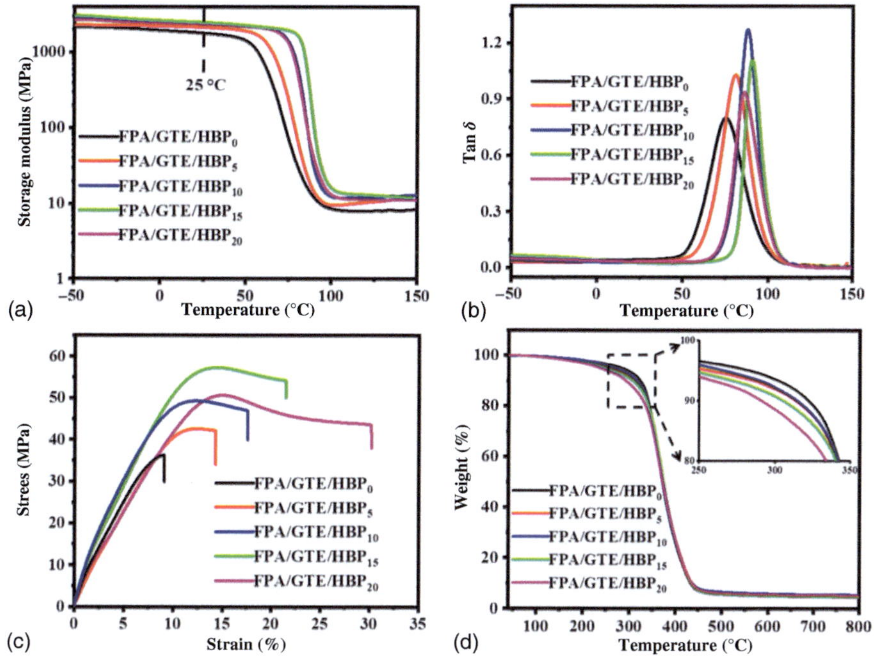

Figure 4.26 Characterization of FPA/GTE/HBP vitrimers. (a) Storage modulus. (b) Tan δ curves. (c) Tensile stress–strain curves. (d) Thermogravimetric analysis (TGA).

Tensile strength and elongation at break also improved with HBP addition up to 15%, reaching 57.2 ± 4.3 MPa and $21.5 \pm 1.2\%$, respectively, compared to 36.3 ± 3.2 MPa and $9.1 \pm 0.6\%$ for the HBP-free vitrimer (Figure 4.26c). Beyond this optimum, strength dropped to 50.6 ± 5.5 MPa for the 20% HBP sample. Although the temperature at 10% weight loss decreased from 323 °C to 291 °C with increasing HBP content—indicating that the oligomeric HBP component is more prone to thermal degradation—all formulations remained stable above 290 °C (Figure 4.26d). This retained thermal stability is ascribed to the robust phenanthrene-ring structure derived from rosin, which compensates for the destabilizing effect of HBP.

To enhance interfacial adhesion, PCFs were functionalized with a thiol-terminated HBP (BTMP-6), which improved fiber wettability and thus strengthened the interfacial interaction between the fibers and the polymer matrix [71]. The thiol-functionalized carbon fibers (TCF) were impregnated with epoxy-terminated HBPs (HTn) of varying molecular weights to produce renewable HTn/TCF composites with competitive properties. At a fixed resin content of ~57 vol%, the influence of HTn molecular weight on the mechanical performance of the cured composites is summarized in Figure 4.27. All HTn/TCF composites (n = 6, 12, 24) exhibited markedly higher mechanical properties than the conventional DGEBA/PCF reference. Strength metrics first increased and then decreased with increasing HTn molecular weight, with HT12 yielding the optimum performance (Figure 4.27a–c). Relative to DGEBA/PCF (tensile strength: 224.60 ± 21.41 MPa; tensile modulus: 21.08 ± 0.85 GPa; flexural strength: 70.28 ± 1.75 MPa; flexural

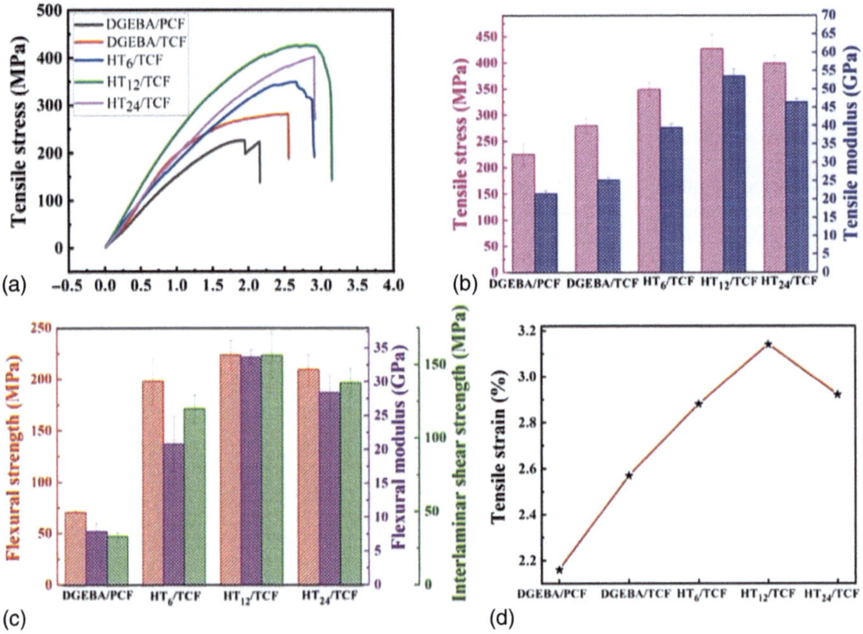

Figure 4.27 The mechanical performance of composites: (a) stress–strain curves, (b) tensile, (c) flexural, and (d) strain.

modulus: 7.85 ± 1.19 GPa; interlaminar shear strength: 32.79 ± 2.97 MPa), the HT12/TCF composite showed increases of 89.84%, 154.55%, 217.89%, 366.62%, and 376.52% in these properties, respectively.

This pronounced enhancement is attributed to the high crosslink density, low free volume, deformable TS of HT12, and strong interfacial interaction between TCF and the matrix. With increasing HTn molecular weight, the polymer adopts a more spherical topology, which improves chain slippage and load transfer, thereby increasing elongation (Figure 4.27d). The HT12 matrix thus provides an optimal balance of crosslink density, topological deformability, and interfacial compatibility, leading to superior mechanical strength. Overall, the synergistic effects of tunable network structure, reduced free volume, flexible topology, and enhanced fiber–matrix adhesion result in HTn/TCF composites that outperform traditional DGEBA/PCF systems across all measured mechanical properties.

Ma et al. first reported double closed-loop highly efficient recycling of resins and carbon fibers without deterioration of performance [72]. TCFs were prepared via a thiol–ene click reaction, while an epoxy-terminated HBP containing a degradable hexahydro-*s*-triazine unit (HT3) was synthesized from sustainable monomers. Single-ply composites were fabricated by impregnating TCF fabric with a mixture of HT3 and diaminodiphenylmethane.

The mechanical properties of the TCF/HT3 composites (≈55 wt% resin content) were evaluated and compared with those of reference composites made from PCF and DGEBA. As shown in Figure 4.28, composites containing TCF outperformed those with PCF, and HT3-based matrices surpassed the DGEBA-based reference.

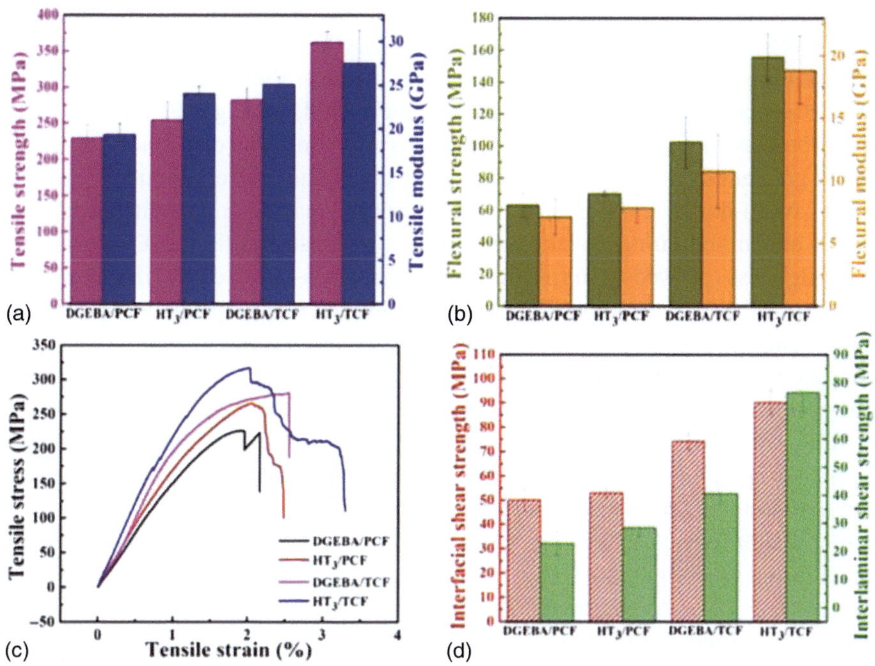

Figure 4.28 Mechanical properties of the carbon fiber composites: (a) tensile strength and modulus; (b) flexural strength and modulus; (c) strain or elongation; (d) the interfacial and interlaminar shear strength.

Specifically, the TCF/HT3 composite exhibited increases of 58.16% in tensile strength, 42.30% in tensile modulus, 146.71% in flexural strength, and 164.98% in flexural modulus relative to the conventional DGEBA/PCF system (Figure 4.28a,b).

4.4.3 Other Composites

A photoluminescent, transparent, and biocompatible hyperbranched epoxy/carbon-dot nanocomposite was synthesized by incorporating carbon dots during the in situ formation of the HER. The nanocomposite was characterized using Fourier transform infrared spectroscopy (FTIR), Nuclear magnetic resonance spectroscopy (NMR), and TEM. After curing with poly(amido-amine), it exhibited a tensile strength of 62.5 MPa, an elongation at break of 45%, thermal stability up to 291 °C, high transparency, wavelength-dependent photoluminescence, and good biocompatibility with skin cells. Its performance was compared with that of pristine hyperbranched epoxy (PHE) and an ex situ prepared nanocomposite. The in situ material demonstrated superior mechanical, optical, and biocompatible properties, highlighting its potential for high-performance transparent fluorescent polymers in optoelectronics. Excellent viability, spreading, and proliferation of skin fibroblasts and keratinocytes further suggest its suitability as a bio-sealant.

Mechanical properties—including tensile strength, elongation at break, toughness, scratch hardness, impact resistance, and bending performance—were evaluated for the PHE, the in situ carbon-dot-incorporated epoxy (CHE), and the physically blended ex situ composite (ECP). The CHE thermoset showed the highest tensile strength, attributed to covalent bonding between the aromatic carbon-dot core and the epoxy matrix (DGEBA and pentaerythritol) during in situ synthesis, which promotes strong interfacial interactions and uniform dispersion. It also exhibited greater elongation at break than PHE, likely due to stress-induced sliding between carbon-dot layers, which increases interlayer spacing and facilitates deformation while maintaining covalent connectivity. Consequently, CHE displayed higher toughness (derived from the area under the stress–strain curve) than PHE.

In contrast, the ECP thermoset, prepared by physical blending, showed reduced tensile strength relative to PHE, owing to predominant physical interactions, carbon-dot agglomeration (observed by TEM), and the absence of strong covalent bonding. However, ECP exhibited increased elongation and toughness compared with PHE. All three materials reached the upper instrument limits for scratch hardness (10 kg) and impact resistance (100 cm), and passed the flexibility test (1-mm mandrel bending) without film damage.

Tian et al. [73] presented a strategy to reprogram an industrial thermoset into a degradable and reconfigurable vitrimer using hyperbranched dynamic crosslinking networks (HDCNs). A comprehensive performance analysis of the resulting materials is summarized in Figure 4.29. The glass-transition temperatures of the neat epoxy (EP) and its vitrimer analogues (EP-3, EP-6, EP-9, EP-12) are 131, 121, 118, 104, and 94 °C, respectively, confirming their glassy state at room temperature. The slight decrease in T_g is ascribed to the amorphous nature of the hyperbranched crosslinker (HBPPB). Below T_g, chain mobility is further restricted by the incorporated HBPPB, leading to an increased loss modulus (Figure 4.29a). Above both T_g and the topology-freezing transition temperature (T_v), dynamic bond exchange reduces chain friction, enabling network flow and consequently lowering the loss modulus.

Notably, despite the introduction of flexible aliphatic chains—which often compromises stiffness—the storage modulus (E') of the vitrimer is enhanced relative to that of neat EP (Figure 4.29b). At 3 wt% HBPPB loading, E' rises from 2.38 GPa (neat EP) to 3.15 GPa, after which it gradually decreases with further HBPPB addition. This reinforcement is attributed to the hyperbranched architecture, which acts as a "supramolecular rivet" that interpenetrates the epoxy network through a combination of hydrogen and covalent bonds, forming a robust interlocked structure (Figure 4.29d). Concurrently, the expanded molecular topology reduces the overall crosslink density (Figure 4.29c). Thus, the hyperbranched crosslinker serves as an effective topological tool for designing strong yet lightly crosslinked polymer networks.

Enhancing the toughness of EPs is crucial for high-performance applications. Lu et al. [74] addressed this by combining a TP (PET) with a hyperbranched ionic liquid (HBIL). Tensile testing results (Figure 4.30) show that neither additive significantly altered the tensile strength, with only the 6P/4H formulation showing a marginal increase.

In contrast, the elastic modulus (Young's modulus) exhibited a pronounced response. The addition of PET (6 phr) and HBIL (4 phr) individually increased the

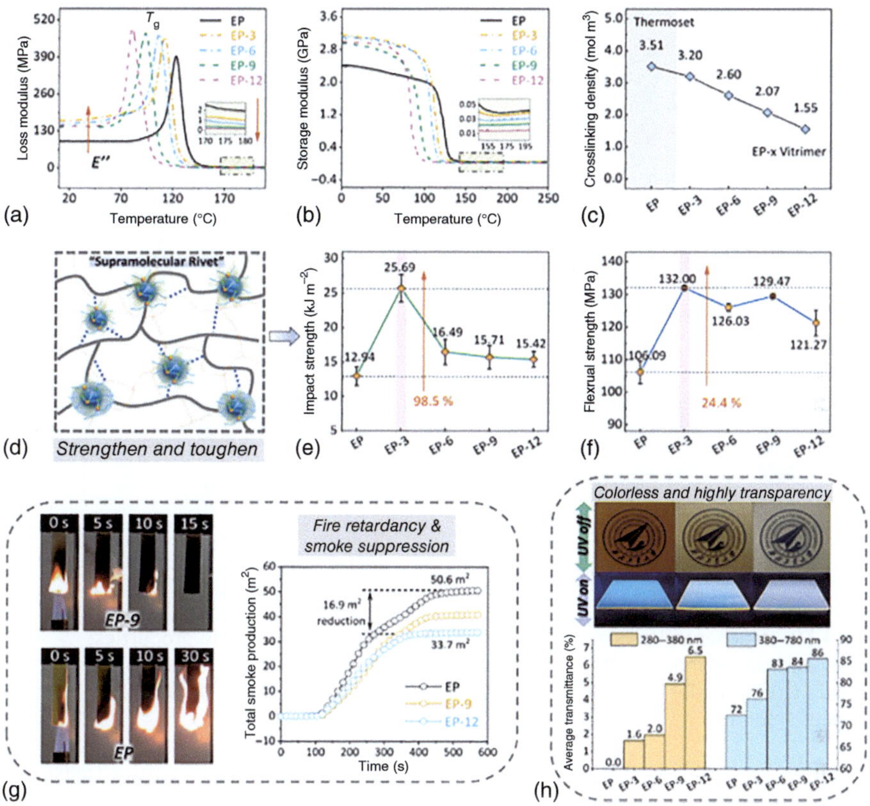

Figure 4.29 Multifunctional performance of the HDCN-modified vitrimers. Dynamic mechanical analysis: temperature-dependent loss modulus (a) and storage modulus (b). (c) Crosslinking density reduction due to expanded network topology. (d) Schematic of simultaneous strengthening and toughening via HDCNs. Enhanced impact toughness (e) and flexural strength (f). (g) Fire-retardant properties: self-extinguishing behavior and suppressed smoke release. (h) Digital images (left to right: EP, EP-9, EP-12) and average transmittance in the UV (220–380 nm) and visible (380–780 nm) regions. *Source:* (g)–(h) [73] / John Wiley & Sons / CC BY 4.0.

modulus by 8.5% and 46.2%, respectively. The hybrid 6P/4H composite achieved the highest modulus in the series at 1.67 GPa—a 28.5% improvement over neat epoxy and 18.4% over the 6PET composite. This enhancement is attributed to the high crosslink density and low free volume of HBIL, which reinforces the three-dimensional network. When both additives are present, the mobility of PET chains moderates the modulus relative to the pure HBIL system, yet the increased overall crosslink density from HBIL ensures the hybrid composite outperforms the PET-only formulation.

4.5 Modifiers for Composite Materials

Hyperbranched epoxy composite modifier, as a new type of polymer material additive, brings significant performance improvement and new application

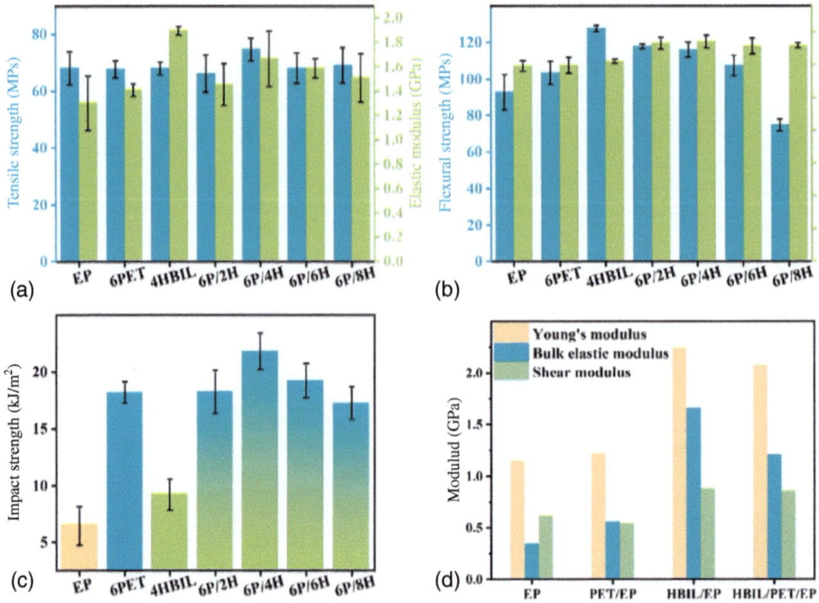

Figure 4.30 Mechanical properties of the epoxy composites. (a) Tensile properties: strength and elastic modulus. (b) Flexural properties: strength and modulus. (c) Impact strength. (d) Molecular dynamics (MD) simulation results for different composite systems.

possibilities for traditional epoxy composite materials (such as temperature-resistant materials, flame-resistant materials, antimicrobial materials, and others) through its unique hyperbranched structure and versatility.

4.5.1 Temperature-Resistant Materials

Heat-resistant composites are indispensable in aerospace (engine components), energy (high-temperature filters), and electronics (thermal management) sectors, demanding long-term stability at >150 °C and excellent mechanical retention under thermal cycling. However, conventional matrix resins (e.g., linear epoxies, phenolics) face critical limitations: linear epoxies exhibit low glass-transition temperatures (T_g < 130 °C) and severe performance degradation at 180 °C; phenolics suffer from brittleness and toxic volatile release during curing.

Emerging demands exacerbate these issues: aerospace components require T_g > 200 °C, ILSS retention ≥80% after 500 hours at 180 °C; energy-related composites need resistance to thermal shocks (−40 °C to 200 °C) without delamination. HERs address these gaps via their high crosslink density (enabled by abundant terminal epoxy groups) for T_g > 180 °C, tunable molecular structures (e.g., siloxane-modified HERs for T_g > 220 °C), and low viscosity (<800 mPa · s at 25 °C) for uniform reinforcement impregnation. Yet, industrial adoption requires optimizing HER thermal oxidative stability and reducing synthesis costs—key to meeting sector-specific heat-resistant demands.

Zhang et al. designed a class of epoxy supramolecular thermosets that enable rapid reprocessing and room temperature closed-loop chemical recycling [75] (Figure 4.31). The synthesis began with vanillin-derived Van-TTE, which was then epoxidized to yield vanillin-based EP (VanEP). A carboxyl-terminated polyetheramine (CBA-TTE) was prepared separately as the curing agent. Catalyst-free curing of VanEP (and its blend with a hyperbranched variant, VanEHBP) with CBA-TTE-produced epoxy networks crosslinked synergistically through transesterification and hydrogen bonding. The resulting supramolecular network (EN-VanEHBP) exhibits excellent mechanical properties, enhanced creep and chemical resistance, and rapid reprocessability. Notably, these thermosets can be fully dissolved into oligomers at room temperature and subsequently regenerated into crosslinked networks without loss of original performance. Tensile curves of the materials are shown in Figure 4.31a. The mechanical properties of the baseline EN-VanEP are comparable to those of conventional EN-DGEBA, owing to their structural similarity. Incorporating VanEHBP simultaneously improves both mechanical performance and glass-transition temperature (T_g). The properties first increase with VanEHBP content, peak at 7 wt%, and then decline. Specifically, EN-VanEHBP7 achieves a tensile strength of 104.5 MPa, toughness of 3.58 MJ/m^3, and impact strength of 57.0 kJ/m^2—representing increases of 43.1%, 109.4%, and 128.0%, respectively, over the cured EN-VanEP (73.0 MPa, 1.71 MJ/m^3, 25.0 kJ/m^2).

Dynamic polyimide adhesives (PI-B_3O_3 DPAs) were fabricated by crosslinking low-molecular-weight PI prepolymers with dynamic boroxines. The boroxine cross-linkers provide a unique combination of covalent stability and dynamic reversibility (activated by Lewis bases or water), which also enhances interfacial adhesion. The resulting adhesives exhibit high tensile strength (78.1–111.3 MPa), Young's modulus (1.2–3.5 GPa), and initial adhesion strength (19.4–26.0 MPa)—values that exceed those of most reported DPAs. They also display outstanding thermal stability ($Td_5\% = 541.6\,°C$) and retain strong adhesion after high-temperature pretreatment (25.8 MPa at 300 °C, 22.6 MPa at 350 °C, 18.3 MPa at 400 °C). The adhesives maintain stable bonding for up to 5 hours at 200 °C and show excellent resistance to a range of solvents, including acids, bases, and organic media. Moreover, the PI-B_3O_3 DPAs can be fully dissolved at room temperature using a mixed solvent (DCM/NMP/HCl), enabling straightforward debonding, recycling, and reuse. This work provides a viable strategy for designing robust, detachable, and reusable adhesives capable of long-term service under extreme conditions.

Zhang et al. synthesized ferulic acid-based hyperbranched EP (FEHBP) to prepare a closed-loop recyclable catalyst-free epoxy glass without affecting its thermal and mechanical properties [41] (Figure 4.32).

The mechanical performance of the CA/FEP/FEHBP vitrimers, presented in Figure 4.33, displayed a non-monotonic dependence on FEHBP content. Optimal properties were achieved at a loading of 10 phr FEHBP, where tensile strength increased from 88.0 to 126.4 MPa, elongation at break from 4.5% to 6.9%, and toughness from 2.26 to 5.29 MJ/m^3. These values compare favorably with previously reported transesterification-based epoxy vitrimers. The enhancement is attributed to the combined catalytic activity and hyperbranched topology of

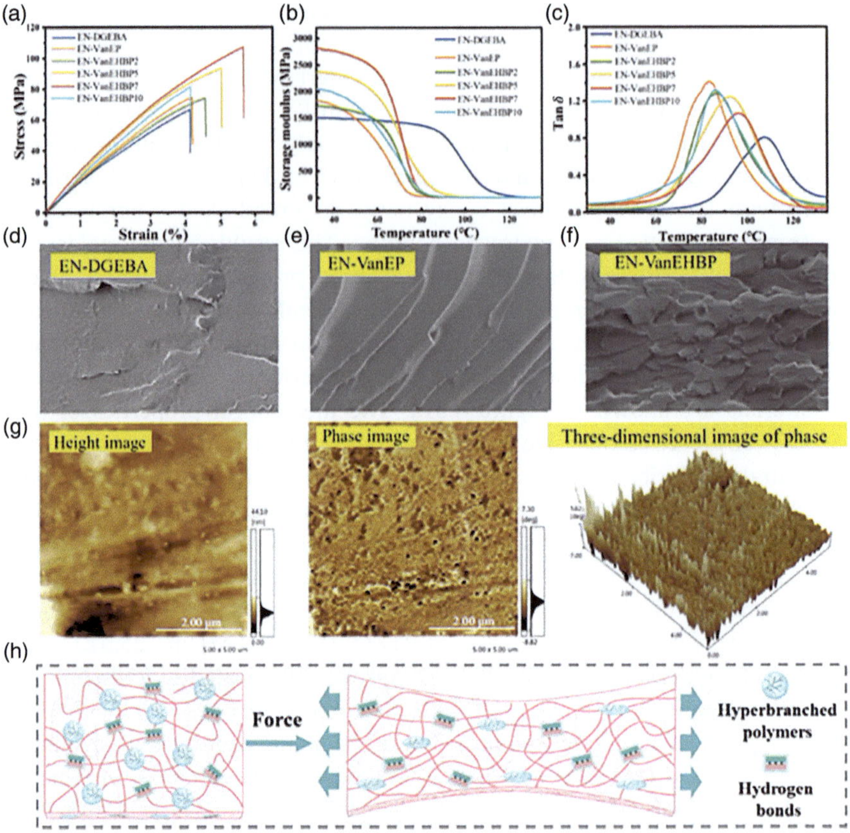

Figure 4.31 Mechanical properties and toughening mechanism of epoxy supramolecular thermosets. (a) Stress–strain curves at room temperature (strain rate: 5 mm/min). (b) Storage modulus and (c) tan δ curves measured from 30 to 130 °C at 5 °C/min and 1 Hz. SEM images of (d) EN-DGEBA, (e) EN-VanEP, and (f) EN-VanEHBP7 (scale bar: 10 μm). (g) AFM image of EN-VanEHBP7 at room temperature (scale bar: 2 μm). (h) Schematic of the supramolecular network deformation mechanism in EN-VanEHBP. Error bars denote s.d. ($n = 3$). Source data are provided. *Source:* [75] / Springer Nature / CC BY 4.0.

FEHBP, which concurrently elevates the crosslinking density and free volume of the network.

While degradable HERs have been reported [76], the efficient preparation of HER-based composites that simultaneously exhibit high strength, elongation, modulus, and low-temperature resistance—along with high degradation efficiency—remains a key challenge. To address this, we synthesized a series of self-curing, degradable hyperbranched EPs (DSEHP-n, $n = 3, 5, 7, 11$) via a rapid thiol–isocyanate click reaction. Incorporating DSEHP-*n* into DGEBA simultaneously enhanced degradability and mechanical performance. The mechanical properties of the DSEHP-n/DGEBA blends showed a non-monotonic dependence on resin structure, with DSEHP-5 providing the most pronounced enhancement. At 12 wt% loading, the DSEHP-5/DGEBA composite achieved a tensile strength of 104.5 MPa (a 64.8% increase over neat DSEHP) and a tensile modulus of

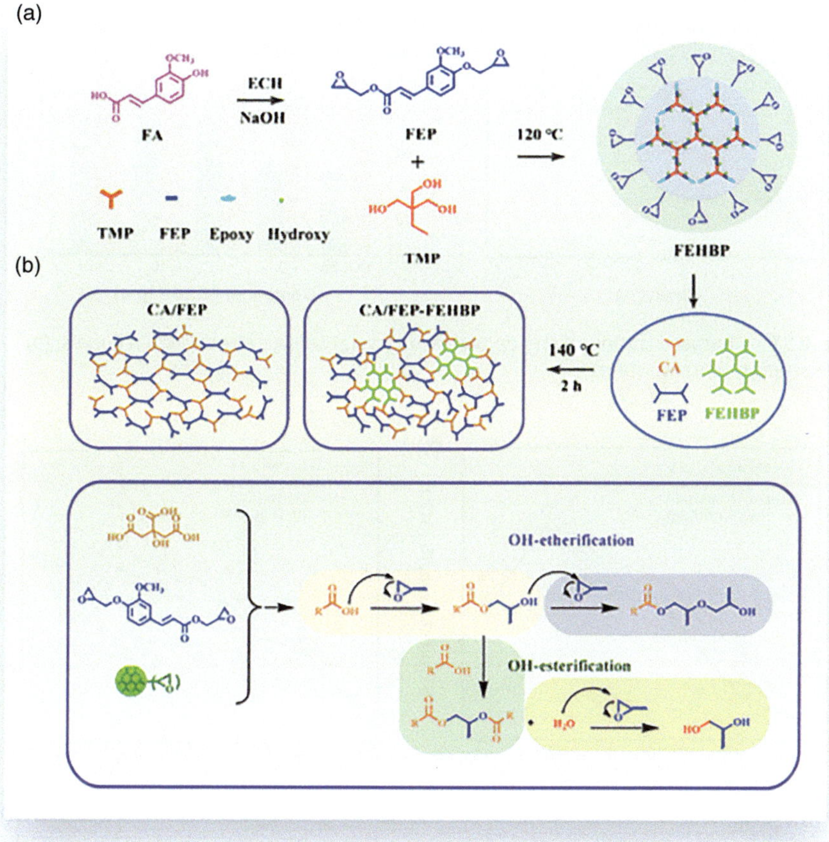

Figure 4.32 Synthetic route to FEP and FEHBP (a) and schematic of the curing process for the epoxy vitrimers (b).

5677 MPa (an 86.4% increase over neat DGEBA) (Figure 4.34a). Its flexural strength and modulus reached 138.8 and 3811 MPa, representing improvements of 38.1% and 64.6%, respectively, relative to neat DGEBA (Figure 4.34b). The composite also exhibited an elongation at break of 4.68% (≈50% higher than neat DGEBA) (Figure 4.34c) and an impact strength increased by 93.7% (Figure 4.34d).

These enhancements arise from the interplay of three structural factors imparted by DSEHP-*n*: increased crosslink density, the presence of intramolecular cavities, and a deformable hyperbranched topology. While crosslink density primarily strengthens tensile and flexural properties, the internal cavities and hyperbranched architecture facilitate energy absorption, chain mobility, and load transfer, thereby boosting impact resistance and elongation.

4.5.2 Flame-Resistant Materials

Teng et al. [77] designed and synthesized a DOPO-containing hyperbranched flame retardant (EHBFR). While the O=P—O bond is generally associated with

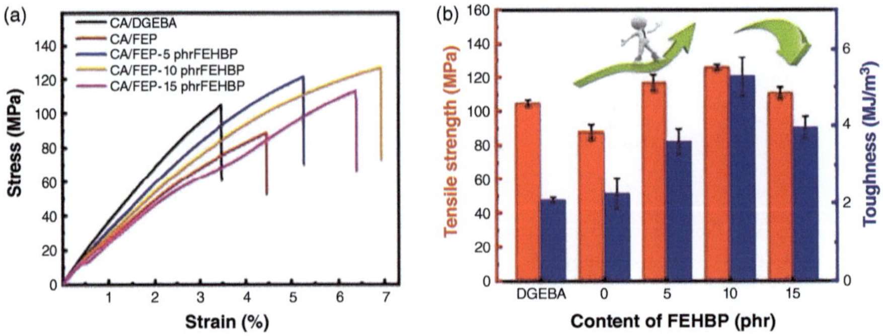

Figure 4.33 Characterization of the epoxy vitrimers: (a) tensile stress–strain curves, (b) tensile strength and toughness.

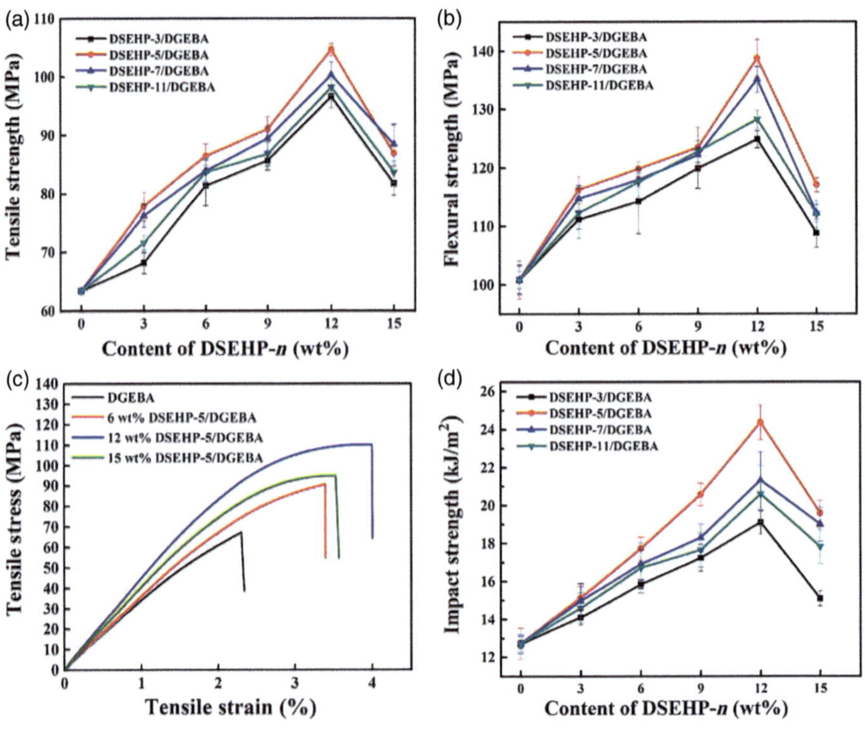

Figure 4.34 Mechanical properties of the cured DSEHP-*n*/DGEBA blends. (a) Tensile strength. (b) Flexural strength. (c) Elongation at break. (d) Impact strength.

limited thermal stability, leading to early decomposition in many DOPO-based polymers, the incorporation of EHBFR significantly enhanced the fire resistance of the epoxy matrix. In the UL-94 vertical burning test, the pristine epoxy (EP) showed no rating, accompanied by sustained flaming drips and heavy smoke, whereas EHBFR-modified formulations achieved improved ratings (Figure 4.35). Microscale combustion calorimetry (MCC) results further quantified

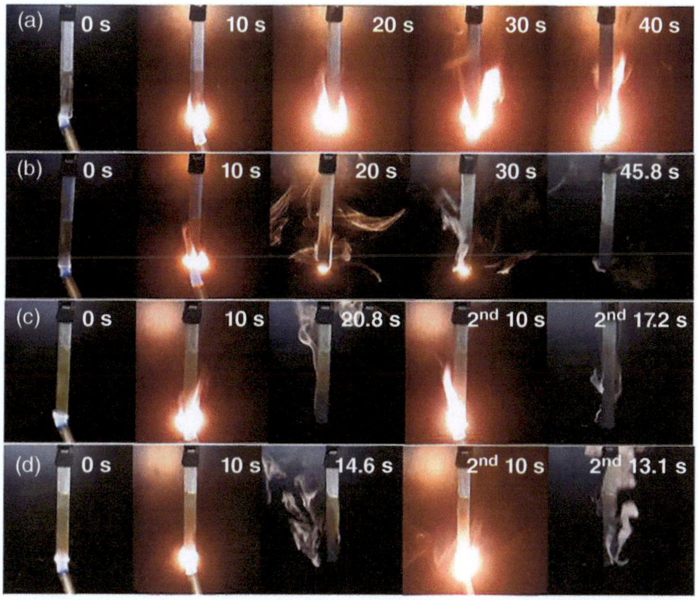

Figure 4.35 Digital photographs of EP (a), EP10 (b), EP20 (c), and EP30 (d) during the vertical burning test. *Source:* [77] / with permission of Elsevier.

this improvement: the peak heat release rate (pHRR) dropped markedly from 625 W/g for EP to 353 W/g for the composite containing 30 wt% EHBFR (EP30). The total heat release (THR) also decreased correspondingly. These data collectively demonstrate that EHBFR effectively suppresses the fire hazard of the EP.

To elucidate the flame-retardant mechanism, thermogravimetric analysis coupled with infrared spectroscopy (TG-IR) was performed (Figure 4.36). The FTIR spectra of pyrolysis products at the maximum decomposition rate are shown in Figure 4.36c (absorbance normalized to sample mass). Both EP and EP30 exhibit similar characteristic peaks corresponding to water (~3650/cm), hydrocarbons (~3034, 2974/cm), CO_2 (~2364/cm), carbonyl compounds (~1767/cm), aromatics (~1603, 1507/cm), and ethers (~1175/cm). A distinct difference is the appearance of a weak but clear peak at ~1061/cm for EP30, assigned to P–O vibration, confirming the release of phosphorus-containing species into the gas phase. Concurrently, the release intensity of carbonyl compounds—mainly generated from free-radical scission of hydroxyl-containing aliphatic chains during combustion—is notably lower for EP30 than for EP (Figure 4.36d), indicating effective suppression of radical-driven degradation pathways in the condensed phase.

A hyperbranched phosphorus/silicon-based flame retardant (TTD) was synthesized from triethylenetetramine, vinyl trichlorosilane, and DOPO [78]. The UV–vis spectra of the epoxy films (Figure 4.37a) show that at 780 nm, transmittance remains high for low TTD loadings: 84.5% (neat EP), 82.1% (EP-TTD3), 83.3% (EP-TTD4), and 73.7% (EP-TTD7), indicating minimal optical loss due to good TTD–matrix compatibility—a trend supported by parallel haze

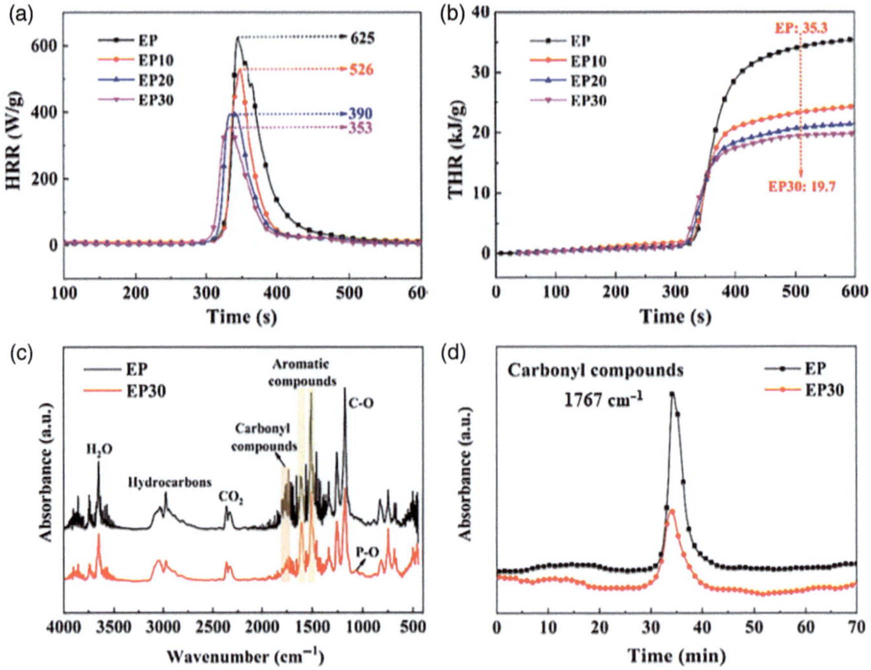

Figure 4.36 HRR (a) and THR (b) curves of epoxy thermosets; FTIR spectra of pyrolysis products for EP and EP30 at maximum decomposition rate (c); intensity of carbonyl compounds versus time for EP and EP30 (d).

measurements (Figure 4.37b). Moreover, all TTD-containing films exhibit reduced transmittance in the 200–400 nm region, confirming enhanced UV-shielding capability. This improvement is attributed to the intrinsic UV absorption of the DOPO group, consistent with earlier reports on DOPO-based additives. XRD patterns (Figure 4.37c) display a single broad peak near 18° for all samples, further evidencing the homogeneous dispersion of TTD in the epoxy matrix. Together, these results demonstrate that the EP-TTD composites maintain high visible-light transparency while offering effective UV-blocking performance.

Furthermore, analyses of both gaseous pyrolysis products and char residues verified the synergistic flame-retardant action of the phosphaphenanthrene, piperazine, and isocyanurate moieties in HPITT. Compared with neat epoxy, the modified thermosets showed markedly enhanced mechanical performance: impact strength, tensile strength, and flexural strength increased by 53.9%, 29.2%, and 58.0%, respectively. These improvements are ascribed to the combined contribution of intramolecular cavities, rigid segments, and a high crosslink density imparted by HPITT. Importantly, the modified systems also maintained excellent thermal properties, with high glass-transition and thermal decomposition temperatures.

The influence of HPITT on the mechanical performance of the epoxy thermosets is summarized in Figure 4.38 [44]. All EP/HPITT composites exhibited higher tensile strength, flexural strength, and modulus than the neat EP system.

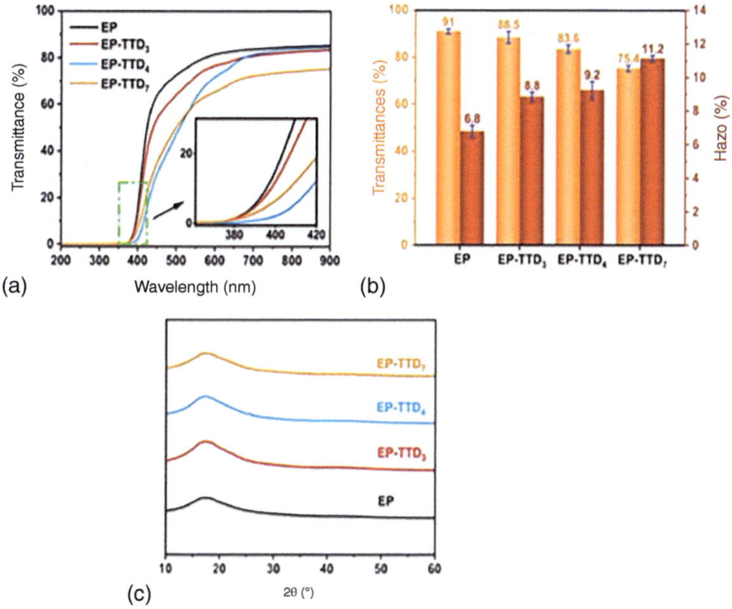

Figure 4.37 Characterization of optical, structural, and mechanical properties. (a) UV–vis spectra. (b) Haze and transmittance. (c) XRD patterns.

These properties initially increased steadily with HPITT content before showing a slight decline at higher loadings, demonstrating an effective reinforcement. For instance, the EP10/HPITT composite achieved a tensile strength of 58.9 MPa (versus 45.6 MPa for neat EP), while maximum flexural strength and modulus reached 128.6 MPa and 3.14 GPa, respectively. This strengthening is attributed to the participation of HPITT's epoxy groups in the curing network, which raises the crosslink density, and to the intrinsic rigidity of the HPITT structure.

Impact strength, fracture toughness (K_{IC}), and fracture energy (G_{IC}) are presented in Figure 4.38b,c. The EP10/HPITT sample exhibited values of 36.8 kJ/m^2, 3.18 MPa m$^{1/2}$, and 4.52 kJ/m^2, corresponding to increases of 53.9%, 50.3%, and 98.8% over neat EP. This toughening effect arises from the flexible segments in the hyperbranched architecture of HPITT, the energy-absorbing free volume provided by its intramolecular cavities, and its integrated role in the cured network, which collectively promote energy dissipation. This modification strategy thus successfully yields epoxy thermosets with an excellent balance of strength and toughness.

Zhang et al. synthesized hyperbranched polysiloxane-functionalized kaolin nanotubes (HBPPSi-HNTs) to prepare flame-retardant and mechanically strong epoxy composites [79].

The mechanical and thermomechanical properties of the epoxy composites modified with HBPPSi-HNTs are presented in Figure 4.39. The tensile strength, toughness, impact strength, and flexural strength of the neat epoxy (EP) were 111.2 MPa, 0.9 MJ/m^3, 12.9 kJ/m^2, and 42.3 MPa, respectively (Figure 4.39a,b). Incorporation of 5.0 phr HBPPSi-HNTs enhanced these properties significantly,

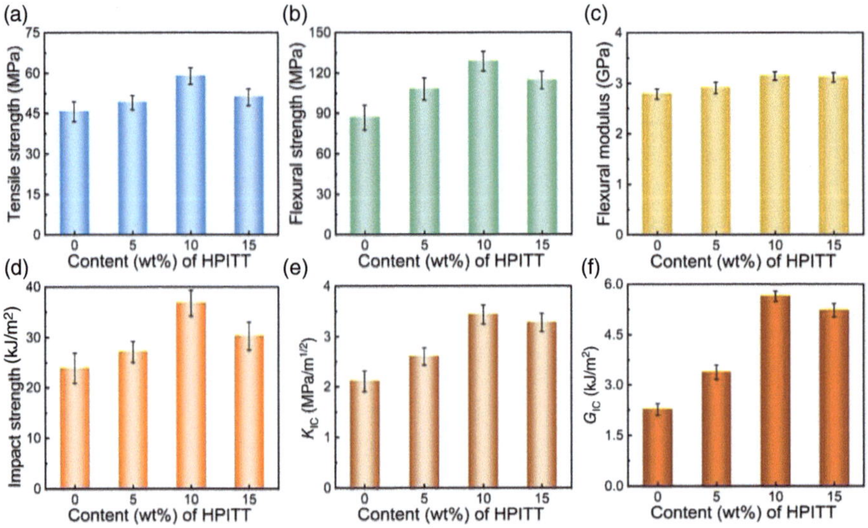

Figure 4.38 (a) Tensile strength, (b) flexural strength, (c) flexural modulus, (d) impact strength, (e) K_{IC}, and (f) G_{IC} of EP and EP/HPITT thermosets. *Source:* [79] / with permission of Elsevier.

with tensile strength reaching 137.6 MPa (a 23.7% increase). The corresponding improvements in toughness, flexural strength, and impact strength were 148.9%, 66.4%, and 48.8%, respectively. These gains are attributed to the uniform dispersion of halloysite nanotubes (HNTs) in the matrix and the abundant intramolecular cavities of the hyperbranched HBPPSi structure, which increase the fractional free volume and promote energy dissipation. For practical applications, further toughening may be pursued by combining HBPPSi-HNTs with other high-strength two-dimensional fillers such as graphene, hexagonal boron nitride, or MXenes.

Dynamic mechanical analysis (Figure 4.39c,d) shows that the storage modulus rises with HNT loading, an effect that is more pronounced with HBPPSi-HNTs due to the rigidity of the nanotubes and their improved dispersion via strong interfacial interactions with the HBP. However, the glass-transition temperature (T_g) shows no significant shift. This can be explained by two competing factors: while better filler dispersion and stronger interfacial adhesion tend to raise T_g, the flexible, hyperbranched topology of HBPPSi enhances chain mobility, thereby offsetting any upward shift—a trend consistent with previous reports on similar epoxy composites.

Zhang et al. [80] reported a bio-based HBP (ITA-HBP) derived from itaconic acid and DOPO, which simultaneously enhances the toughness and flame retardancy of EP at low phosphorus loadings. Cone calorimetry results (Figure 4.40) show that the neat epoxy ignites rapidly upon sparking, reaching a pHRR of 678.7 kW/m² and a THR of 157.9 MJ/m². In contrast, ITA-HBP incorporation markedly reduces the pHRR, THR, and total smoke release (TSR), confirming its strong flame-retardant effect.

With increasing ITA-HBP content (5–20 phr), the time to ignition shortens relative to neat epoxy. This is attributed to the early thermal decomposition of ITA-HBP,

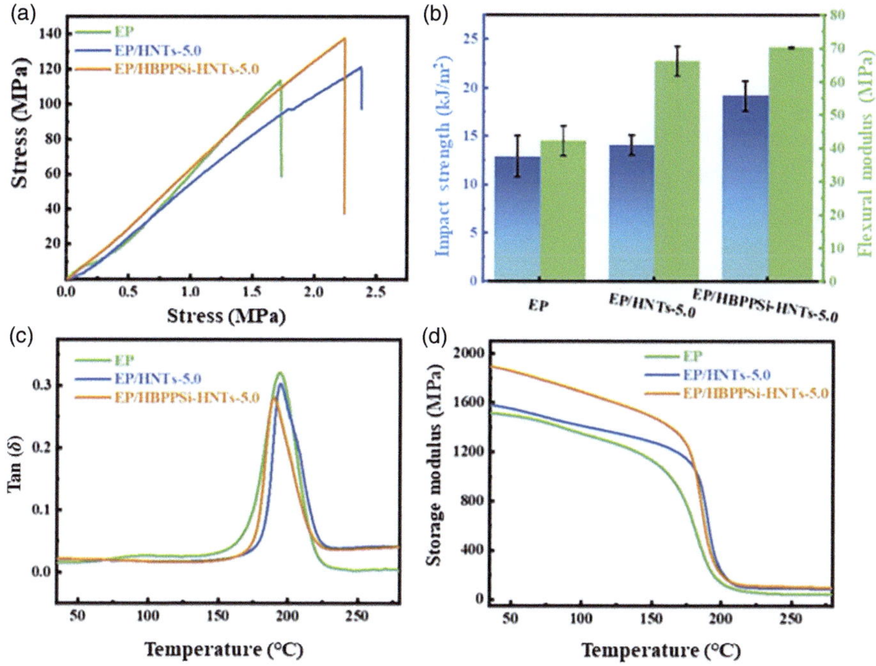

Figure 4.39 Mechanical and dynamic mechanical properties of the epoxy (EP) and its composites. (a) Tensile stress–strain curves. (b) Impact strength and flexural properties. Dynamic mechanical analysis showing storage modulus (c) and tan δ (d) curves. *Source:* [79] / with permission of Elsevier.

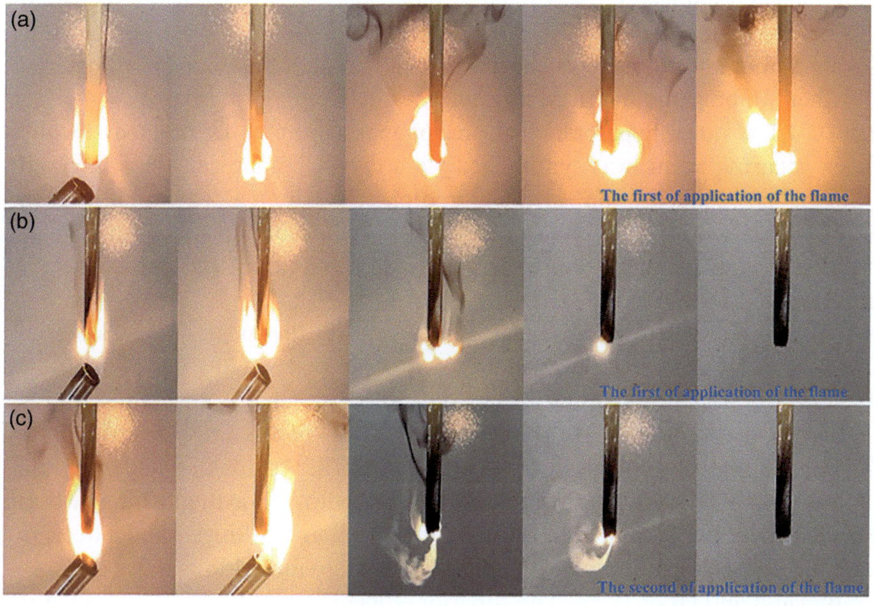

Figure 4.40 Photographs of the samples captured during the UL-94 burning tests: (a) EP, (b and c) EP20. *Source:* [77] / with permission of Elsevier.

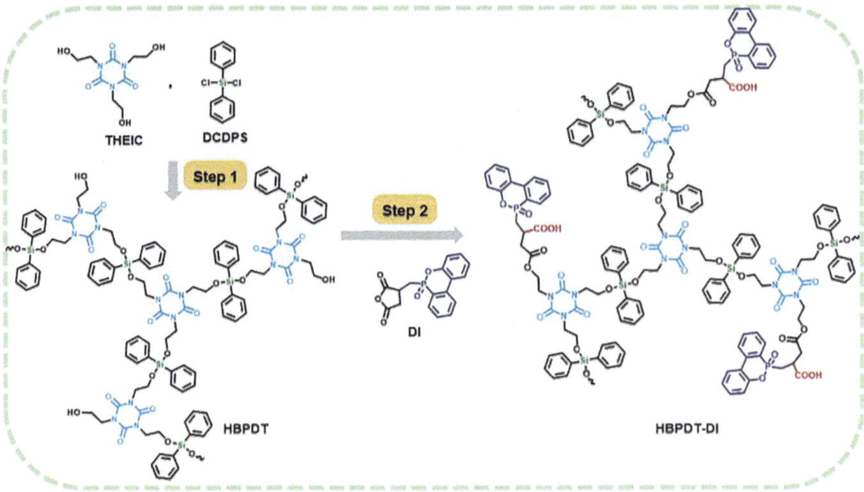

Figure 4.41 Synthesis processes of HBPDT and HBPDT-DI.

which releases radicals that promote premature pyrolysis of the epoxy matrix. Despite earlier ignition, the combustion process is strongly suppressed: at 20 phr loading, the pHRR, THR, and average effective heat of combustion (av-EHC) drop to 468.0 kW/m^2, 110.2 MJ/m^2, and 19.2 MJ/m^2, respectively, while TSR falls from 14,441 to 11,159 m^2/m^2. The reduction in heat release is due to gas-phase activity: volatile phosphorus-containing species generated from ITA-HBP dilute oxygen, capture radicals, and lead to incomplete combustion, as reflected by the increased CO/CO_2 yield ratio. Concurrently, in the condensed phase, these volatiles promote the formation of an intumescent char layer that insulates the underlying material and further inhibits heat and mass transfer.

HBPDT-DI, a novel hyperbranched phosphaphenanthrene-terminated silicon–nitrogen cocuring agent, was synthesized via a two-step approach. This involved the initial polycondensation of dichlorodiphenylsilane (A_2 monomer) with tris(2-hydroxyethyl) isocyanurate (B_3 monomer), followed by a substitution reaction with a phosphaphenanthrene-functionalized itaconic anhydride derivative [81] (Figure 4.41).

The flame-retardant performance of the epoxy thermosets was assessed via LOI and UL-94 vertical burning tests. Neat epoxy (EP) showed a low LOI of 23.5%, produced flaming drips, and failed to achieve any UL-94 rating (Figure 4.42). Incorporating HBPDT-SA or HBPDT-DI significantly improved both self-extinguishing behavior and LOI. With 12 wt% HBPDT-SA, the LOI increased by 30.9%, and a UL-94 V-0 rating was attained, with self-extinguishing within 8 seconds—an effect attributed to the synergistic action of nitrogen and silicon in the additive. In contrast, only 6 wt% HBPDT-DI was required to achieve a UL-94 V-0 rating and an LOI of 31.2%. The superior efficiency of HBPDT-DI is likely due to the additional contribution of its phosphaphenanthrene groups, which further enhance gas-phase and condensed-phase flame inhibition.

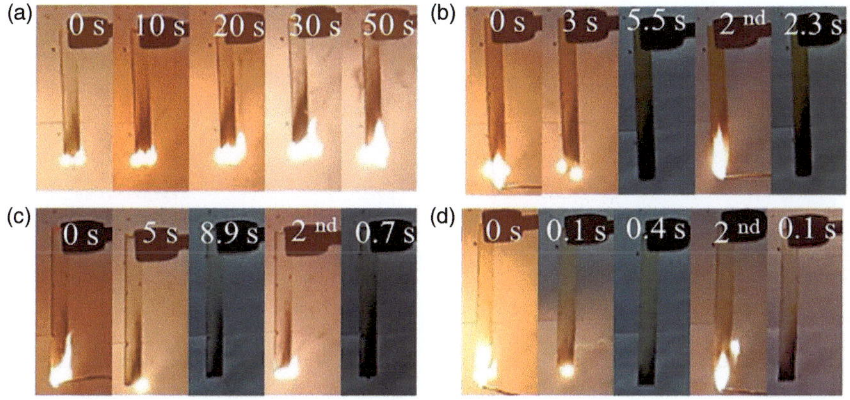

Figure 4.42 Combustion behavior in the UL-94 test. Images captured for (a) neat EP, (b) EP12/HBPDT-SA, (c) EP6/HBPDT-DI, and (d) EP12/HBPDT-DI. *Source:* [81] / with permission of Elsevier.

The combustion behavior of the epoxy thermosets was evaluated by cone calorimetry (Figure 4.43). Compared with neat EP, the EP9/HBPDT-DI composite showed marked reductions in key combustion parameters: the pHRR dropped by 36.1% to 993.5 kW/m^2, the THR by 20.8% to 84.3 MJ/m^2, and the av-EHC by 12.5% to 30.2 MJ/kg (Figure 4.43a,b). The fire-growth rate (FIGRA) decreased from 14.5 to 10.4 $kW/m^2/s$, while the fire-performance index (FPI) increased from 0.046 to 0.065 m^2 s/kW, confirming suppressed flame spread.

Smoke production metrics also improved. The total smoke production (TSP) and peak smoke production rate (PSPR) for EP9/HBPDT-DI were 13.6 m^2 and 0.20 m^2/s, respectively—21.8% and 31% lower than those of neat EP (Figure 4.43c,d). The char residue increased nearly 9.5-fold to 10.5 wt%, consistent with TGA data and indicative of a strong condensed-phase char-forming effect. The average CO yield (av-COY) rose slightly (0.42 versus 0.37 kg/kg for neat EP), whereas the average CO_2 yield (av-CO_2Y) decreased (4.79 versus 5.06 kg/kg), reflecting more incomplete combustion due to gas-phase radical scavenging by phosphorus species. Overall, HBPDT-DI imparts superior flame retardancy compared with HBPDT-SA, owing to the synergistic activity of its phosphorus, nitrogen, and silicon constituents in both the gas and condensed phases.

4.5.3 Antimicrobial Materials

Antibacterial EPs hold significant promise for medical and electronic applications. Here, artemisia naphtha (AN), a waste byproduct from artemisinin extraction, is introduced as a novel bio-based antibacterial agent due to its intrinsic bioactive components. The properties of epoxy composites with varying AN loadings were systematically evaluated. The AN/EP composites exhibited strong antibacterial activity, achieving a 100% sterilization ratio against *Escherichia coli* and 99.96% against *S. aureus*. Meanwhile, the thermal properties—including curing behavior and glass-transition temperature—were well preserved, and mechanical performance was notably enhanced. Specifically, flexural strength increased

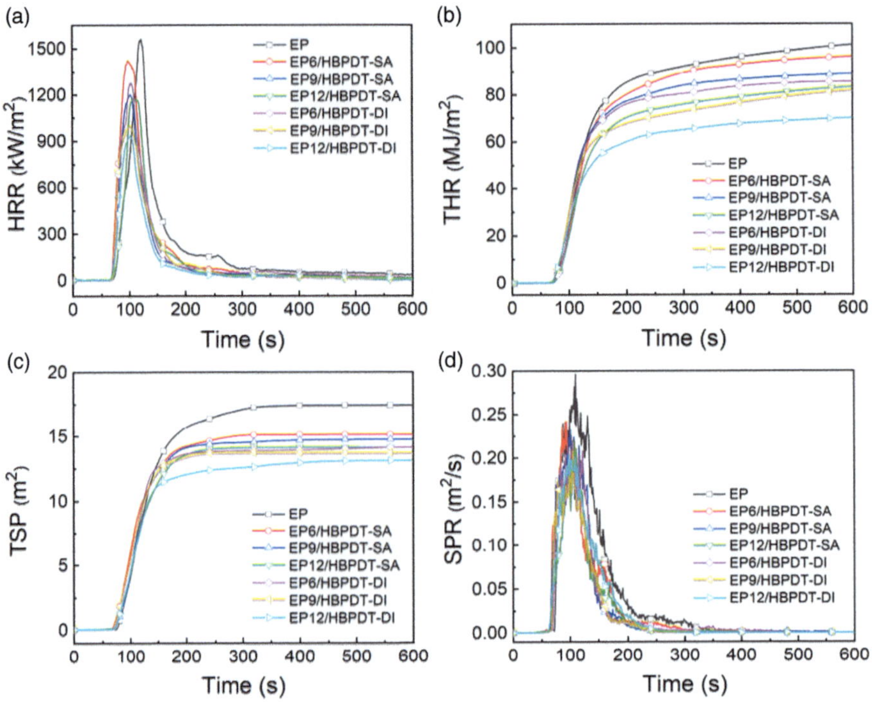

Figure 4.43 Cone calorimetry data of the epoxy thermosets. (a) Heat release rate (HRR). (b) Total heat release (THR). (c) Total smoke production (TSP). (d) Smoke production rate (SPR).

by 62.9% at an AN content of 5 wt%. For comparison, epoxy composites with directly milled artemisia annua powder (AAP) also showed excellent antibacterial performance. These results demonstrate that waste-derived AN can serve as an effective natural antibacterial filler, yielding epoxy composites with simultaneously improved antimicrobial and mechanical properties for potential use in biomedical and healthcare materials.

Nanocomposite NCMES2.5 showed the highest and constant hemolysis prevention from harmful free radicals [82]. Nevertheless, the ability of nanocomposite films to prevent hemolysis exhibited a gradual decline over time, a phenomenon that could be ascribed to the saturation of the films following their reaction with free radicals. For NCMES2.5, its strong capacity to inhibit hemolysis may stem from two key factors: first, the presence of polyphenolic aromatic segments within the hyperbranched epoxy matrix; second, the adequate dispersion and exfoliation of clay layers at this specific clay loading concentration, which results in a large exposed surface area of the silicate clay layers. Therefore, the dispersion state and morphological characteristics of the clay layers are likely to play a vital role in safeguarding red blood cells (RBCs) against hemolysis. In conclusion, the anti-hemolytic assay results demonstrate that the nanocomposites possess a certain degree of antioxidant activity and the ability to scavenge oxygen-free radicals, while having no impact on cell lysis.

4.5.4 Other Materials

Hyperbranched modifiers can also effectively disperse and absorb external forces by forming microphase separation structures or acting as stress concentration points, thereby improving the toughness and strength of composite materials. HBPs have emerged as promising toughening agents for EPs, owing to their highly branched, three-dimensional architecture, functional peripheral groups, and low melt viscosity. Their compact, spherical structure minimizes intermolecular chain entanglements, resulting in a favorable viscosity-to-molecular-weight ratio that enhances toughness without compromising processability. The intrinsic free volume of HBPs further contributes to toughening. While commercial hyperbranched polyesters™ (e.g., Boltron™) have been extensively studied and demonstrate effective epoxy toughening, numerous tailored HBPs with diverse backbones and terminal functionalities (e.g., hydroxyl, epoxy, amino) have since been synthesized via methods such as self-condensing vinyl polymerization and step-growth polycondensation for this purpose. A Common challenge, however, is that toughness improvement often comes at the expense of other critical properties, including glass-transition temperature (T_g), modulus, and strength. For instance, when an epoxidized HBP was used to toughen an epoxy-anhydride system, the fracture toughness (K_{IC}) increased with HBP content, reaching a maximum of 1.21 MPa·m$^{1/2}$ at 20 wt% loading (Figure 4.44a,b). The toughening mechanism was identified as particle cavitation, supported by SEM observations of spherical HBP particles (1–5 μm) dispersed in the epoxy matrix.

To enhance fracture toughness while preserving key mechanical properties, the use of miscible hyperbranched modifiers offers a viable route. An epoxy-functional hyperbranched polymer (HBE) with high epoxide content was synthesized and incorporated into an aromatic epoxy-amine system. As shown in Fig. 4.44c, the fracture toughness (K_{IC}) increased substantially upon HBE addition, reaching a 175% improvement at 10 wt% loading. Tensile modulus likewise rose with HBE content. Dynamic mechanical analysis revealed a single tan δ peak for all blends, confirming a homogeneous, phase-mixed network after curing.

These outcomes have directed attention to molecular-level design, aiming to correlate HBP architecture with blend performance. Five HBPs—HBP40, HBP60, BE1, HBP95, and BE2—with decreasing epoxy equivalent weights (EEW: 1050, 841, 563, 408, and 373 g/mol, respectively) were used to toughen EPs. The glass-transition temperature (T_g) of the blends decreased markedly with the addition of BE2 (lowest EEW), as low-EEW HBPs remain miscible, forming a continuous phase. As HBP functionality decreases (EEW increases), T_g gradually rises toward that of neat epoxy, and phase separation emerges, with the HBP segregating into discrete particles.

The influence of EEW on toughening efficiency is summarized in Figure 4.44d. While all HBPs improved fracture toughness, K_{IC} exhibited a non-monotonic dependence on EEW, first increasing and then decreasing. At low EEW, the miscible, continuous phase promotes plastic deformation via shear yielding. As EEW rises and phase separation occurs, the compatibility between matrix and

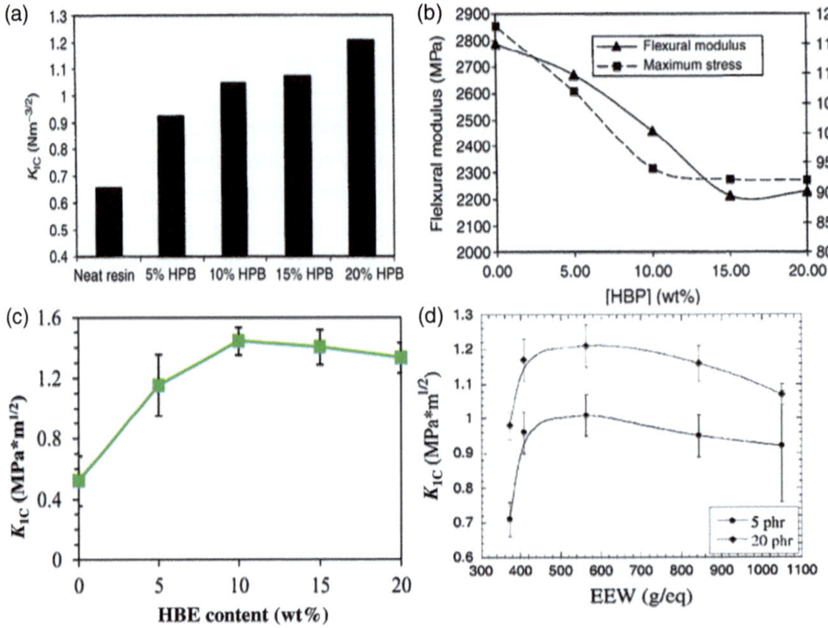

Figure 4.44 Toughening effects of epoxy by different HBPs: (a) K_{IC} of EP/HPB, (b) flexural modulus and maximum stress of EP/HPB, (c) K_{IC} of HBE system, (d) K_{IC} plotted versus HBP EEW for two different concentrations: 5 and 20 phr.

particles declines, reducing the plastic deformation capacity of the continuous phase. Moreover, stress concentration around particles reaches an optimum at intermediate EEW, beyond which it diminishes, linked to the volume of matrix that can yield plastically below its intrinsic yielding stress. These findings underscore that tuning HBP microstructure provides a unique and powerful means to tailor composite properties—an advantage inherent to hyperbranched architectures.

Zhang et al. reported degradable HERs (HER-HTn) containing hydrolytically labile HT bonds, prepared via slow condensation, which exhibit high degradability [76] (Figure 4.45). Nevertheless, the sustainable development of thermosets still faces a key challenge: the efficient preparation of degradable hyperbranched epoxies (HERs) and their composites that simultaneously offer high strength, elongation, modulus, and low-temperature resistance, while also achieving a high degree of degradation. Inspired by our earlier work, we synthesized a series of self-curing, degradable hyperbranched EPs (DSEHP-n, n = 3, 5, 7, 11) through an efficient thiol–isocyanate click reaction. Incorporating DSEHP-n into diglycidyl ether of bisphenol A (DGEBA) enhanced both the degradability and the mechanical performance of the matrix, yielding DSEHP-n/DGEBA composites with excellent strength, elongation, modulus, and low-temperature resistance.

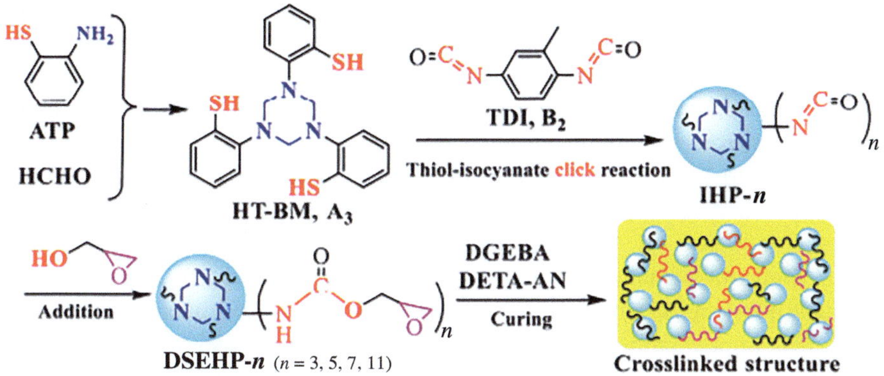

Figure 4.45 Synthesis and curing process of DSEHP-*n*.

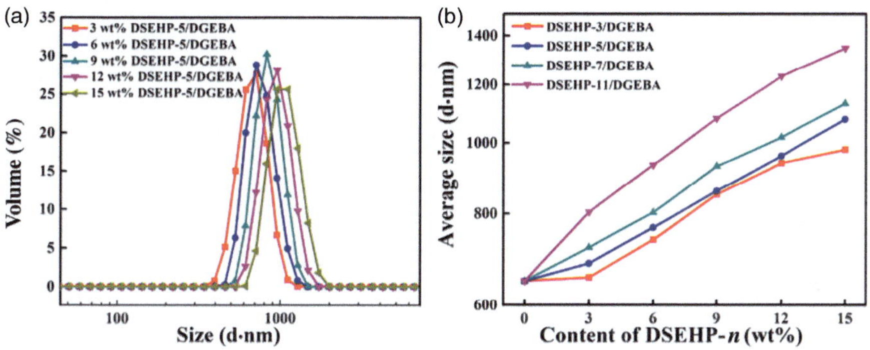

Figure 4.46 Particle size distribution of DSEHP-*n*/DGEBA blends. (a) As a function of DSEHP-5 content. (b) As a function of hyperbranched epoxy molecular weight (n = 3, 5, 7, 11).

The particle size distribution of the hyperbranched epoxy/DGEBA blends was characterized by dynamic light scattering (DLS). As shown in Figure 4.46, the average particle size of the DSEHP-5/DGEBA blends increases systematically with DSEHP-5 content—from 648 nm for neat DGEBA to 685, 766, 859, 960, and 1077 nm at 0–15 wt% loadings (Figure 4.46). This growth in particle dimensions is attributed to the high molecular weight and polarity of DSEHP-5, which promote chain entanglement and strong intermolecular interactions. The trend is consistent with the corresponding rise in blend viscosity, reflecting enhanced molecular packing and reduced mobility as the hyperbranched modifier content increases.

4.6 Conclusions and Perspectives

HERs have emerged as a transformative class of advanced thermosetting materials, distinguished by their unique three-dimensional dendritic architecture, high functionality, low viscosity, and excellent processability. Over the past two decades, HERs have demonstrated significant promise across diverse application domains—particularly in adhesives, coatings, and composite matrices—where they address long-standing limitations of conventional EPs, such as brittleness, high internal stress, and poor interfacial adhesion.

Compared to classical heterogeneous toughening approaches—including rubber modification, TP blending, and NP incorporation—HERs offer several distinct advantages: (i) Homogeneous toughening without phase separation: Due to their molecular compatibility with epoxy networks, HERs integrate seamlessly into the matrix. (ii) Low-viscosity enhancement: Enables easier processing and impregnation of fibers, especially in resin transfer molding (RTM) or vacuum-assisted processes. (iii) Multifunctional end groups: Provide reactive sites for crosslinking, functionalization, or interaction with fillers and substrates. (iv) Balanced property profiles: Improve toughness without sacrificing modulus, thermal stability, or chemical resistance—an elusive goal with many conventional modifiers. Despite these advances, research on homogeneous toughening using HBPs remains relatively limited compared to second-phase methods.

While HERs have made impressive strides, several challenges remain to be addressed for broader industrial adoption:

(1) Scalable and Sustainable Synthesis
Although synthetic routes for HBPs are more straightforward than for perfect dendrimers, achieving precise control over molecular weight distribution, degree of branching, and end-group fidelity at scale is still challenging. Moreover, there is a growing demand for bio-based HERs derived from renewable resources (e.g., soy protein, lignin, cardanol) to meet sustainability goals. While some progress has been made (e.g., bio-based soy protein adhesives with flame retardancy and antimicrobial activity), large-scale production and cost competitiveness remain barriers.

(2) Structure–Property Relationship Optimization
The performance of HERs is highly sensitive to structural parameters such as generation level, core type, terminal group chemistry, and compatibility with curing agents. A deeper understanding of how these factors influence network topology, crosslinking dynamics, and final material properties is needed to enable rational design. Although HBPs improve compatibility, optimizing interfacial interactions between the hyperbranched modifier and the epoxy matrix—especially in hybrid or nanocomposite systems—requires deeper mechanistic understanding.

(3) Long-Term Durability and Environmental Resistance
Limited data exist on the long-term aging behavior of HER-based materials under extreme conditions (e.g., UV exposure, hydrothermal environments, cyclic loading). Further studies on hydrolytic stability, creep resistance, and environmental degradation mechanisms are essential for qualification in critical engineering applications.

(4) Multifunctionality and Smart Responsiveness
Future developments should focus on designing stimuli-responsive HERs capable of self-healing, shape memory, or sensing capabilities through dynamic covalent bonds (e.g., Diels–Alder adducts, disulfide exchange) or supramolecular interactions. Such "smart" epoxy systems could revolutionize maintenance and safety in infrastructure and transportation sectors.
(5) Digital Modeling and AI-Driven Formulation
Utilize computational modeling and machine learning to predict the impact of branching degree, end-group type, and filler interactions on macroscopic properties. Accelerate material discovery and formulation optimization through high-throughput screening and digital twins.
(6) Integration into Industrial Processing Protocols
While HERs improve processability, adapting them to existing industrial workflows—especially in aerospace prepreg manufacturing or high-throughput coating lines—requires validation of cure kinetics, shelf life, and reproducibility. Establish standardized testing protocols for HER-based materials, particularly in emerging areas like bioadhesives and sustainable composites. Foster collaboration between academia, industry, and regulatory bodies to facilitate technology transfer and market entry.

In summary, HERs have evolved from academic curiosities into viable solutions for high-performance applications in adhesives, coatings, and composite materials. They uniquely combine enhanced toughness, excellent processability, and multifunctionality, overcoming key drawbacks of conventional epoxies. Notably, their success in bioadhesives and sustainable formulations highlights their adaptability to emerging societal needs in healthcare and environmental stewardship.

Looking ahead, continued innovation in molecular design, green synthesis, multiscale modeling, and industrial integration will determine the extent to which HERs can displace legacy materials. With growing global demand for EPs—projected to exceed 4.69 million tons in 2024—and increasing emphasis on durability, safety, and sustainability, hyperbranched epoxy systems are poised to play a central role in shaping the future of advanced polymer technologies.

References

1 Xu, Z., Liang, Y., Ma, X. et al. (2019). Recyclable thermoset hyperbranched polymers containing reversible hexahydro-*s*-triazine. *Nat. Sustainability* 3 (1): 29–34. https://doi.org/10.1038/s41893-019-0444-6.

2 Mi, X., Wu, J., Jiang, Y. et al. (2022). Toughness and its mechanisms in epoxy resins. *Prog. Mater. Sci.* 130: 100977. https://doi.org/10.1016/j.pmatsci.2022.100977.

3 Chen, B., Wu, Q., Li, J. et al. (2020). A novel and green method to synthesize a epoxidized biomass Eucommia gum as the nanofiller in the epoxy composite coating with excellent anticorrosive performance. *Chem. Eng. J.* 379: 122323. https://doi.org/10.1016/j.cej.2019.122323.

4 Chowaniec, A., Sadowski, Ł., and Żak, A. (2020). The chemical and microstructural analysis of the adhesive properties of epoxy resin coatings modified using waste glass powder. *Appl. Surf. Sci.* 504: 144373. https://doi.org/10.1016/j.apsusc.2019.144373.

5 Kim, Y.-J., Chun, H., Park, S.-Y. et al. (2018). Preparation and curing chemistry of ultra-low CTE epoxy composite based on the newly-designed triethoxysilyl-functionalized ortho-cresol novolac epoxy. *Polymer* 147: 81–94. https://doi.org/10.1016/j.polymer.2018.05.073.

6 Wu, Y., Ye, K., Liu, Z. et al. (2019). Cotton candy-templated fabrication of three-dimensional ceramic pathway within polymer composite for enhanced thermal conductivity. *ACS Appl. Mater. Interfaces* 11 (47): 44700–44707. https://doi.org/10.1021/acsami.9b15758.

7 Wang, S., Ma, S., Xu, C. et al. (2017). Vanillin-derived high-performance flame retardant epoxy resins: facile synthesis and properties. *Macromolecules* 50 (5): 1892–1901. https://doi.org/10.1021/acs.macromol.7b00097.

8 Zhang, X., Si, Y., Mo, J., and Guo, Z. (2017). Robust micro-nanoscale flowerlike ZnO/epoxy resin superhydrophobic coating with rapid healing ability. *Chem. Eng. J.* 313: 1152–1159. https://doi.org/10.1016/j.cej.2016.11.014.

9 Han, J., Du, G., Gao, W., and Bai, H. (2019). An Anisotropically high thermal conductive boron nitride/epoxy composite based on nacre-mimetic 3D network. *Adv. Funct. Mater.* 29 (13): 1900412. https://doi.org/10.1002/adfm.201900412.

10 Yang, G., Wu, W.-H., Wang, Y.-H. et al. (2019). Synthesis of a novel phosphazene-based flame retardant with active amine groups and its application in reducing the fire hazard of epoxy resin. *J. Hazard. Mater.* 366: 78–87. https://doi.org/10.1016/j.jhazmat.2018.11.093.

11 Qiu, Y., Qian, L., Chen, Y., and Hao, J. (2019). Improving the fracture toughness and flame retardant properties of epoxy thermosets by phosphaphenanthrene/siloxane cluster-like molecules with multiple reactive groups. *Compos. Part B Eng.* 178: 107481. https://doi.org/10.1016/j.compositesb.2019.107481.

12 Seo, J., Yui, N., and Seo, J.-H. (2019). Development of a supramolecular accelerator simultaneously to increase the cross-linking density and ductility of an epoxy resin. *Chem. Eng. J.* 356: 303–311. https://doi.org/10.1016/j.cej.2018.09.020.

13 Cao, Q., Li, J., Qi, Y. et al. (2022). Engineering double load-sharing network in thermosetting: much more than just toughening. *Macromolecules* 55 (21): 9502–9512. https://doi.org/10.1021/acs.macromol.2c01599.

14 Wang, W., Li, Y., Ma, Z. et al. (2025). Ultra-tough poly(urea-urethane) plastics with superior impact resistance for cryogenic applications. *Adv. Mater.* 37: e2509421. https://doi.org/10.1002/adma.202509421.

15 Lu, X., Guo, L., Wang, Y., and Sun, J. (2025). Reversibly cross-linked plastics meet carbon fibers: closed-loop recyclable composites with enhanced mechanical strength and stability. *Adv. Funct. Mater.* 35: 2503106. https://doi.org/10.1002/adfm.202503106.

16 Cheng, Q., Huang, C., and Tomsia, A.P. (2017). Freeze casting for assembling bioinspired structural materials. *Adv. Mater.* 29 (45): 1703155. https://doi.org/10.1002/adma.201703155.

17 Wu, W., Feng, H., Xie, L. et al. (2024). Reprocessable and ultratough epoxy thermosetting plastic. *Nat. Sustainability* 7 (6): 804–811. https://doi.org/10.1038/s41893-024-01331-9.

18 Zheng, N., Xu, Y., Zhao, Q., and Xie, T. (2021). Dynamic covalent polymer networks: a molecular platform for designing functions beyond chemical recycling and self-healing. *Chem. Rev.* 121 (3): 1716–1745. https://doi.org/10.1021/acs.chemrev.0c00938.

19 Yang, B., Ni, T., Wu, J. et al. (2025). Circular 3D printing of high-performance photopolymers through dissociative network design. *Science* 388 (6743): 170–175.

20 Li, S.-C., Hu, B.-C., Ding, Y.-W. et al. (2018). Wood-derived ultrathin carbon nanofiber aerogels. *Angew. Chem. Int. Ed.* 57: 7085–7090.

21 Yue, X., Yang, H., Han, Z. et al. (2024). Tough and moldable sustainable cellulose-based structural materials via multiscale interface engineering. *Adv. Mater.* 36 (7): e2306451. https://doi.org/10.1002/adma.202306451.

22 Li, Y., Lian, W., and Cheng, Q. (2025). High-performance nacre-inspired 2D carbon-based nanocomposites. *Adv. Mater.* 37, 2501932.

23 Liu, L., Liu, X., Li, X. et al. (2024). An integrative chemical recycling approach for catalytic oxidation of epoxy resin and in situ separation of degraded products. *Angew. Chem. Int. Ed.* 63 (26): e202405912. https://doi.org/10.1002/anie.202405912.

24 Li, X., Liu, L., Yu, J. et al. (2025). Full recovery of epoxy resin wastes into bisphenol A and epoxy monomers via Lewis acid. *Angew. Chem. Int. Ed.* 64: e202422472. https://doi.org/10.1002/anie.202422472.

25 Chen, H., Liu, X., He, Q. et al. (2024). Upcycling waste thermosetting polyimide resins into high-performance and sustainable low-temperature-resistance adhesives. *Adv. Mater.* 36 (4): e2310779. https://doi.org/10.1002/adma.202310779.

26 Zhao, X., Long, Y., Xu, S. et al. (2023). Recovery of epoxy thermosets and their composites. *Mater. Today* 64: 72–97. https://doi.org/10.1016/j.mattod.2022.12.005.

27 Wang, H., Lu, R., Yan, J. et al. (2023). Tough and conductive nacre-inspired MXene/epoxy layered bulk nanocomposites. *Angew. Chem. Int. Ed.* 62 (9): e202216874. https://doi.org/10.1002/anie.202216874.

28 Peng, J., Huang, C., Cao, C. et al. (2020). Inverse nacre-like epoxy-graphene layered nanocomposites with integration of high toughness and self-monitoring. *Matter* 2 (1): 220–232. https://doi.org/10.1016/j.matt.2019.08.013.

29 Wan, S., Chen, Y., Huang, C. et al. (2024). Scalable ultrastrong MXene films with superior osteogenesis. *Nature* 634 (8036): 1103–1110. https://doi.org/10.1038/s41586-024-08067-8.

30 Zhou, T., Wu, C., Wang, Y. et al. (2020). Super-tough MXene-functionalized graphene sheets. *Nat. Commun.* 11 (1): 2077. https://doi.org/10.1038/s41467-020-15991-6.

31 Zhou, Y., Zhao, Z., Zeng, Y. et al. (2025). Ultra-robust and tough epoxy resin enabled by pulley mechanism-based curing agent for strong and reversible adhesion. *Adv. Funct. Mater.* 35: e17052. https://doi.org/10.1002/adfm.202517052.

32 Park, Y.T., Qian, Y., Chan, C. et al. (2015). Epoxy toughening with low graphene loading. *Adv. Funct. Mater.* 25 (4): 575–585. https://doi.org/10.1002/adfm.201402553.

33 Chen, S., Zhang, J., Zhou, J. et al. (2018). Dramatic toughness enhancement of benzoxazine/epoxy thermosets with a novel hyperbranched polymeric ionic liquid. *Chem. Eng. J.* 334: 1371–1382. https://doi.org/10.1016/j.cej.2017.11.104.

34 Wang, S., Xue, H., Guo, S. et al. (2021). Graphene/nanorubber reinforced electrically conductive epoxy composites with enhanced toughness. *J. Appl. Polym. Sci.* 138 (14): 50163. https://doi.org/10.1002/app.50163.

35 Pruksawan, S., Samitsu, S., Yokoyama, H., and Naito, M. (2019). Homogeneously dispersed polyrotaxane in epoxy adhesive and its improvement in the fracture toughness. *Macromolecules* 52 (6): 2464–2475. https://doi.org/10.1021/acs.macromol.8b02450.

36 Xu, C., Qu, T., Zhang, X. et al. (2019). Enhanced toughness and thermal conductivity for epoxy resin with a core–shell structured polyacrylic modifier and modified boron nitride. *RSC Adv.* 9 (15): 8654–8663. https://doi.org/10.1039/C8RA10645B.

37 Zhao, X., Li, Y., Chen, W. et al. (2019). Improved fracture toughness of epoxy resin reinforced with polyamide 6/graphene oxide nanocomposites prepared via in situ polymerization. *Compos. Sci. Technol.* 171: 180–189. https://doi.org/10.1016/j.compscitech.2018.12.023.

38 Zainol Abidin, M.S., Herceg, T., Greenhalgh, E.S. et al. (2019). Enhanced fracture toughness of hierarchical carbon nanotube reinforced carbon fibre epoxy composites with engineered matrix microstructure. *Compos. Sci. Technol.* 170: 85–92. https://doi.org/10.1016/j.compscitech.2018.11.017.

39 Kim, S.H., Heo, Y.-J., Park, M. et al. (2018). Effect of hydrophilic graphite flake on thermal conductivity and fracture toughness of basalt fibers/epoxy composites. *Compos. Part B Eng.* 153: 9–16. https://doi.org/10.1016/j.compositesb.2018.07.022.

40 Liu, X., Wu, H., Xu, W. et al. (2023). Ultrastrong and high-tough thermoset epoxy resins from hyperbranched topological structure and subnanoscaled free volume. *Adv. Mater.* 36 (9): e2308434. https://doi.org/10.1002/adma.202308434.

41 Zhong, L., Hao, Y., Zhang, J. et al. (2022). Closed-loop recyclable fully bio-based epoxy vitrimers from ferulic acid-derived hyperbranched epoxy resin. *Macromolecules* 55 (2): 595–607. https://doi.org/10.1021/acs.macromol.1c02247.

42 Teng, N., Dai, J., Wang, S. et al. (2022). Hyperbranched flame retardant for epoxy resin modification: simultaneously improved flame retardancy, toughness and strength as well as glass transition temperature. *Chem. Eng. J.* 428: 131226. https://doi.org/10.1016/j.cej.2021.131226.

43 Liu, H., Wang, C., Liu, S. et al. (2025). Enhancing interfacial properties of epoxy coatings via hyperbranched modification of SiC fillers: experimental and simulation insights. *Chem. Eng. J.* 511: 161841. https://doi.org/10.1016/j.cej.2025.161841.

44 Tang, Y., Zhang, X., Liu, X. et al. (2024). Facile synthesis of hyperbranched epoxy resin for combined flame retarding and toughening modification of epoxy thermosets. *Chem. Eng. J.* 482: 148992. https://doi.org/10.1016/j.cej.2024.148992.

45 Liang, X., Li, X., Tang, Y. et al. (2022). Hyperbranched epoxy resin-grafted graphene oxide for efficient and all-purpose epoxy resin modification. *J. Colloid Interface Sci.* 611: 105–117. https://doi.org/10.1016/j.jcis.2021.12.068.

46 Dong, T., Tian, N., Xu, B. et al. (2022). Biomass poplar catkin fiber-based superhydrophobic aerogel with tubular-lamellar interweaved neurons-like structure. *J. Hazard. Mater.* 429: 128290. https://doi.org/10.1016/j.jhazmat.2022.128290.

47 Mezzenga, R., Boogh, L., Månson, J.-A.E., and Pettersson, B. (2000). Effects of the branching architecture on the reactivity of epoxy–amine groups. *Macromolecules* 33 (12): 4373–4379. https://doi.org/10.1021/ma991906w.

48 Li, D., Zhou, S., Wu, Y. et al. (2025). Bifunctional hyperbranched modifier enabled interlocking interfaces for ultra-strong recyclable carbon fiber/bio-epoxy composites. *Chem. Eng. J.* 521: 166591. https://doi.org/10.1016/j.cej.2025.166591.

49 Han, X., Wu, Y., Han, H. et al. (2023). Super-strong reusable hot melt adhesives prepared from hyperbranched epoxy resin. *Mater. Chem. Front.* 7 (13): 2676–2682. https://doi.org/10.1039/D3QM00090G.

50 Feng, W., Cao, H., Jin, T. et al. (2023). Air oxidation and hyperbranched crosslinking approach for a boiling water-resistant sucrose-derived wood adhesive. *Ind. Crop. Prod.* 202: 117052. https://doi.org/10.1016/j.indcrop.2023.117052.

51 Cui, C., Liu, B., Wu, T. et al. (2022). A hyperbranched polymer elastomer-based pressure sensitive adhesive. *J. Mater. Chem. A* 10 (3): 1257–1269. https://doi.org/10.1039/D1TA08091A.

52 Shao, J., Li, X., Liu, T. et al. (2024). Multiple function hyperbranched polysiloxane nanoclusters for controlling a cross-linking structure to convert soy meal into a strong, tough, and multifunctional adhesive. *ACS Sustain. Chem. Eng.* 12 (10): 3966–3976. https://doi.org/10.1021/acssuschemeng.3c06212.

53 Chen, H., Lin, H., Sun, Z. et al. (2023). An ion-terminated hyperbranched polymer towards multipurpose adhesive with record-high bonding strength and sensitive stress-sensing. *J. Mater. Chem. A* 11 (5): 2443–2451. https://doi.org/10.1039/D2TA08457K.

54 Moazami, S., Kharaziha, M., Emadi, R., and Dinari, M. (2023). Multifunctional bioinspired bredigite-modified adhesive for bone fracture healing. *ACS Appl. Mater. Interfaces* 15 (5): 6499–6513. https://doi.org/10.1021/acsami.2c20038.

55 Zhang, Y., Li, X., Wei, W., and Liu, X. (2021). A strong dual-component bioadhesive based on solventless thiol-isocyanate click chemistry. *ACS Biomater. Sci. Eng.* 7 (7): 3389–3398. https://doi.org/10.1021/acsbiomaterials.1c00504.

56 Wu, Y., Huang, S., Liu, X. et al. (2023). Preparation of high-performance polyurethane-polyacrylate coatings based on hyperbranched structure and application in anti-ultraviolet film. *Prog. Org. Coat.* 182: 107614. https://doi.org/10.1016/j.porgcoat.2023.107614.

57 Fu, J., Yu, H., Wang, L. et al. (2020). Preparation and properties of UV-curable hyperbranched polyurethane acrylate hard coatings. *Prog. Org. Coat.* 144: 105635. https://doi.org/10.1016/j.porgcoat.2020.105635.

58 Xie, R., Ai, X., Xie, Q. et al. (2023). Non-silicone elastic coating with fouling resistance and fouling release abilities based on degradable hyperbranched polymer. *Prog. Org. Coat.* 175: 107350. https://doi.org/10.1016/j.porgcoat.2022.107350.

59 Liu, C., Jiao, S., Sun, Z. et al. (2024). Development of a hyperbranched oxazolidinone dynamic omniphobic liquid-like coating with high hardness and flexibility. *Appl. Surf. Sci.* 654: 159351. https://doi.org/10.1016/j.apsusc.2024.159351.

60 Li, Z., Wu, J., Chen, P. et al. (2022). A new metallic composite cathode originated from hyperbranched polymer coated MOF for high-performance lithium-sulfur batteries. *Chem. Eng. J.* 435: 135125. https://doi.org/10.1016/j.cej.2022.135125.

61 Dağlar, Ö., Eisenreich, F., and Tomović, Ž. (2024). Chemical and solvent-based recycling of high-performance epoxy thermoset utilizing imine-containing secondary amine hardener. *Adv. Funct. Mater.* 34: 2408299. https://doi.org/10.1002/adfm.202408299.

62 Li, Y., Xiong, T., Ge, L. et al. (2025). Highly-effective thermoelectric cooling for power semiconductor devices packed with thermal-expansion offset and flame retardancy epoxy resin. *Adv. Funct. Mater.* 35: 2420944. https://doi.org/10.1002/adfm.202420944.

63 Wang, S., Ruan, K., Guo, Y. et al. (2025). Thermally conductive naphthalene epoxy resin by tailoring flexible chain length and liquid crystal structure. *Angew. Chem. Int. Ed.* 64: e202501459. https://doi.org/10.1002/anie.202501459.

64 Liang, N., Liu, X., Hu, J. et al. (2023). Influence of topological structure on mechanical property of recyclable bio-based hyperbranched epoxy/carbon fiber fabric composites. *Chem. Eng. J.* 471: 144329. https://doi.org/10.1016/j.cej.2023.144329.

65 Yu, L., Hu, Q., Li, T. et al. (2023). Ultrahigh flowability and excellent mechanical performance of glass fiber/PA6 composites prepared by hyperbranched polymers. *Macromol. Mater. Eng.* 308 (8): 2300012. https://doi.org/10.1002/mame.202300012.

66 Cao, Z., Hao, T., Wang, P. et al. (2017). Surface modified glass fiber membranes with superior chemical and thermal resistance for O/W separation. *Chem. Eng. J.* 309: 30–40. https://doi.org/10.1016/j.cej.2016.10.013.

67 Liang, X., Hu, Q., Cheng, K. et al. (2024). Glass fiber reinforced PA6 composites with high strength and high flowability by hyperbranched polyesters. *J. Polym. Sci.* 62 (24): 5584–5597. https://doi.org/10.1002/pol.20240711.

68 Zhuo, D., Gu, A., Liang, G. et al. (2011). Flame retardancy and flame retarding mechanism of high performance hyperbranched polysiloxane modified bismaleimide/cyanate ester resin. *Polym. Degrad. Stab.* 96 (4): 505–514. https://doi.org/10.1016/j.polymdegradstab. 2011.01.006.

69 Xu, H., Hu, J., Liu, X. et al. (2023). Interface strengthening and high-value recycling of epoxy resin/carbon fiber fabric composites. *Chem. Eng. J.* 465: 142998. https://doi.org/10.1016/j.cej.2023.142998.

70 Xu, Y., Zhang, H., Dai, S. et al. (2022). Hyperbranched polyester catalyzed self-healing bio-based vitrimer for closed-loop recyclable carbon fiber-reinforced polymers. *Compos. Sci. Technol.* 228: 109676. https://doi.org/10.1016/j.compscitech.2022.109676.

71 Ma, X., Wu, Y., Liang, N. et al. (2023). High-efficiently renewable hyperbranched epoxy resin/carbon fiber composites with both long service life and high performance. *Compos. Commun.* 40: 101630. https://doi.org/10.1016/j.coco.2023.101630.

72 Ma, X., Xu, H., Xu, Z. et al. (2021). Closed-loop recycling of both resin and fiber from high-performance thermoset epoxy/carbon fiber composites. *ACS Macro Lett.* 10 (9): 1113–1118. https://doi.org/10.1021/acsmacrolett.1c00437.

73 Zhang, Y., Yan, H., Yu, R. et al. (2024). Hyperbranched dynamic crosslinking networks enable degradable, reconfigurable, and multifunctional epoxy vitrimer. *Adv. Sci.* 11 (2): 2306350. https://doi.org/10.1002/advs.202306350.

74 Huang, Y., Ye, L., Wei, Q. et al. (2025). Synergistic effect of recycled PET and hyperbranched ionic liquids for toughening epoxy resin. *Compos. Commun.* 54: 102286. https://doi.org/10.1016/j.coco.2025.102286.

75 Jiang, C., Deng, G., Luo, M. et al. (2024). Closed-loop recycling of tough epoxy supramolecular thermosets constructed with hyperbranched topological structure. *Nat. Commun.* 15 (1): 4869. https://doi.org/10.1038/s41467-024-49272-3.

76 Ma, X., Guo, W., Xu, Z. et al. (2020). Synthesis of degradable hyperbranched epoxy resins with high tensile, elongation, modulus and low-temperature resistance. *Compos. Part B Eng.* 192: 108005. https://doi.org/10.1016/j.compositesb.2020.108005.

77 Teng, N., Dai, J., Wang, S. et al. (2021). Hyperbranched flame retardant to simultaneously improve the fire-safety, toughness and glass transition temperature of epoxy resin. *Eur. Polym. J.* 157: 110638. https://doi.org/10.1016/j.eurpolymj.2021.110638.

78 Shi, Q., Huo, S., Wang, C. et al. (2022). A phosphorus/silicon-based, hyperbranched polymer for high-performance, fire-safe, transparent epoxy resins. *Polym. Degrad. Stab.* 203: 110065. https://doi.org/10.1016/j.polymdegradstab.2022.110065.

79 Zhong, L., Li, T., Zhang, J. et al. (2022). Simultaneously improving the flame retardancy and toughness of epoxy composites with hyperbranched phosphorus-containing polysiloxane functionalized halloysite nanotubes. *Eur. Polym. J.* 179: 111564. https://doi.org/10.1016/j.eurpolymj.2022.111564.

80 Zhang, J., Mi, X., Chen, S. et al. (2020). A bio-based hyperbranched flame retardant for epoxy resins. *Chem. Eng. J.* 381: 122719. https://doi.org/10.1016/j.cej.2019.122719.

81 Tang, Y., Zhang, X., Ge, A. et al. (2024). Novel hyperbranched phosphaphenanthrene-terminaled silicon-nitrogen cocuring agent towards high-performance epoxy thermosets. *Prog. Org. Coat.* 197: 108828. https://doi.org/10.1016/j.porgcoat.2024.108828.

82 Roy, B. (2013). Silver-embedded modified hyperbranched epoxy/clay nanocomposites as antibacterial materials. *Bioresour. Technol.* 127: 175–180.

5

Reinforcing and Toughening and Mechanism of Hyperbranched Epoxy Resins

Junheng Zhang and Daohong Zhang

South-Central Minzu University, Hubei R&D Center of Hyperbranched Polymers Synthesis and Applications, School of Chemistry and Materials Science, No.182, Minzu Ave. Hongshan District, Wuhan Hubei Provinces, 430074, P.R.China

5.1 Introduction

Recognized as an exceptionally versatile class of materials, epoxy resins currently occupy a preeminent position in a diverse spectrum of engineering applications. These span from composite components for wind turbine blades in the renewable energy sector to sophisticated structural parts employed in aerospace systems [1, 2]. Structurally, epoxy resins are characterized by the presence of oxirane rings—also designated as epoxy groups—within their molecular backbones. They can form rigid three-dimensional cross-linked networks through cross-polymerization with a secondary component, which is commonly referred to as a hardener or curing agent. Alternatively, epoxy chains are also capable of undergoing homopolymerization with one another via either anionic or cationic reaction mechanisms. A wide array of conventional hardeners includes multifunctional amines, acid anhydrides, carboxylic acids, thiols, and polyols. A major group of epoxy resins are synthesized by reaction of epichlorohyrdin (1-chloroprene 2-oxide) and bisphenol-A (bis(4-hydroxy phenylene)-2,2 propane) in the presence of sodium hydroxide.

Furthermore, epoxy resins find extensive utilization as high-performance engineering adhesives and matrix materials for fiber-reinforced plastics (FRPs). This widespread adoption stems from their appealing synergy of robust adhesive properties and superior processability [3, 4]. Nevertheless, virgin (uncured) epoxy resin precursors exhibit notable deficiencies in mechanical robustness, chemical stability, and thermal tolerance. Through the cross-linking reaction between linear epoxy resins and appropriately selected curing agents, it is possible to construct three-dimensionally cross-linked thermosetting networks. Such cured structures endow the resulting materials with high elastic modulus, excellent fracture strength, and reliable bonding performance, rendering them well-suited for a broad spectrum of industrial applications [5, 6].

Despite these favorable properties, conventional cured epoxy resins possess a highly cross-linked architecture that affords minimal plastic deformation. Furthermore, volumetric shrinkage that occurs during curing can induce residual internal stress, which retards crack initiation and propagation—consequently confining the application of epoxy resins in high-performance composite materials. To mitigate these toughness-related limitations, both homogeneous and heterogeneous toughening tactics have been developed [7–9]. Typically, enhancing the toughness of cured epoxies entails incorporating a second phase into the epoxy matrix, thereby forming a two-phase heterogeneous system composed of the epoxy matrix phase and a toughening phase. Strictly defined, the addition of any component with a dissimilar chemical constitution to the matrix resin can be considered the introduction of a second phase.

A diverse array of toughening methodologies has been devised to address this challenge, encompassing the modification of epoxy resins with rubber modifiers (REs, i.e., elastomers) [10–12], thermoplastic additives (TPs) [13, 14], core–shell polymeric particles (CSPs) [15, 16], liquid crystalline polymers (LCPs) [17, 18], block copolymers (BCPs) [19, 20], nanomaterial (NM) fillers [21, 22], hyperbranched polymers (HBPs) [23, 24], and bio-based modifiers (BBMs) [25, 26]. In particular, heterogeneous toughening strategies commonly entail two primary pathways: the in-situ formation of interpenetrating polymer networks (IPNs) or semi-interpenetrating polymer networks (SIPNs) during the curing process, or the direct incorporation of functional modifiers (e.g., REs, TPs, CSPs, LCPs, BCPs, or NMs). These modifiers tend to phase-separate into a discrete second phase within the matrix due to thermodynamically unfavorable free energy conditions during curing [27, 28]. The cured epoxy composites obtained via such approaches typically possess a multiphase microstructure featuring submicron- or micron-sized domains, which effectively imparts enhanced toughness to the material [29, 30]. Nevertheless, heterogeneous toughening is not without drawbacks: it may induce inherent immiscibility between the matrix and the modifiers, resulting in undesirable side effects such as elevated system viscosity, diminished fluidity, and reduced optical transparency. These issues consequently impede the deployment of modified epoxies in high-end applications including electronic packaging and insulating coatings for wind power equipment.

In sharp contrast, homogeneous toughening systems are characterized by modifiers that achieve molecular-level uniform dispersion throughout the epoxy matrix. These systems form single- or multi-molecular interpenetrating networks within the cured cross-linked framework, with no discernible phase separation observable at either the microscale or the nanoscale. Such single-phase structures can be realized via the incorporation of either flexible molecular chain segments or HBPs into the cross-linked epoxy network [31–33]. Notably, these modification strategies not only enhance the material's toughness but also preserve, or even further improve, its mechanical strength. Epoxy-terminated HBPs (EHBPs), in particular, possess a remarkable ability to drastically reduce the viscosity of epoxy resin systems—a key advantage that renders them exceptionally well-suited for high-end advanced material applications.

Alongside the advancement of polymer materials science and state-of-the-art analytical methodologies, the toughening mechanisms of epoxy resins have now been more comprehensively elucidated and systematically categorized. To illustrate,

REs and CSPs are effective in enhancing epoxy toughness, yet they frequently compromise the material's elastic modulus, mechanical strength, and thermal stability, thereby restricting their applicability in high-performance engineering scenarios. BCPs, on the other hand, tend to boost toughness at the expense of mechanical strength. LCPs exhibit a notable advantage in that they can drastically improve toughness while preserving the original elastic modulus and thermal resistance of epoxies; however, their inherent drawbacks—including poor solubility in epoxy matrices, weak interfacial compatibility, and low melt fluidity—often lead to cumbersome processing procedures and the formation of undesirable phase-separated microstructures [34, 35]. NM fillers hold great potential for simultaneously toughening and strengthening epoxy resins, yet their strong propensity for agglomeration poses a major hurdle to achieving uniform dispersion within the matrix, which frequently results in inconsistent batch-to-batch material performance. Additionally, renewable or biodegradable modifiers can be incorporated into epoxy matrices to improve toughness; this approach, however, typically comes at the cost of diminished mechanical and thermal performance. While IPNs are capable of enhancing both impact strength and overall toughness of epoxies, HBPs have emerged as one of the most promising and effective toughening modifiers to date [36, 37]. They not only elevate the toughness of epoxy matrices but also contribute to a certain degree of improvement in mechanical strength. More importantly, HBPs demonstrate exceptional interfacial compatibility with epoxy matrices and can effectively reduce the overall system viscosity, which facilitates the formation of homogeneous microstructures and expands their application scope—especially in high-end advanced material fields such as precision electronic packaging and high-performance composite fabrication.

5.2 Heterogeneous Mechanism

5.2.1 Nano-inorganic and Carbon-Based Nanofillers

First synthesized and commercialized in the 1960s, bisphenol-A diglycidyl ether (DGEBA) rapidly established itself as one of the most prevalent epoxy resin varieties in industrial practice. This resin finds utility not only in the formulation of high-performance engineering adhesives but, more crucially, serves as the predominant matrix material for fiber-reinforced polymers (FRPs). Such widespread adoption is attributed to its exceptional adhesive properties, robust mechanical strength, and favorable processability. Despite these outstanding attributes, conventional DGEBA-based epoxy resins are plagued by inherent brittleness—a critical limitation characterized by facile crack initiation and severe suppression of plastic void expansion during mechanical deformation. Among the various strategies proposed to mitigate this defect, the incorporation of nanofillers to construct a discrete secondary microphase within the epoxy matrix has emerged as a particularly promising approach [38, 39]. Epoxy–nanofiller composite systems have thus garnered considerable research and industrial attention, as they enable the simultaneous enhancement of mechanical strength and toughness while preserving the inherent merits of thermosetting polymers. These advantages include high structural stiffness, superior specific strength, and minimal volumetric shrinkage during the curing process. Furthermore, by precisely tailoring the species, morphology, and loading content of nanofillers, the performance of epoxy composites can be finely tuned to meet the stringent requirements of diverse targeted applications.

Numerous investigations have confirmed that the incorporation of nanofillers—including graphene oxide (GO) [40, 41], nanoclays [42, 43], graphene nanoplatelets [41, 44], carbon nanotubes (CNTs) [45, 46], MoS_2 [47], SiO_2 [48], and TiO_2 [49, 50],—into epoxy matrices can significantly enhance the toughness of epoxy resins. However, nanoparticles (NPs) strongly tend to agglomerate owing to their lack of reactive surface functional groups; this issue leads to inadequate dispersion and weakens interfacial adhesion between the fillers and the epoxy matrix, thereby impeding efficient stress transfer within the composite system. Furthermore, excessive nanofiller content dramatically elevates the system viscosity, which compromises the processability of epoxy resins and induces structural defects within the matrix.

Although crack pinning and crack deflection were once regarded as the predominant toughening mechanisms in epoxy/alumina and epoxy/titanium oxide composites, distinct core mechanisms govern the toughening behavior of other epoxy nanocomposite systems. For epoxy matrices reinforced with conventional spherical NPs, the critical toughening pathways involve localized shear band-induced plastic deformation, as well as NP debonding coupled with subsequent void nucleation and growth. Meanwhile, investigations into epoxy composites modified with nanofibrous reinforcements have revealed that the fiber bridging mechanism effectively suppresses crack propagation and void expansion, thereby yielding a notable enhancement in material toughness. With respect to CNT-reinforced epoxies, numerous studies have documented that the pullout of nanotubes exerts a pivotal role in improving fracture toughness. In the case of epoxy systems strengthened by layered NMs (e.g., nanoclays), crack deflection and the formation of microvoids have been identified as the primary toughening mechanisms. Beyond these filler-specific mechanisms, the efficacy of a given nanoadditive in tailoring epoxy performance is governed by a multifaceted set of parameters. These include the nanoadditive's intrinsic properties (size, volume fraction, morphological features, aspect ratio, dispersion uniformity, and surface modification status), as well as the epoxy matrix's inherent characteristics (cross-link density and glass transition temperature, T_g). It has been consistently established that increasing the concentration of NPs generally translates to enhanced fracture toughness of the composite. In contrast, the existing literature presents contradictory findings regarding the influence of particle size on epoxy fracture toughness: some studies report that particle size exerts no statistically significant effect on this property, while others demonstrate that reducing particle size can induce either positive or negative impacts on fracture toughness. In general, disentangling the complex, individual contributions of all these intertwined parameters remains a formidable challenge in this research field. Notably, the surface modification of NPs can exert a profound influence on the curing kinetics of epoxy resins, thereby altering the matrix's cross-link density and the resulting composite's fracture toughness. However, a clear, systematic correlation between these interconnected parameters has not yet been elucidated in published studies. Additionally, while void nucleation and propagation have been widely recognized as one of the dominant toughening mechanisms, research has indicated that relatively weak interfacial adhesion between NPs and the epoxy matrix is conducive to improved toughness. Interestingly, this conclusion stands in contrast to findings for CNT-reinforced systems: it has been well-documented that stronger interfacial adhesion between CNTs and the epoxy matrix enhances fracture toughness, with amine-functionalized CNTs yielding significantly higher toughness values compared to their pristine and carboxylated counterparts.

To realize efficient nanoscale reinforcement in epoxy composites, two critical prerequisites must be satisfied: the homogeneous dispersion of nanofillers within the matrix and the formation of robust interfacial interactions between the nanofillers and the polymer matrix. Surface functionalization of NPs—encompassing both chemical and physical modification approaches—has emerged as one of the most viable strategies to address these challenges, as it simultaneously ameliorates nanofiller dispersion stability and strengthens filler–matrix interfacial adhesion [51, 52]. Among the diverse functionalization tactics reported to date, the synergistic integration of HBPs with nanofillers has been validated as a particularly potent and versatile method [53]. For example, epoxy composites modified with a dual system of carboxyl-terminated hyperbranched aromatic polyester (CHBPE) and silica (SiO_2) NPs exhibit remarkable enhancements in their overall comprehensive performance. The unique hyperbranched architecture of CHBPE facilitates strong electrostatic interactions with SiO_2 NPs, thereby effectively inhibiting NP agglomeration. Furthermore, the abundant carboxyl (—COOH) groups present on the CHBPE chains can undergo chemical reactions with the epoxy groups in the matrix resin, which, in turn, constructs strong covalent bonds at the filler–matrix interface [54].

A novel surface modification strategy was devised, wherein amino-terminated HBPs serve as molecular linkers to anchor nano-silica (SiO_2) onto GO nanosheets. The primary objective of this design is to simultaneously ameliorate the dispersion uniformity of GO within the epoxy matrix and fortify the interfacial interactions between the reinforcement phase and the resin. The multi-step modification process is executed as follows: first, the surface of SiO_2 NPs is functionalized with γ-aminopropyltriethoxysilane (KH-550) and amino-terminated HBPs via two sequential Michael addition reactions. Subsequently, the resulting SiO_2–HBP hybrid NPs are covalently grafted onto the GO surface to form a stable GO–SiO_2–HBP ternary nanostructure.

The HBPs immobilized on the SiO_2 surface play a dual pivotal role in the composite system: on the one hand, they facilitate robust chemical bonding with the functional groups on the GO surface; on the other hand, they actively participate in the curing reaction of the epoxy resin, thus effectively enhancing the interfacial compatibility and interaction between the GO–SiO_2 hybrid filler and the epoxy matrix. Furthermore, the SiO_2–HBP conjugates construct a rigid-flexible hierarchical structure on the GO surface, which enables efficient load transfer throughout the composite. Benefiting from these synergistic effects, the epoxy composite incorporating a mere 0.1 wt% of the GO–SiO_2–HBP hybrid filler exhibits remarkable enhancements in mechanical performance compared to the neat epoxy resin. Specifically, the storage modulus, tensile strength, flexural strength, and impact strength of the composite are improved by 46.6%, 37.81%, 42.09%, and 86.86%, respectively.

The exceptional mechanical reinforcement achieved at such a low filler loading can be attributed to four synergistic mechanisms: (i) the immobilization of SiO_2 NPs on the GO surface effectively inhibits the re-agglomeration of both nanofillers, thereby significantly improving their dispersibility in the epoxy matrix; (ii) the grafting of amino-terminated HBPs onto the SiO_2 surface promotes the formation of strong covalent bonds and a mechanical interlocking effect with the epoxy resin, which, in turn, enhances the interfacial adhesion between the GO–SiO_2 hybrid filler and the matrix; (iii) the unique three-dimensional point-plane nanostructure of the GO–SiO_2–HBP hybrid facilitates the formation of a gradient modulus interface with the epoxy matrix, enabling efficient and uniform load transfer under external stress; and (iv) the rigid-flexible structure of the SiO_2–HBP conjugates can effectively absorb additional fracture energy during material failure, while simultaneously increasing the complexity and tortuosity of the crack propagation path, which further dissipates fracture energy. In conclusion, the rational integration of HBPs, SiO_2 NPs, and GO nanosheets enables a pronounced synergistic reinforcement and toughening effect, which substantially elevates the overall mechanical properties of the epoxy composite [55].

Empirical findings have demonstrated that epoxy (EP)/carboxyl-terminated hyperbranched aromatic polyester-silica (CHBPE–SiO_2) composites exhibit pronounced improvements in both mechanical performance and thermal stability. Specifically, upon the incorporation of 16 wt% CHBPE–SiO_2 hybrid modifier into the epoxy matrix, the resultant composite achieved an impact strength of 46 kJ/m^2, a flexural strength of 165 MPa, a tensile strength of 71 MPa, and a glass transition temperature (T_g) of 126 °C. All these performance metrics are substantially superior to those of the pristine epoxy resin, which validates the robust synergistic effect of HBPs and nanofillers in simultaneously enhancing the composite's toughness, mechanical strength, and thermal resistance.

To summarize, silica NPs act as potent toughening modifiers for epoxy resins, as they can induce plastic deformation in the matrix—an effect that has been validated in silica/epoxy nanocomposites. Crack propagation in these nanocomposites originates from the debonding of silica NPs; the primary energy dissipation mechanisms underlying toughening involve shear yielding of the epoxy matrix and void nucleation/growth promoted by this debonding behavior. As evidenced by micrographs and research findings from the aforementioned studies, the homogeneous incorporation of silica particles (either nanoscale or microscale) into pristine epoxy resins enhances fracture toughness (K_1C), normalized fracture energy (G_1C), and overall toughness. Notably, nanosized silica fillers outperform microscale counterparts as toughening agents across the entire filler loading range (up to 60 wt%). A highly promising strategy for epoxy resin modification involves the synergistic integration of GO with hyperbranched epoxy resins (HBPEs), leveraging the complementary properties of the two components. On the one hand, hyperbranched epoxy resins can form covalent bonds with GO nanosheets, which effectively inhibits GO agglomeration and enhances filler–matrix interfacial compatibility. On the other hand, the inherently high modulus of GO compensates for the modulus reduction and viscosity increase that are often associated with hyperbranched epoxy resin-based modifications [56, 57].

For instance, GO covalently modified with a hyperbranched epoxy resin (HBPEE-epoxy) affords a hybrid NM denoted as HPE–GO, which achieves homogeneous dispersion within the epoxy matrix. In comparison with pristine GO, this HPE–GO hybrid exhibits markedly improved dispersion uniformity and enhanced filler–matrix interfacial bonding, thus conferring dramatic enhancements in the composite's key performance metrics: tensile strength is elevated by 65.0%, elastic modulus by 36.2%, elongation at break (EBA) by an impressive 259.1%, and thermal conductivity by 178.9%. In a similar vein, benzoxazine (BOZ) composites reinforced with GO-hyperbranched epoxy resin (GO–HE) hybrids also demonstrate significant mechanical and thermal reinforcement effects. Specifically, the incorporation of a mere 0.05 wt% GO–HE into the BOZ matrix increases the material's flexural strength and modulus by 14.4% and 10.0%, respectively, relative to neat BOZ resin. Furthermore, both the glass transition temperature (T_g) and overall thermal stability of the composite are remarkably enhanced. These improvements are attributed to the homogeneous dispersion of GO–HE hybrids and the robust interfacial interactions established between the hybrid fillers and the BOZ matrix [58].

The enhanced toughness of such modified composites can be ascribed to the composite system's increased capacity for energy absorption and expanded fracture surface area. Nevertheless, microstructural characterizations reveal certain limitations in some formulations: discrete aggregates were observed within the fracture surface dimples, indicating the occurrence of uneven GO dispersion in the matrix. More notably, high-magnification scanning electron microscopy (SEM) images further unveiled distinct gaps at the interface between GO nanosheets and the polymer matrix, which signifies relatively weak interfacial interactions between the filler and the matrix in these unoptimized systems.

By sharp contrast, the fracture surface of HPE–GO/epoxy (EP) composites exhibits a significantly rougher morphology compared to that of pristine GO/EP composites with the same filler loading. Additionally, no obvious GO agglomerates were detected in the HPE–GO/EP system, a morphological feature that implies greater energy dissipation during the fracture process. Moreover, detailed SEM observations confirm that the edges of HPE–GO nanosheets are firmly embedded in the epoxy matrix, with the individual sheets achieving excellent dispersion throughout the system. Collectively, these results indicate that surface modification of GO with hyperbranched epoxy resins (HERs) effectively ameliorates GO dispersion and establishes covalent linkages between GO nanosheets and the epoxy matrix.

In this synergistic structure, HER moieties act as flexible domains to enhance energy dissipation during deformation, while the GO nanosheets serve as rigid domains to facilitate efficient load transfer within the composite.

A novel modification strategy was proposed, wherein hyperbranched polyamide-functionalized graphene oxide (HPA-GO) is incorporated into epoxy matrices to simultaneously optimize interfacial compatibility and mechanical performance. Owing to the enhanced dispersion uniformity of the filler and reinforced interfacial interactions, HPA-GO/epoxy composites exhibit remarkable improvements in both mechanical and thermomechanical properties even at ultra-low GO loading levels. Systematic investigations into the HPA-GO/epoxy interface were subsequently conducted, which confirmed the formation of robust covalent chemical bonds between the functionalized filler and the resin matrix. Such strong interfacial bonding facilitates efficient stress transfer and uniform stress distribution across the HPA-GO/epoxy interface. Consequently, the overall mechanical strength of the epoxy composites is effectively enhanced, which can be attributed to two key factors: the homogeneous dispersion of HPA-GO within the matrix and the specific chemical interactions established at the HPA-GO/epoxy interface. In quantitative terms, the incorporation of a mere 0.10 wt% HPA-GO into the epoxy matrix results in impressive increases of 310.5% and 37.2% in impact strength and tensile strength, respectively, when compared with the pristine epoxy resin [59].

Following functionalization with HPA, the water contact angle of HPA-GO increased markedly from 51.5° to 76.2°, while its glycerol contact angle rose from 48.8° to 78.6°—both trends are indicative of a significant enhancement in hydrophobic characteristics. In contrast, the contact angle of HPA-GO with DGEBA decreased notably from 83.5° to 51.2° after HPA grafting. The improved wetting behavior between HPA-GO and epoxy resin can be attributed to the introduction of HPA moieties into the interfacial region. This enhancement in wettability effectively promotes the formation of both covalent linkages and mechanical interlock effects at the HPA-GO/epoxy interface.

The surface free energy values of pristine GO and HPA-GO were determined to be 51.28 and 30.10 mJ/m^2, respectively. The reduction in both the dispersion component (γ^d) and polar component (γ^p) of surface free energy stems from the high density of amine groups introduced by the grafted HPA. Consequently, HPA functionalization fosters more pronounced hydrophobicity and diminishes the agglomeration tendency of GO nanosheets by lowering their surface free energy. Concomitantly, the adhesion work between the filler and epoxy matrix increased from 52.32 to 76.45 mJ/m^2 after functionalization. This elevated adhesion work substantially facilitates the subsequent formation of covalent bonds and physical interactions at the interface, thereby significantly improving the interfacial strength of the resulting composites.

Notably, the adhesion work between HPA-GO and epoxy resin is considerably higher than that between pristine GO and epoxy. This finding demonstrates that the HPA-GO/epoxy interface enables more stable interfacial adhesion and a more efficient stress transfer mechanism from the filler to the matrix. The enhanced interfacial strength of the composites arises from three synergistic factors: (i) robust π–π stacking interactions between HPA-GO and epoxy resin, which are promoted by planetary ball milling treatment; (ii) improved interfacial compatibility induced by HPA functionalization; and (iii) sufficient covalent reactions at the interface. During the curing process, it is postulated that a chemical reaction occurs between HPA-GO and the epoxy polymer chains, forming strong interfacial bonds that facilitate the effective separation and exfoliation of HPA-GO nanosheets.

Pristine GO exhibits poorer wetting by epoxy resin compared to HPA-GO, rendering it more prone to agglomeration in the matrix. In contrast, HPA-GO can be dispersed more uniformly in the epoxy matrix due to thermodynamic favorability, as evidenced by a highly negative Gibbs free energy change (ΔG_i) for dispersion. This favorable interfacial performance originates from specific interactions between the amino groups on HPA-GO and the epoxy groups in the matrix resin. In summary, HPA grafting onto GO nanosheets enables their homogeneous dispersion and strong interfacial adhesion within the composite, which are critical factors governing the reinforcement efficiency of the filler.

A schematic illustration depicting the fabrication of HPA-GO/epoxy composites is presented in Figure 5.1. The amine-functionalized HPA-GO exhibits excellent compatibility with the epoxy matrix; specifically, the amine groups on the HPA-GO surface participate in the curing reaction to form a cross-linked network at the filler–matrix interface. This unique interphase structure enhances the mobility of HPA-GO within the epoxy matrix and facilitates efficient load transfer and fracture energy dissipation during mechanical deformation (Figure 5.1).

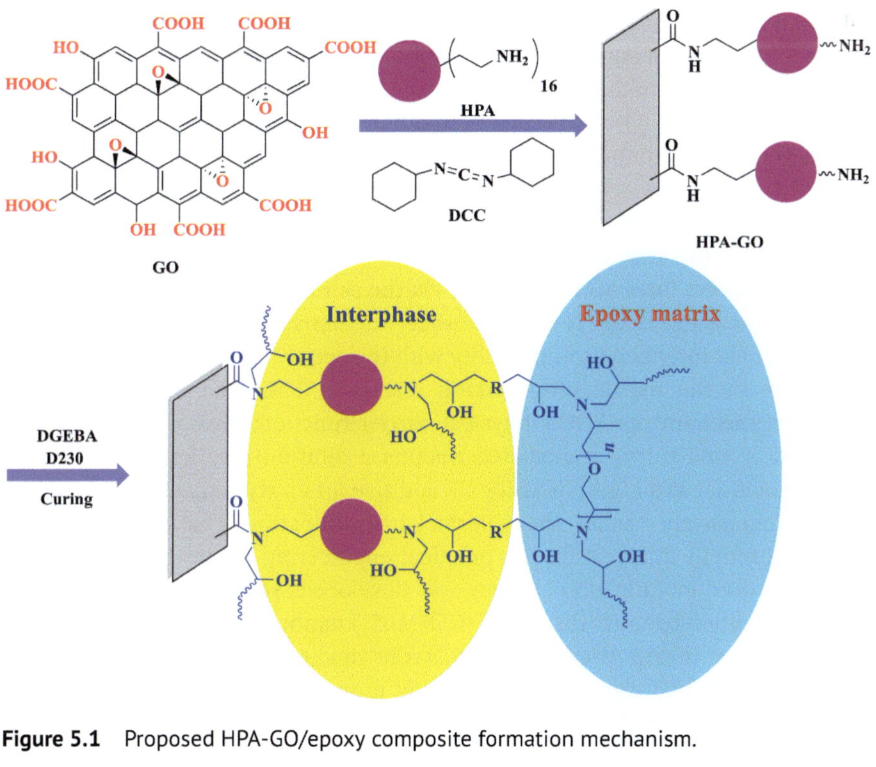

Figure 5.1 Proposed HPA-GO/epoxy composite formation mechanism.

Hyperbranched epoxy resin (HER)-modified GO has also been investigated as a high-efficiency compatibilizer for asphalt–epoxy composite systems. Specifically, EHBPs were covalently grafted onto GO nanosheets to fabricate a hybrid functional filler (denoted as GO–EHBP), which was subsequently incorporated into cold-mixed epoxy asphalt (CEA) formulations. This dual-modification strategy yielded remarkable enhancements in the composite's mechanical and thermomechanical performance: EBA was elevated by 74.5%, and the glass transition temperature (T_g) was increased to 62.6 °C. This performance improvement can be attributed to the following mechanism: a tailored dosage of EHBPs introduces an optimal concentration of epoxy functional groups into the CEA system, which actively participate in the cross-linking network formation during the curing process, thereby effectively improving the toughness of the CEA composite. Furthermore, at elevated service temperatures (≥40 °C), the storage modulus (E') of GO–EHBP/CEA composites surpassed that of unmodified CEA. This result confirms that the incorporation of GO–EHBP significantly enhances the material's resistance to high-temperature deformation. Collectively, these findings underscore the dual functionality of GO–EHBP: it not only improves the interfacial compatibility between the asphalt and epoxy phases but also simultaneously imparts toughening effects to the CEA composite [60].

Commercially sourced GO, leveraging its intrinsic two-dimensional layered architecture, is regarded as a promising candidate material for large-scale flame-retardant (FR) applications. However, pristine GO exhibits suboptimal FR efficiency in epoxy matrices due to strong π–π stacking interactions and van der Waals forces, which impede homogeneous dispersion. Furthermore, GO inherently lacks FR elements, rendering it incapable of suppressing heat transfer or effectively facilitating matrix carbonization during combustion. To mitigate these limitations, hyperbranched polyamide–amine (HPAA)—a functional modifier rich in amide groups—has been employed. During the combustion process, HPAA can decompose to generate nitrogen and other inert gases, which dilute the concentration of oxygen and flammable volatiles in the combustion zone, thereby endowing it with potential as an efficient FR. Inspired by this mechanism, phosphorus–nitrogen co-functionalized GO (PN-GO) nanosheets have been fabricated via the self-assembly of reactive HPAA onto the GO surface. These PN-GO nanosheets demonstrate excellent dispersion uniformity, robust interfacial compatibility with the epoxy matrix, and the capacity to participate in the matrix cross-linking reactions. The synergistic effects of three distinct mechanisms—GO's physical barrier function, phosphorus-induced char formation, and nitrogen-mediated gas-phase dilution—confer PN-GO with significant potential as a high-efficiency FR additive for epoxy resin systems [61].

To comprehensively tackle the interfacial incompatibility and insufficient mechanical strength that restrict the practical engineering applications of epoxy composites, a dual-modification strategy was developed. Specifically, the reduced graphene oxide/molybdenum disulfide (rGO/MoS_2) nanocomposite was fabricated via a solvothermal in-situ growth method, while vinyl-terminated hyperbranched polysiloxane (HBPSi-V) was synthesized using a one-pot synthetic strategy [47]. Notably, both materials exhibit excellent compatibility with the epoxy resin matrix and can directly optimize interfacial properties without the need for additional surface modification. These two components were then incorporated into the epoxy resin in both synergistic (joint) and individual (separate) manners to prepare epoxy (EP) composites. A systematic investigation was subsequently conducted to elucidate their distinct individual contributions to the composite's mechanical strength, thermal properties, and friction-reducing/anti-wear performance. The tribological mechanism underlying the friction-reducing and anti-wear properties of this soft-hard NM hybrid system can be divided into two key aspects. First, with regard to anti-wear performance, the "soft" HBPSi-V—characterized by its flexible Si—O—C molecular backbones and highly branched molecular structure—not only significantly toughens and strengthens the epoxy matrix but also modulates the dispersion of the "hard" rGO/MoS_2 nanocomposite. This modulation effect stems from HBPSi-V's extremely low surface energy and functional end groups, which facilitate the formation of an excellent interface between the epoxy matrix and NPs through strong coupling effects. During the friction process, the toughened epoxy matrix enables the generation of a robust tribofilm and transfer film. Meanwhile, the "hard" rGO/MoS_2 nanocomposite, which forms a strong interfacial bond with the epoxy matrix, remarkably enhances the composite's load-bearing capacity. Collectively, the improved toughness, enhanced load tolerance, and excellent interfacial strength of the composite synergistically result in a significantly lower volumetric wear rate. Second, in terms of lubrication capability, the rGO/MoS_2 composite NM exhibits superior lubricating performance compared to individual rGO or MoS_2. In both abrasive and adhesive wear regimes, the transfer films formed by the nanocomposite and HBPSi-V-reinforced epoxy resin are far more durable and self-lubricating than those of neat epoxy resin. Within these films, the hard rGO/MoS_2 nanocomposite primarily plays a role in lubrication and polishing action. Furthermore, silicon-containing polymers—particularly Si—O-based polymers—are recognized as inherently smooth and soft materials. This is because the Si—O bond exhibits a larger bond angle and longer bond length than the C—O bond. During the friction process, such soft materials tend to slide easily due to their extremely low surface energy, a phenomenon widely utilized in silicone-based lubricants such as shampoos and greases. Additionally, the flexibility of the Si—O molecular chains can mitigate frictional vibration, stabilize the friction coefficient, and transfer frictional forces via its dendritic molecular structure. In light of this synergistic mechanism, both the soft and hard NMs contribute to enhancing the tribological performance of the epoxy resin, yielding superior overall performance compared to the incorporation of either component individually.

To attain homogeneous dispersion of aluminum oxide (Al_2O_3) NPs within the epoxy matrix and optimize the interfacial interactions between Al_2O_3 fillers and the epoxy resin—thereby enhancing the thermal performance of the resultant nanocomposites—a two-step surface modification strategy was implemented for the Al_2O_3 NPs. Initially, the NPs were functionalized with a silane-coupling agent to graft amine groups onto their surfaces [62]. Subsequently, a hyperbranched aromatic polyamide was covalently immobilized onto the premodified Al_2O_3 surface. This HBP modifier self-assembled into an ultrathin polymeric coating on the Al_2O_3 NPs, which not only facilitated the uniform dispersion of the fillers in the epoxy matrix but also significantly strengthened the interfacial bonding between the Al_2O_3 NPs and the epoxy resin.

CNTs rank among the most promising multifunctional NMs, boasting excellent electrical conductivity, high mechanical strength, and other distinctive properties. However, fabricating homogeneously dispersed CNT/epoxy nanocomposites that simultaneously exhibit high strength and elongation remains a significant challenge. To address this issue, thiol-terminated hyperbranched polymers (THBPs) were employed to functionalize CNTs, yielding thiol-functionalized CNTs with superior dispersibility and good compatibility with the epoxy matrix [63]. The grafted THBPs act as molecular bridging agents between the epoxy resin and CNTs. Consequently, the thiol-functionalized CNT/epoxy nanocomposites exhibit substantial enhancements in key properties: thermal conductivity increased by 23.33%, elongation by 37.46%, tensile strength by 60.74%, flexural strength by 26.72%, and impact strength by 86.71%.

Mechanistically, the flexible, ellipsoidal HBP chains bonded to the CNT surface can alter their conformation under external loading, enabling efficient stress transfer and thereby synergistically enhancing the mechanical strength and elongation of the composites. The reinforcement of mechanical properties in the cured CNT/DGEBA nanocomposites is attributable to two core factors: the excellent dispersion of CNTs within the DGEBA matrix and the robust interfacial interactions established between them. The curing process of the nanocomposites, involving DETA-AN (diethylenetriamine–acetonitrile adduct) and DGEBA, is illustrated in Figure 5.3.

Notably, CNTs play a pivotal role in improving the composite's mechanical performance owing to their intrinsic high mechanical robustness—they effectively share and transfer applied loads/stresses, while inhibiting crack propagation to enhance the material's strength. However, with increasing CNT content, excessive CNTs fail to disperse completely and tend to agglomerate, which, in turn, leads to a reduction in mechanical strength. Furthermore, the thiol groups grafted onto CNTs can undergo a thiol-epoxy click reaction with the epoxy groups of DGEBA, resulting in increased cross-linking density and tensile strength of the cured nanocomposites—supplementing the primary cross-linking reaction between DETA-AN and DGEBA. Additionally, the π–π stacking interactions between the THBP chains and CNTs suppress CNT agglomeration and improve their dispersibility, which further enhances the interfacial interactions between CNTs and the DGEBA matrix, thereby contributing to the improved mechanical properties of the nanocomposites. Moreover, the flexible, ellipsoidal HBP chains on the CNT surface can deform under stress and facilitate load transfer, ultimately boosting the impact strength of the nanocomposites (Figure 5.2).

HBPs have also been utilized in the design of epoxy composites with integrated FR and mechanically robust properties. To this end, hyperbranched poly(phosphorus–silicon)-modified halloysite nanotubes (HBPPSi–HNTs) were incorporated into epoxy resins [64]. At a filler loading of 5.0 phr, the tensile strength and impact strength of the resultant composite were enhanced by 23.7% and 48.8%, respectively, compared to the pristine epoxy resin. The significant improvement

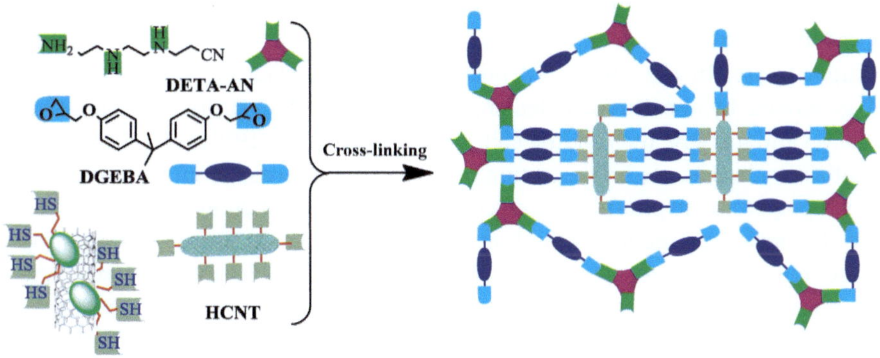

Figure 5.2 Curing process of CNTs, DETA-AN, and DGEBA.

in the composite's mechanical properties is attributable to two key factors: first, the homogeneous dispersion of HNTs within the epoxy matrix and second, the highly branched architecture of HBPPSi, which possesses abundant intramolecular cavities. These cavities increase the fractional free volume of the composite, thereby further contributing to the effective enhancement of mechanical performance.

Additionally, the HBPPSi–HNT/epoxy composites exhibited excellent FR properties, benefiting from the synergistic FR effects of HBPPSi–HNTs in both the gas and condensed phases. When the filler loading was increased to 10.0 phr, the composite achieved a UL-94 V-0 classification and a limiting oxygen index (LOI) of 42.6%. Relative to the pristine epoxy resin, the peak heat release rate (PHRR), total heat release (THR), total smoke production (TSP), and carbon monoxide yield (COY) of the composite were reduced by 50.1%, 32.6%, 30.1%, and 39.5%, respectively (Figure 5.3).

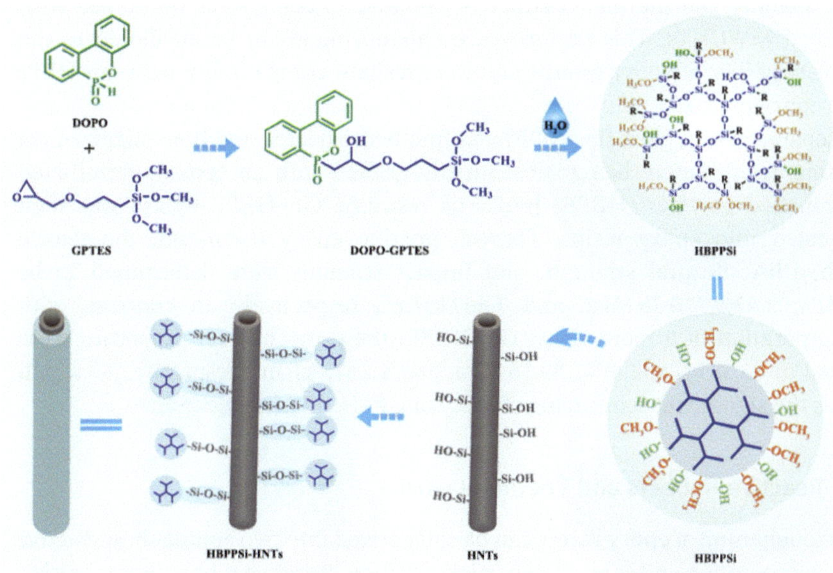

Figure 5.3 Synthesis route of HBPPSi–HNTs.

During combustion, HBPPSi–HNTs sustainably release PO·/HPO· radicals, which suppress combustion via quenching highly reactive free radicals (e.g., ·H, ·OH) in the vapor phase. In the condensed phase, phosphoric acid and pyrophosphoric acid generated by the thermal decomposition of HBPPSi–HNTs facilitate char formation and increase the residual char yield. Meanwhile, HBPPSi–HNTs activate aromatization, accelerate isomerization reactions, and thereby promote the formation of a polyaromatic phosphorus/silicon/carbon (P/Si/C) hybrid protective layer. These char residues function as binding agents to connect adjacent HNTs, forming a dense and robust protective barrier that effectively inhibits the erosion of the epoxy matrix by heat and oxygen.

Additionally, the release of crystal water from HNTs dilutes combustible volatiles and absorbs substantial amounts of heat. Furthermore, the tubular structure of HNTs aids in trapping gases emitted during combustion, reducing smoke release. In summary, the efficient barrier effect of the dense char residue, the dilution effect on combustible gases, and the quenching effect of phosphorus-containing radicals synergistically enhance the FR of epoxy resins.

Xiong et al. developed an efficient and rapid strategy for the covalent conjugation of multiwalled carbon nanotubes (MWCNTs) with hyperbranched poly(thiol ether-ester) (HPTEE) via thiol–yne click chemistry [65]. Their findings confirmed that HBPs were successfully covalently grafted onto the MWCNT surface, which significantly enhanced both the dispersion uniformity of MWCNTs in the epoxy matrix and the interfacial interactions between MWCNTs and the matrix. At a filler loading of 0.6 wt% MWCNTs–HPTEE, the composite achieved optimal mechanical performance: the impact strength and flexural strength were increased by 180% and 72.3%, respectively, compared to the pristine epoxy resin. Notably, at the same filler content, the MWCNTs–HPTEE/epoxy composites exhibited superior thermal stability and thermal conductivity relative to composites reinforced with unmodified MWCNTs. This performance enhancement can be attributed to the high polarizability of epoxy groups and the excellent compatibility between HBPs and the epoxy matrix.

In a separate study, glass fiber (GF) was first hydroxylated and then silanized via a coupling agent, before being subsequently grafted with an epoxide-terminated hyperbranched polymer (HBPE) [66]. The resulting GF–HBPE hybrid was then incorporated into epoxy resins. For the pristine epoxy thermoset, the tensile strength, EBA, flexural strength, and impact strength were determined to be 37.69 MPa, 2.43%, 70.78 MPa, and 7.667 kJ/m^2, respectively. In contrast, with the incorporation of a mere 1 wt% GF–HBPE, these mechanical properties were enhanced to 46.58 MPa, 5.47%, 89.16 MPa, and 13.375 kJ/m^2, respectively—which confirms the significant reinforcing efficacy of GF–HBPE modification.

5.2.2 Reactive Rubbers and Thermoplastics

Rubber toughening of epoxy resins can be categorized into two approaches: reactive rubber incorporation and core–shell particle addition. Reactive rubber incorporation represents the most established strategy for epoxy toughening. In this method, functionalized acrylonitrile-butadiene rubbers—such as carboxyl-terminated butadiene-acrylonitrile (CTBN), amino-terminated butadiene-acrylonitrile (ATBN), epoxy-terminated butadiene-acrylonitrile (ETBN), and vinyl-terminated butadiene-acrylonitrile (VTBN)—are dissolved in epoxy resins. A key prerequisite for a rubber to qualify as an epoxy toughener is its miscibility with the epoxy matrix and subsequent phase separation during curing (i.e., reaction-induced phase separation). Additionally, the functional groups of the rubber must react with the epoxy groups of the matrix. Notably, CTBN is the most widely used RE in this category. However, these REs suffer from drawbacks, including cumbersome handling and challenges in controlling the size of dispersed particles.

Core–shell rubber particles are regarded as the next generation of reactive REs. These preformed rubber particles are encapsulated by a thin glassy shell. The incorporation of core–shell modifiers facilitates precise control over the size of the dispersed phase. The glassy shell prevents agglomeration of the rubber particles by mitigating their inherent stickiness, resulting in superior thermo-mechanical properties compared to reactive rubber toughening. Core–shell toughening agents are typically synthesized via emulsion polymerization, with particle size tunable based on polymerization conditions. Common shell materials include epoxy-compatible polymers, among which poly(methyl methacrylate) (PMMA) is the most prevalent. Core materials are generally composed of butadiene-based and acrylate-polyurethane rubbers.

While rubber modification typically enhances the fracture toughness of epoxy resins by several orders of magnitude, it concurrently diminishes the matrix's strength and stiffness. In fact, a linear correlation exists between the dosage of REs and the reduction in strength and stiffness. Incomplete phase separation of the rubber may lead to a decrease in glass transition temperature (T_g), an increase in coefficient of linear thermal expansion (CLTE), and compromised transparency—a critical property for coating applications.

The synergistic incorporation of rubbers and HBPs into epoxy resins has also garnered significant research interest. Experimental data demonstrate that this approach achieves a remarkable improvement in fracture toughness and ductility. Nevertheless, the stiffness of the cured polymer decreases, as the presence of low-modulus rubber particles typically impairs the stiffness and thermal stability of polymer-based materials.

Thermoplastic-modified epoxies were first proposed by Bucknall and Partridge in 1983 to address the limitations of rubber-toughened epoxies—specifically the reduction in strength and modulus, as well as the ineffectiveness of REs in highly cross-linked epoxy systems. Owing to their high glass transition temperature (T_g), high modulus, rigidity, and ductility, thermoplastics are regarded as promising candidates for epoxy toughening. Typically, the thermoplastic is dissolved in the epoxy resin (with a solvent added if necessary); similar to REs, phase separation occurs during the curing process.

A key criterion for selecting a thermoplastic for epoxy toughening is the similarity in chemical structure between the epoxy resin and the modifier. Common thermoplastic tougheners include polyethersulfone (PES), polyether ether ketone (PEEK), polyphenylene oxide (PPO), polystyrene (PS), PMMA, and poly(ether imide) (PEI). Phase separation of the thermoplastic phase is another critical prerequisite for effective epoxy toughening; thus, thermoplastics with high miscibility with epoxy resins are not ideal toughening modifiers. Generally, the presence of numerous reactive end groups—which enhance miscibility—can be considered a disadvantage for epoxy toughening applications.

Key factors influencing toughening efficiency include thermoplastic content, chemical structure (e.g., molecular weight and functional groups), ductility of the epoxy matrix, interfacial adhesion between the thermoplastic and epoxy matrix, as well as curing kinetics and system morphology. Notably, while rubber toughening is more effective for epoxy matrices with low cross-link density, thermoplastics exhibit superior performance in highly cross-linked epoxy resins.

Vinyl esters (VEs) are typically synthesized by reacting bisphenol-A epoxy resins with acrylic or methacrylic acid. For the purpose of enhancing the toughness of bisphenol-A-based vinyl ester–urethane hybrid (VEUH) resins, liquid nitrile rubbers, hyperbranched polyesters, and core–shell rubber particles bearing different functional groups (vinyl, carboxyl, and epoxy) were incorporated. Specifically, 10 wt% liquid nitrile rubbers with vinyl (VTBN), carboxyl (CTBN), and epoxy (ETBN) functionalities, as well as 10 wt% hyperbranched polyesters with vinyl (VHBP) and epoxy (EHBP) functionalities, were added to VEUH. The toughening mechanism of the CTBN/VHBP combination was attributed to cavitation and crack pinning [67]. Adding additional NPI to the VEUH containing 10 wt% CTBN results in a very coarse salami-type distribution of CTBN particles. The improvement in toughness is governed by the same mechanisms mentioned earlier (cavitation, matrix yielding, and crack pinning). Notably, the crack pinning process was accompanied by significant shear deformation of the matrix between particles. Instead of forming inclusions via phase separation, the rubber acts as a "binding agent" for the morphological entities of VE. It should be recalled that VE exhibits a two-phase morphology: a nodular (micrometer-sized) VE network dispersed in a polystyrene phase. However, additional NPI may restore the phase-separated structure. A similar phenomenon was observed for the CTBN + EHBP combinations. The dominant fracture mechanisms in these systems are likely crack tip blunting and crack branching/bifurcation—an interesting finding, as these crack bifurcation mechanisms are generally considered markedly less efficient than cavitation/shear deformation-type mechanisms.

The toughening enhancement by the CTBN + VHBP modifier combination stems from cavitation and crack pinning. It is worth noting that although the fracture energy (G_c) values of the modifier combinations were comparable, the underlying fracture mechanisms—as discussed earlier—were significantly different. Consequently, deducing general rules for the toughening strategy of VEUH resins is not a straightforward task.

Based on this study, which aimed to improve the toughness of VEUH resins by incorporating functional polymers that either affect the heteropolymerization of VE with styrene (St, via vinyl functionality) or the polyurethane reaction (via epoxy and carboxyl functionalities), the following conclusions can be drawn: (i) The most effective toughness modifiers are combinations that participate in both reactions (i.e., free radical polymerization and polyaddition); (ii) the stoichiometric ratio, which is influenced by the functionality of the modifier (in this case, the carboxyl groups of CTBN), exerts a significant impact on fracture energy; and (iii) the phase structure governs the fracture mechanisms, although different fracture mechanisms may yield identical fracture mechanical responses.

To explore high-value utilization approaches for waste tire rubber (WTR), hyperbranched poly(amidoamine) (PAMAM) was grafted onto the surface of WTR powders to endow them with chemical reactivity. Adopting a divergent synthesis strategy, numerous amine groups were anchored on the WTR surface, yielding PAMAM-functionalized WTR powders. The grafting of PAMAM onto the WTR surface and the subsequent chain propagation were accomplished via Tomalia's divergent growth method. The synthetic process consists of two alternating sequential chain-formation reactions: exhaustive amidation and exhaustive Michael addition, which are repeated cyclically. Specifically, the exhaustive amidation of ester-terminated PAMAM with a large excess of ethylenediamine (EDA) in methanol results in fully generation PAMAM with amine-terminal groups. Conversely, the Michael addition of methyl acrylate (MA) to amine-terminated PAMAM in methanol produces half-generation PAMAM with ester terminal groups. By repeating these Michael addition and amidation reactions, PAMAM of higher generations can be obtained. These modified powders integrate the elastomeric properties of WTR with abundant surface amine groups, enabling them to serve as curing agents and potential toughening modifiers for epoxy resins [68].

A new reactive core–shell structure (DAS@HBPA/EP) has been successfully constructed, with dialdehyde starch (DAS) serving as the core and hyperbranched epoxy (HBPA/EP) as the shell [69]. This core–shell system was integrated into soy protein-based adhesives (SM) to enhance their water resistance, bonding strength, and toughness. The reactive HBPA/EP shell is capable of forming covalent bonds with protein molecules in SM, which facilitates the formation of a highly interconnected network and thereby improves adhesion performance and water resistance. Owing to the synergistic effect of the DAS core and the flexible hyperbranched shell, the SM adhesives modified with DAS@HBPA/EP exhibited enhanced toughness through microphase separation. Experimental results indicated that plywood bonded with the SM/DAS@HBPA/EP adhesive achieved an outstanding dry shear strength of 2.14 MPa, representing a 589% increase compared to the pure SM adhesive. Its wet shear strength reached 1.24 MPa, which is comparable to that of commercial MUF resins (E0 grade). Additionally, the SM/DAS@HBPA/EP adhesive showed a high residual rate (82.3%), extremely low water absorption (0.94%), and a homogeneous and tightly interwoven microstructure. The excellent water-resistant bonding strength of the SM/DAS@HBPA/EP adhesive can be ascribed to the formation of a hyperbranched cross-linked network and the robust core–shell architecture. On the one hand, the HBPA/EP shell interacts with protein molecules in SM, leading to the formation of a hyperbranched cross-linked structure. This interaction increases the cross-linking density, thus simultaneously enhancing the bonding strength and water resistance of the modified adhesive. On the other hand, the core–shell structure of DAS@HBPA/EP promotes microphase separation within the SM adhesive matrix, which effectively suppresses crack propagation under external stress. Furthermore, the abundant cavities in the hyperbranched shell of DAS@HBPA/EP can easily deform to absorb impact energy, contributing to the improved toughness of the SM/DAS@HBPA/EP adhesive through a synergistic effect. The enhanced toughness of the SM adhesive plays a crucial role in boosting its bonding strength. In summary, the SM/DAS@HBPA/EP adhesive demonstrates significant commercial potential and promising prospects for practical applications.

Additionally, dumbbell-shaped SEBS-g-PEG copolymers, which represent a transitional molecular structure between block and hyperbranched copolymers, were synthesized by grafting methoxy polyethylene glycol (PEG) onto the end blocks of polystyrene-*b*-poly(ethylene-*ran*-butylene)-*b*-polystyrene (SEBS) [70]. These copolymers exhibited good compatibility with epoxy thermosets without inducing plasticization, while concurrently forming microphase-separated morphologies that effectively toughened the epoxy matrices. At higher branching densities, the incorporation of 10 wt% SEBS-g-PEG enhanced the fracture toughness (K) and impact strength by approximately 230%, without compromising thermal stability, glass transition temperature (T), flexural strength, or modulus. Morphological analysis indicated that increased branching density enlarged the soft nanostructural domains and blurred the phase interfaces. Thus, the toughening effect is attributed to stronger interfacial bonding coupled with larger, softer interphase regions. The structures and properties of the toughened epoxy thermosets undergo the following changes: (i) increased size of the soft secondary phases; (ii) monotonic enhancement of toughness without sacrificing other mechanical and thermal properties; and (iii) significant inhibition of the plasticizing effect in the toughened epoxy thermosets. Dumbbell-shaped SEBS-g-PEG with appropriate molecular structures is expected to be a desirable transitional molecular chain between linear BCPs and hyperbranched copolymers for enhancing the toughness of epoxy thermosets. This work paves the way for the design of highly efficient toughened epoxy materials using dumbbell-shaped copolymers.

5.2.3 Fiber Materials

Thermoset epoxy resin/carbon fiber (EP/CF) composites are extensively utilized in the automotive, aerospace, and wind energy sectors. Nevertheless, two key challenges hinder their wider application in advanced sustainable materials: the difficulty in interfacial enhancement and the high-value recycling of CFs. Owing to their smooth, highly graphitized surfaces, CFs exhibit chemical inertness that leads to weak interactions with epoxy resins, severely compromising the performance and practical applications of these composites. CF-reinforced epoxy resin composite laminates (CF/EP) are highly sought after in the field of high-strength, lightweight, and durable equipment components, attributed to their exceptional strength and modulus, low thermal expansion coefficient, and tunable properties. Despite these advantages, such components are prone to damage, degradation, and failure under various thermos-mechanical and impact loads (e.g., bird strikes and lightning strikes). Typically, damage initiates at the micron scale as microvoids, progresses to microcracking and interfacial fracture, and ultimately culminates in catastrophic component failure, thereby reducing their service life and economic feasibility. In CF/EP composites, significant defects are inherently present due to the high chemical inertness of the untreated CF surface. Microcracks at the composite interface cause stress concentration under loading, leading to long-term degradation and eventual failure. It is well-established that CF/EP composites develop microcracks during prolonged service; if left unaddressed, these microcracks can deteriorate material performance. Existing research confirms that the presence of matrix microcracks impairs the key properties of CF/EP composites, making healing essential to enhance their durability and reliability. Three main strategies are commonly employed to expand the application scope of CF/EP composites and extend component service life. First, CF surface modification can be performed to improve CF/EP composite properties. Currently, research on CF surface modification primarily focuses on two aspects: on the one hand, introducing grafts onto the CF surface increases the number of covalent bonds available, which enhances the interfacial properties of CF/EP composites through strong chemical bonding during resin curing.

The incorporation of natural fibers is also one of the most appealing approaches for epoxy resin modification. Over the past few years, interest in and applications of natural fibers as polymer reinforcements have increased significantly in both engineering and research fields. Driven by concerns regarding environmental sustainability, waste management, and the depletion of nonrenewable resources, researchers and scientists have also placed greater emphasis on the utilization of renewable materials. Numerous studies have focused on natural fiber reinforcement, including fibers derived from date palms, coir, oil palm empty fruit bunches (OPEFB), ramie, bamboo, hemp, flax, sisal, jute, and kenaf. These natural fibers hold potential as substitutes or alternatives to synthetic fibers (e.g., glass and CFs) in composite applications. In the composite industry, epoxy serves as a binder to secure reinforcing materials in place. Typically, natural fibers are employed in advanced polymer composites in three forms, selected based on their aspect ratio. First, natural fibers used as powder fillers can enhance the desired properties of composites compared to pristine resins. Second, short natural fibers are randomly distributed at a designed volume fraction. Third, continuous biofibers are incorporated into polymer composites. Studies have also been conducted to explore the feasibility of introducing new natural fibers as fillers in polymer matrices for developing cost-effective and lightweight structural materials. High-performance composites with enhanced properties based on thermosetting polymers can be achieved by simultaneously optimizing fiber properties, resin performance, and their interfacial adhesion. Epoxy composites reinforced with natural fibers offer a unique combination of versatility, high performance, and processing advantages at a more competitive cost compared to thermoplastic-based composites. Researchers have reported that the incorporation of natural fibers (e.g., jute, kenaf,

OPEFB) improves the mechanical and physical properties of polymer matrices in most cases. Additionally, studies have indicated that most epoxy-based thermoset composites are fabricated via manual lay-up and spray-up techniques [71].

Beyond introducing polar functional groups, the topological structure of grafted polymers also regulates interfacial interactions during CF surface modification. HBPs, characterized by flexible and deformable topological structures, exhibit distinct advantages over linear, cyclic, or trapezoidal polymers. They not only enhance the wettability of CFs with epoxy matrices but also facilitate interfacial stress transfer and load distribution, thereby improving the interfacial performance of composites. Furthermore, HBPs possess abundant functional groups (e.g., amine, epoxy, hydroxyl), which provide numerous active chemical sites for forming cross-linked networks or transition layers at the CF/epoxy (CF/EP) interface, enabling robust mechanical interlocking. Currently, chemical grafting of HBPs onto CF surfaces to strengthen interfacial interactions is a major research focus. Hyperbranched poly(thioether-yne) (HPTEY) has been developed as a rapid, efficient, and versatile bridging agent for grafting CNTs and CFs via thiol-alkyne click chemistry [72, 73]. The synergistic integration of CNTs and HPTEY significantly enhances CF surface wettability and surface energy. Microdroplet tests demonstrated that the interfacial shear strength (IFSS) of CF–HPTEY–CNT/epoxy composites increased by 106.4% compared to pristine CF/epoxy composites. Similarly, the superior interfacial properties of CF–HPTEY–GO/EP composites (corrected "HPTY" to "HPTEY" for consistency) are primarily attributed to the elevated surface energy of CF–HPTEY–GO, which enhances fiber–matrix interfacial adhesion. Additionally, polar functional groups in HPTEY and GO improve compatibility with the epoxy matrix and provide reactive sites for chemical bonding, further reinforcing interfacial adhesion. Grafting HPTEY and GO also imparts a rougher surface and highly branched structure to CFs, acting as "anchors" to enhance mechanical interlocking. GO sheets, tightly attached to CF surfaces via HPTEY, fill surface defects and exert a reinforcing effect, further boosting interfacial bonding strength. Overall, the significant improvement in interfacial properties stems from the synergistic effects of enhanced compatibility, mechanical interlocking, reinforcement, and defect filling of the GO–HPTEY graft layer. The bending strength of CF–HPTEY–CNT/epoxy composites was also improved by 40.4%, mainly due to more efficient interfacial stress transfer. Moreover, the monofilament tensile strength of CF–HPTEY–CNT increased by 3.3% relative to pristine CF, as grafting CNTs and HPTEY repaired surface defects. The enhanced tensile and bending strengths are attributed to four key factors: (i) HPTEY and GO introduction increases CF surface functional groups and wettability, promoting the formation of chemical and hydrogen bonds between fibers and the epoxy matrix—these bonds dissipate substantial energy when broken under external loads; (ii) the GO–HPTEY graft layer fills CF surface defects, and active groups react with defect sites, reducing interfacial stress concentration; (iii) GO grafting increases CF surface roughness, enhancing mechanical interlocking between resin and fibers and increasing energy consumption during composite failure; and (iv) GO sheets covering CF surfaces bridge surface microcracks, exerting a strengthening effect. Collectively, the synergistic effect of HPTEY and GO significantly enhances the mechanical properties of the composites.

Gu et al. grafted hyperbranched poly(2,2-bis(hydroxymethyl)propionic acid) (HBP-32-OH) onto CF surfaces to fabricate modified CF/epoxy composites [74]. The uniform distribution of HBP-32-OH on CF surfaces enhanced fiber–epoxy wettability, mechanical interlocking, and chemical interactions. Compared with unmodified CF composites, the IFSS, interlaminar shear strength (ILSS), and impact strength of the modified composites increased by 78.2%, 40.9%, and 39.2%, respectively, with concurrent improvements in thermal stability. Detailed fracture interface microstructure analysis validated the interfacial strengthening mechanisms. Similarly, grafting hyperbranched polyurethane (HBPU) with multiple terminal hydroxyl groups (HBPU–OH) onto CF surfaces reinforced the fiber–matrix interface, significantly improving ILSS via enhanced chemical bonding and mechanical interlocking. Additionally, a simple and eco-friendly one-step method was developed to graft hydroxyl-terminated hyperbranched polymers (HTHBPs) onto CF surfaces through the condensation of isophorone diisocyanate (IPDI) and tris(hydroxymethyl) aminomethane (TOAM) [75]. The covalently grafted HTHBPs increased CF surface polarity, roughness, and wettability, thereby enhancing interfacial properties. After modification, the ILSS and IFSS of CF/epoxy composites increased from 58.6 to 81.8 MPa and 48.8 to 87.8 MPa, respectively, while the impact strength rose from 55.7 to 76.9 kJ/m^2. Notably, the single-fiber tensile strength was retained.

In an alternative design, octa(γ-chloropropyl) polyhedral oligomeric silsesquioxane (POSS) was synthesized and further converted into amino-terminated hyperbranched structures (Gn-POSS–NH_2, where $n = 1, 2$, or 3), which were grafted onto CF surfaces via a sequential chemical approach [76]. By tuning the dendrimer generation (n), the polarity, roughness, and surface morphology of CFs could be precisely regulated. Increasing the generation significantly enhanced the IFSS, ILSS, and dynamic mechanical properties of the composites, attributed to strengthened interfacial interactions, improved mechanical interlocking, and better resin wetting. The incorporation of POSS cages and additional hexamethylenediamine expanded the interface area and restricted the mobility of the epoxy resin matrix. Furthermore, POSS–NH_2 mitigated stress concentration between CFs and the matrix resin, altered crack propagation paths, and prevented crack tips from directly contacting the fiber surface. For CF–G3–POSS–NH_2, the higher loading of POSS–NH_2 at the fiber interface and the formed hyperbranched structure induced more cracks and hindered stress propagation. In summary, hyperbranched POSS–NH_2 functionalization played a crucial role in significantly enhancing the interfacial microstructure and interactions between POSS–NH_2-modified CFs and the resin matrix.

To concurrently improve both aesthetic and mechanical properties, a novel approach was developed to graft GO onto CF surfaces using hyperbranched poly(thioether-yne) (HPTY) as an intermediary via thiol–yne click chemistry [73]. This modification tailored the fiber surface morphology while boosting wettability and surface energy, leading to a 94.4% increase in the IFSS of CF–HPTY–GO/epoxy composites. Moreover, compared with pristine CF/epoxy composites, the tensile and bending strengths of the modified composites improved by 38.2% and 49.1%, respectively, owing to more efficient interfacial stress transfer. This strategy exhibits substantial potential for fabricating advanced structural composites with superior interfacial and mechanical performances.

The development of high-performance epoxy composites based on carbon NMs has attracted considerable attention in recent years, though several challenges remain unresolved. Current research focuses on the mechanical, electrical, magnetic, thermal conductivity, and FR properties of carbon nanofiller/epoxy composites. However, key issues persist: the performance enhancement of carbon NM-reinforced epoxy composites primarily relies on three core factors: the intrinsic properties of nanofillers, their dispersion within the matrix, and interfacial interactions between fillers and epoxy. Nevertheless, further research is needed to improve the filler–epoxy interfacial interaction and control filler orientation in the epoxy matrix. Molding is the most widely used manufacturing technology for epoxy matrices, which requires composite materials with low viscosity to ensure product quality. While carbon NMs can enhance multiple epoxy properties, they significantly increase composite viscosity, potentially impairing processability. Other critical research focuses include the effects of carbon NMs on the glass transition temperature (T_g), curing reaction, and curing degree of epoxy composites. Investigations into flexible, semi-flexible, and rigid cross-linked epoxy resins are also meaningful, as is the influence of carbon NM dimensions (length, width, and thickness). Driven by growing environmental concerns, developing bio-based epoxy monomers or oligomers from renewable resources to replace traditional epoxy resins is a major development trend. Another eco-friendly direction is recyclable epoxy resins, whose polymer chains can be dissociated from the epoxy network under specific conditions, enabling thermoplastic-like recycling. Additionally, waterborne epoxy is an environmentally benign material, particularly in the coating field. Further studies are required to explore the properties of these eco-friendly epoxies integrated with carbon NMs. To advance the application of epoxy composites, especially in high-tech fields, the development of multifunctional epoxy composites modified with carbon NMs represents the next research focus. Such composites are in demand for aerospace and automotive applications, which require not only high mechanical properties and thermal stability but also excellent antidegradation, FR, tribological, and biodegradable performance. Recent emerging research areas for carbon nanofiller/epoxy composites include coatings, biocompatibility and biodegradability, gas barrier applications, and shape memory materials. Carbon NM-functionalized epoxy composites are particularly prominent in multifunctional coatings, such as impact- or blast-resistant coatings, tribological coatings, marine anticorrosive coatings, and microwave-absorbing coatings for aerospace, civil, and electrical industries.

5.3 Homogeneous Mechanism

Over the past few decades, hyperbranched topologies—characterized by branched backbones, peripheral functional groups, and inherently low viscosities—have been integrated into epoxides to prepare hyperbranched epoxy resins (HERs). HERs can act as hardeners for fabricating high-performance epoxy thermosets. Unlike conventional toughening strategies that utilize rubbers, liquid crystal polymers, or

NMs as second-phase modifiers, HER-based epoxy systems adopt a homogeneous toughening mechanism. In this mechanism, modifiers disperse uniformly within the epoxy matrix and form interpenetrating networks consisting of single- or multi-molecular aggregates. These aggregates connect into continuous networks without generating a distinct second phase or inducing micro- or nanoscale phase separation. Consequently, both the toughness and strength of the epoxy are significantly enhanced, while the matrix viscosity is reduced, resulting in low-viscosity epoxy systems. Owing to their spherical architecture and excellent solubility, HBPs exhibit no intermolecular entanglement and possess much lower viscosities than linear polymers. Their spherical geometry provides a small free volume and abundant functional groups, which enable energy absorption through deformation under impact. Additionally, interfacial interactions can be further strengthened via chemical reactions at the matrix interface. Based on this platform, we have designed structures capable of simultaneously improving both the strength and toughness of epoxy resins (EPs).

The single toughening mechanism of rubbers and low-reactivity HBPs mentioned earlier is primarily attributed to phase separation, which forms a sea-island or two-phase structure. When used to toughen DGEBA, low-reactivity HBPs (e.g., Boltorn E1) reduce mechanical strength—including tensile and flexural strength—to varying degrees. This is due to the low cross-linking density and weak interactions between Boltorn E1 (with a high epoxy equivalent weight [EEW] of 875 g/eq) and DGEBA. To enhance interfacial interactions between hyperbranched epoxy polymers (EHPs) and DGEBA as well as improve the mechanical strength of DGEBA, we synthesized three types of aromatic EHPs (HTDE, HTME, and HTBE) [77]. These EHPs [78–80] not only significantly enhanced DGEBA's toughness but also remarkably improved its tensile and flexural strengths (Figure 5.4).

Notably, a large number of protonema were detected on the fractured surfaces of cured HTDE-2/DGEBA composites, with the protonema density rising as the HTDE-2 content increased. Intriguingly, dynamic mechanical analysis (DMA) curves of the cured HTDE-2/DGEBA blends exhibited only a single glass transition temperature, suggesting a homogeneous structure free of phase separation between HTDE-2 and the DGEBA matrix. On this basis, a putative in-situ homogeneous reinforcement and toughening mechanism is proposed herein. Owing to its small molecular size (several nanometers), HTDE-2 was able to disperse uniformly within the DGEBA matrix. The epoxy groups of both HTDE-2 and DGEBA underwent a reaction with the amine groups of the curing agent. Post-curing, the external cross-linked structure formed by HTDE-2 restricted the mobility of its end groups. During impact loading, numerous non-cross-linked intramolecular cavities within the system could absorb energy, undergo deformation, and generate protonema—an observation indicative of high toughness. Meanwhile, benefiting from its small hydrodynamic volume and ellipsoidal structure, HTDE-2 was capable of dispersing and permeating into the molecular chains of the DGEBA matrix. This enhanced the intermolecular interactions during the curing process, ultimately improving the tensile and flexural strength of the composites. Consequently, HTDE-2 demonstrated a simultaneous reinforcement and toughening effect in the modification of DGEBA.

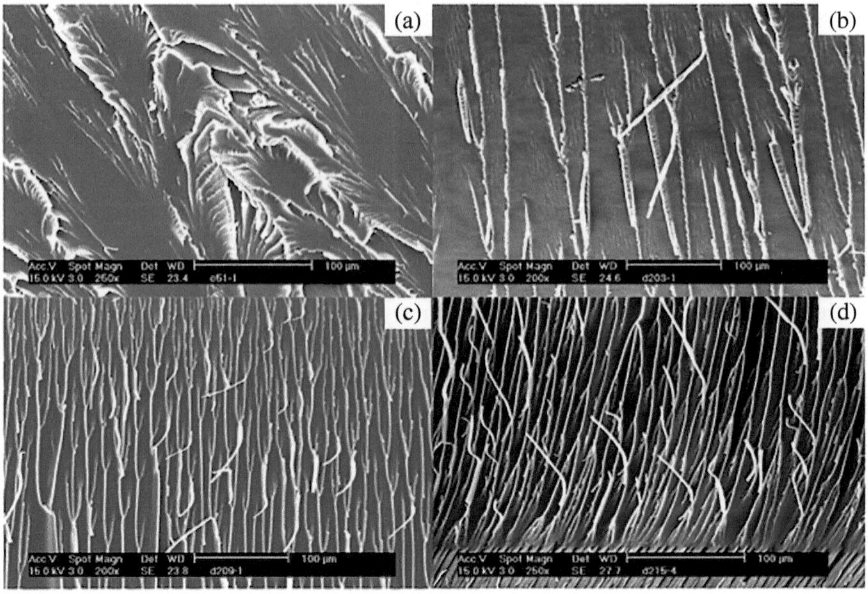

Figure 5.4 SEM micrographs of cured HTDE-2/DGEBA with various HTDE-2 contents: (a) 0 wt%, (b) 3 wt%, (c) 9 wt%, and (d) 15 wt%. *Source:* [77] / with permission of John Wiley & Sons.

In contrast to the chemical structure of HTDE, ethylene glycol and butanediol were employed as substitutes for diethylene glycol to synthesize two other epoxy hardeners (EHPs), namely, HTME [81] and HTBE [82], respectively. Both HTBE [80] and HTME exhibited remarkable reinforcement and toughening effects on DGEBA. Additionally, a substantial number of protonema were observed in the SEM micrographs of the fractured surfaces of cured HTME/DGEBA and HTBE/DGEBA composites. This phenomenon further validated the in-situ reinforcement and toughening mechanism proposed previously (Figure 5.5).

By modifying the dihydric alcohol structure to 1,4-cyclohexanedimethanol, which contains a rigid cyclohexane ring, epoxy hardener (EHP) [83] (designated as HEHP-*n*) could also be successfully prepared. With the increase in molecular weight and content of HEHP-*n*, the mechanical strength of HEHP-*n*/DGEBA composites initially increased and subsequently decreased. Initially, the reinforcing effect of EHPs (including HTDE, HTME, and HTBE) on DGEBA was attributed to the positive contribution of their aromatic structural rigidity, in contrast to the aliphatic structural flexibility of Boltorn E1—which lacks reinforcing capability. To verify this initial hypothesis, we synthesized low-viscosity aliphatic EHPs (HTPE) [84]. HTPE shared a similar chemical structure with Boltorn E1, as the pentaerythritol pentaethoxylate used in Boltorn E1 synthesis was substituted with tris(2-hydroxyethyl) isocyanurate (THEIC) in HTPE preparation (Figure 5.6).

As presented in Figure 5.7, numerous protonema are visible in the SEM micrograph of the impact-fractured surface of cured HTPE-2/DGEBA. This phenomenon is also observed in other cured systems, including unsaturated

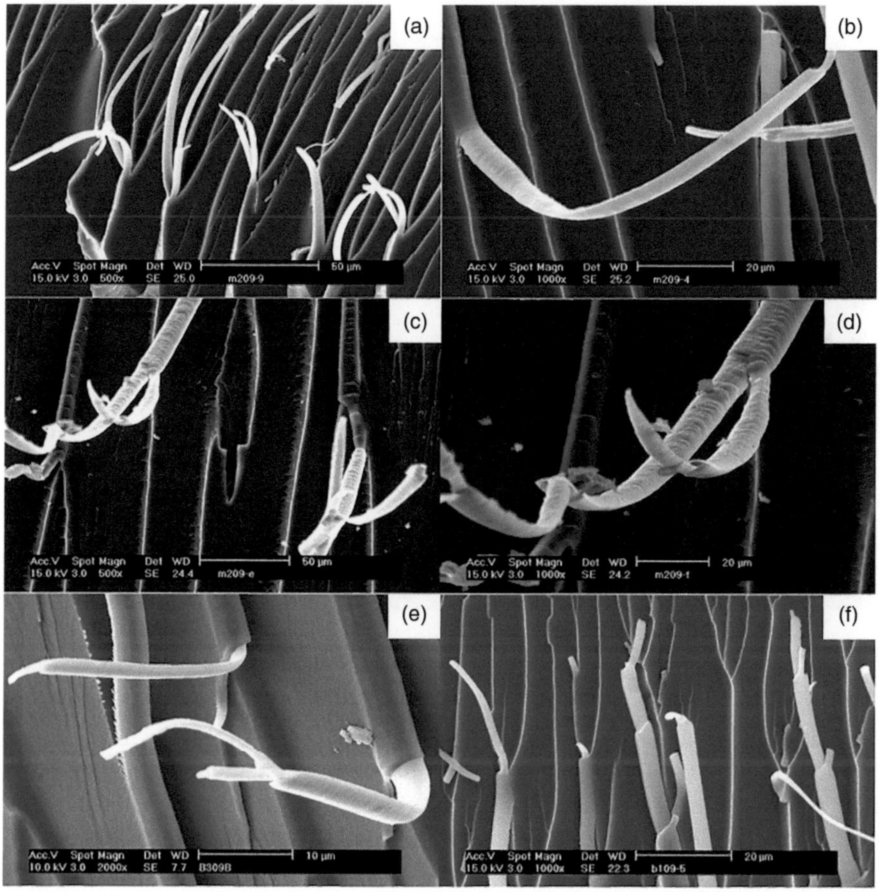

Figure 5.5 SEM micrographs of cured (a–d) 9 wt% HTME-2/DGEBA, (e) 9 wt% HTBE-3/DGEBA, and (f) 9 wt% HTBE-1/DGEBA.

hyperbranched polyesters/linear unsaturated polyesters (UHPR-2/UP-191) [85], hydroxyl-terminated hyperbranched polyphosphate ester (HPPE-3)/DGEBA [86], and degradable aliphatic epoxy-terminated hyperbranched polyesters (DEHP-2)/DGEBA [87]. Furthermore, these four types of HBPs exhibited significant reinforcing effects in either DGEBA or linear unsaturated polyester matrices. These findings further indicate that reactive HBPs—such as hydroxyl-containing HPPE-3, double bond-bearing UHPR-2, and epoxy-functionalized DEHP-2—can enhance the tensile strength of their respective matrices while improving toughness. The primary underlying reason is that these reactive, ellipsoid-shaped HBPs with small hydrodynamic volumes are capable of dispersing and permeating into the molecular chains of the matrix. This process enhances intermolecular interactions, densifies the composite blends, and ultimately improves tensile strength. Notably, the simultaneous reinforcement and toughening mechanism has also been corroborated in the EHBP/DGEBA system [88].

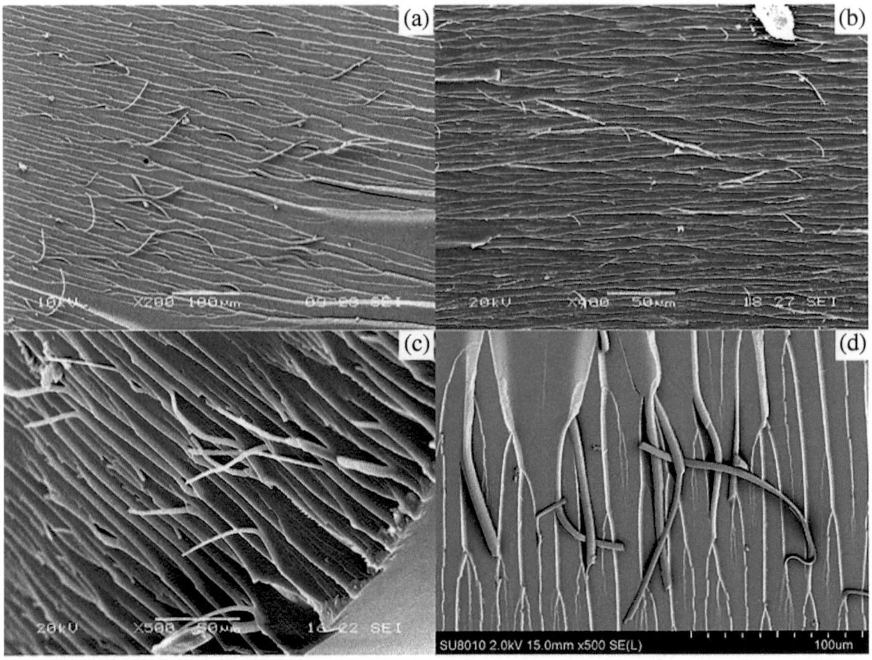

Figure 5.6 SEM micrographs of impact-fractured surfaces of cured samples. (a) HTPE-3/DGEBA, (b) UHPR-2/UP-191, (c) HPPE-3/DGEBA, and (d) DEHP-2/DGEBA.

Degradable citric acid was selected as a monomer to synthesize epoxy resins with different molecular weights, and their performance was systematically assessed in combination with hyperbranched resins. As the molecular weight of degradable aliphatic epoxy-ended hyperbranched polyesters (DEHP) increased, the mechanical properties of DEHP/DGEBA composites first enhanced and then decreased. For example, compared with pure DGEBA (17.1 kJ/m^2), the impact strength of cured composites containing 12 wt% DEHP-1, DEHP-2, and DEHP-3 increased by approximately 117.9%, 137.8%, and 126.1%, respectively. Correspondingly, the tensile strength rose by about 34.2%, 43.4%, and 40.0%, while the flexural strength increased by 35.6%, 48.1%, and 40.7% [87]. The decreased toughness observed at higher DEHP molecular weights is ascribed to the formation of intramolecular cavities within the non-cross-linked hyperbranched structures. These cavities facilitate energy absorption, thereby improving toughness, but concurrently reduce the tensile and flexural strengths of the composites. In contrast, an increase in the cross-linking density of the composites generally leads to enhanced tensile and flexural strengths. Consequently, intermolecular voids generated during cross-linking and the elevated cross-linking density exert competing influences on the mechanical properties of the composites. Overall, the incorporation of 12 wt% medium-molecular-weight DEHP achieves an optimal balance, endowing the composites with high toughness while preserving excellent tensile and flexural strengths (Figure 5.7).

EEW exerts a critical effect on the properties of hyperbranched epoxy resins (HERs) by regulating their cross-linking density. A correlation has been established

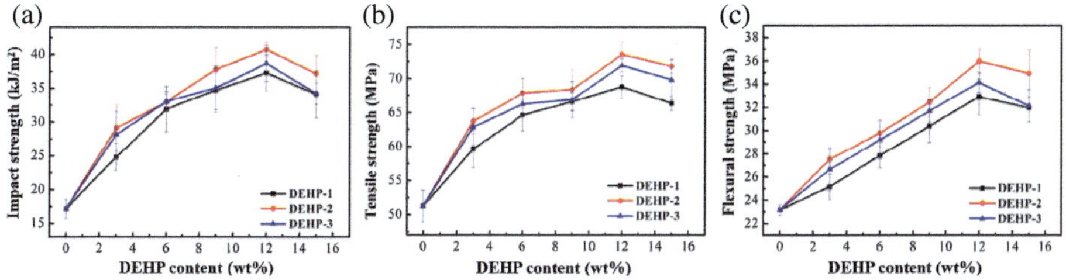

Figure 5.7 Effects of the molecular weight and content of DEHP on the (a) impact, (b) tensile, and (c) flexural strengths of DEHP/DGEBA composites.

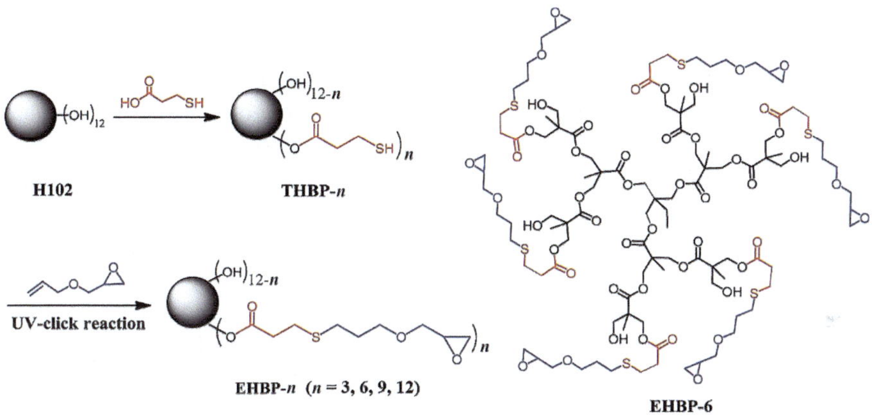

Figure 5.8 Synthesis schemes of EHBP-n ($n = 3, 6, 9, 12$) and chemical structure of EHBP-6.

between the EEW of HERs and the extent of property enhancement. To further explore the relationship between EEW and HER properties—including mechanical strength, adhesive strength, thermal performance, and rheological behavior—a series of hyperbranched epoxy resins (EHBP-*n*, $n = 3, 6, 9, 12$) with varying EEWs were synthesized. This synthesis was accomplished via a thiol–ene click reaction between thiol-terminated hyperbranched polyesters and allylglycidyl ether [89] (Figure 5.8).

When the content of EHBP-*n* increases from 0 to 15 wt% and the EEW decreases from 595.24 to 300.48 g/eq, the viscosity of the blends decreases from 5.93 to 3.46 Pa·s and from 4.71 to 3.94 Pa·s, respectively—indicating improved flowability of EHBP-*n*. This viscosity reduction is attributable to the branched —OH groups in EHBP-*n*, which creates extra free space for the movement of epoxy chains. Generally, hyperbranched epoxy resins possess low viscosity due to their globular structure and the absence of molecular entanglement.

The viscosity of EHBP-*n*/DGEBA blends decreases further as the EEW of EHBP-*n* declines, since fewer hydroxyl groups weaken intermolecular interactions between hyperbranched chains. With the increase in EHBP-*n* content and decrease in EEW, the number of hydrogen bonds decreases while the proportion of flexible segments increases, rendering the blend more liquid-like. Correspondingly, both the storage modulus (G') and loss modulus (G'') of the blends decrease with increasing EHBP-*n* content and decreasing EEW, as the nonentangled EHBP-*n* chains exhibit greater mobility compared to DGEBA chains. As the EEW of EHBP-*n* decreases from 595.24 to 300.48 g/eq, the mechanical properties of EHBP-*n*/DGEBA composites first increase and then decrease. The optimal performance was achieved with EHBP-6 (EEW = 398.72 g/eq), where the tensile, flexural, impact, and adhesive strengths were enhanced by 38.9%, 35.3%, 158.6%, and 96.2%, respectively. This trend stems from the competing effects of intramolecular cavities and cross-linking: intramolecular cavities enhance toughness by absorbing energy but reduce strength, while cross-links improve tensile and flexural strength. The primary mechanism underlying toughness enhancement in EHBP-*n*/DGEBA composites is the formation of cavities within non-cross-linkable hyperbranched structures, which efficiently absorb impact energy.

The degree of branching (DB) is a pivotal parameter for HBPs, as it regulates the transition of molecular topologies from linear chains to three-dimensional spherical structures and thereby exerts a profound influence on physicochemical properties. Nevertheless, precise control of DB and the quantitative correlation between DB and material properties remain unresolved challenges. To tackle this issue, we synthesized nitrogen–phosphorus backbone epoxy-terminated hyperbranched polymers (NPEHP-*n*) with tunable DB levels [90].

In NPEHP-*n*/DGEBA blends, particle size, viscosity, storage modulus (G'), and loss modulus (G'') all decreased with increasing DB and NPEHP-*n* content. This confirms that a higher DB enhances blend flowability through a molecular disentanglement effect. Mechanically, NPEHP-*n*/DGEBA composites attained maximum tensile, flexural, and impact strengths at an NPEHP-*n* loading of approximately 9 wt%, with respective improvements ranging from 31.7% to 42.8%, 56.4% to 76.4%, and 172.0% to 198.3%. As DB increases, the molecular structure becomes more spherical, and the number of intramolecular cavities increases. These cavities undergo deformation under impact loading, absorb energy, and form protonema-like structures. A higher DB leads to the formation of denser protonema networks, endowing the composites with superior toughness compared to pure DGEBA (Figure 5.9).

Furthermore, a series of hyperbranched epoxy resins (NMHEP-*n*) with precisely regulated DBs were synthesized via thiol–ene click chemistry [91]. Their topological characteristics, including hydrodynamic radius (R_h), root mean square radius of gyration (R_g), and shape factor ($\rho = R_g/R_h$), were investigated. The shape factor ρ is indicative of molecular topology in dilute solutions: $\rho \approx 0.774$ for solid spheres, ~1.0 for vesicles, 1.5–1.8 for random coils, and >2 for rigid chains. As DB increases, chain density elevates, intermolecular forces intensify, and both R_h and R_g decrease. For NMHEP-0, NMHEP-0.5, and NMHEP-1, the shape factors were 2.57, 1.52, and 1.26, respectively—suggesting a rigid rod conformation for NMHEP-0 and random coil conformations for NMHEP-0.5 and NMHEP-1. In terms of performance, pure DGEBA exhibited a tensile strength of 63.87 MPa, flexural strength of 99.04 MPa, and impact strength of 17.73 kJ/m^2. The incorporation of 12 wt% NMHEP-0.5 increased these mechanical properties to 87.13 MPa, 134.27 MPa, and 43.41 kJ/m^2, corresponding to improvements of 36.42%, 35.57%, and 144.84%. These findings

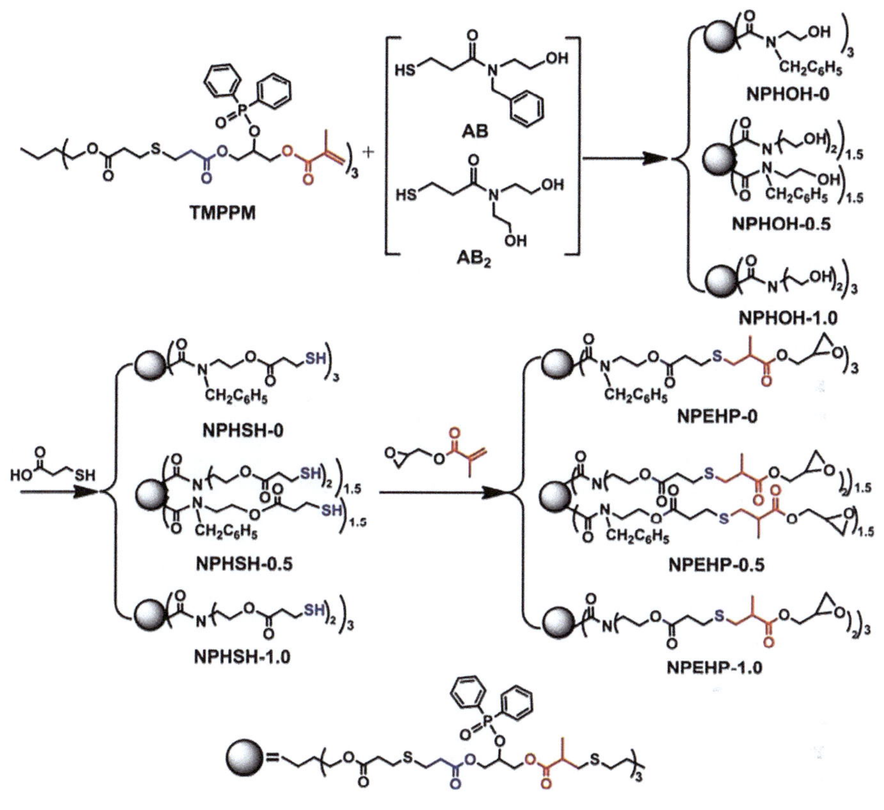

Figure 5.9 The synthesis scheme of NPEHP-*n* with different degrees of branching.

confirm that the distinctive hyperbranched topology, coupled with optimized cross-link density and free volume, enables the simultaneous enhancement of toughness, strength, and elongation in epoxy-based systems (Figure 5.10).

Generally, the size or fraction of free volume exhibits an inverse correlation with mechanical properties (e.g., tensile strength, flexural strength, and fracture toughness [KIC]) and is also associated with the storage modulus. Positron annihilation lifetime spectroscopy (PALS) combined with molecular simulations has confirmed a negative relationship between topological defects and free volume in cured DGEBA/amine systems. These studies suggest that relatively large free-volume holes and high free-volume fractions lead to inferior mechanical strength in pure epoxy systems. To mitigate this issue, a hydroxy-terminated hyperbranched polymer (H204) with an ellipsoidal topology was incorporated into DGEBA. The addition of H204 significantly enhanced the impact and compression properties of the resulting DGEBA/H204 composites by reducing both the size and quantity of free-volume holes. Similarly, in EHPI-*n*/DGEBA systems, the free-volume fraction exerted a positive effect on tensile strength and toughness. However, the relationship between mechanical performance and free volume in epoxy resins remains controversial, with both positive and negative effects documented in the literature. This inconsistency is likely attributed to the presence

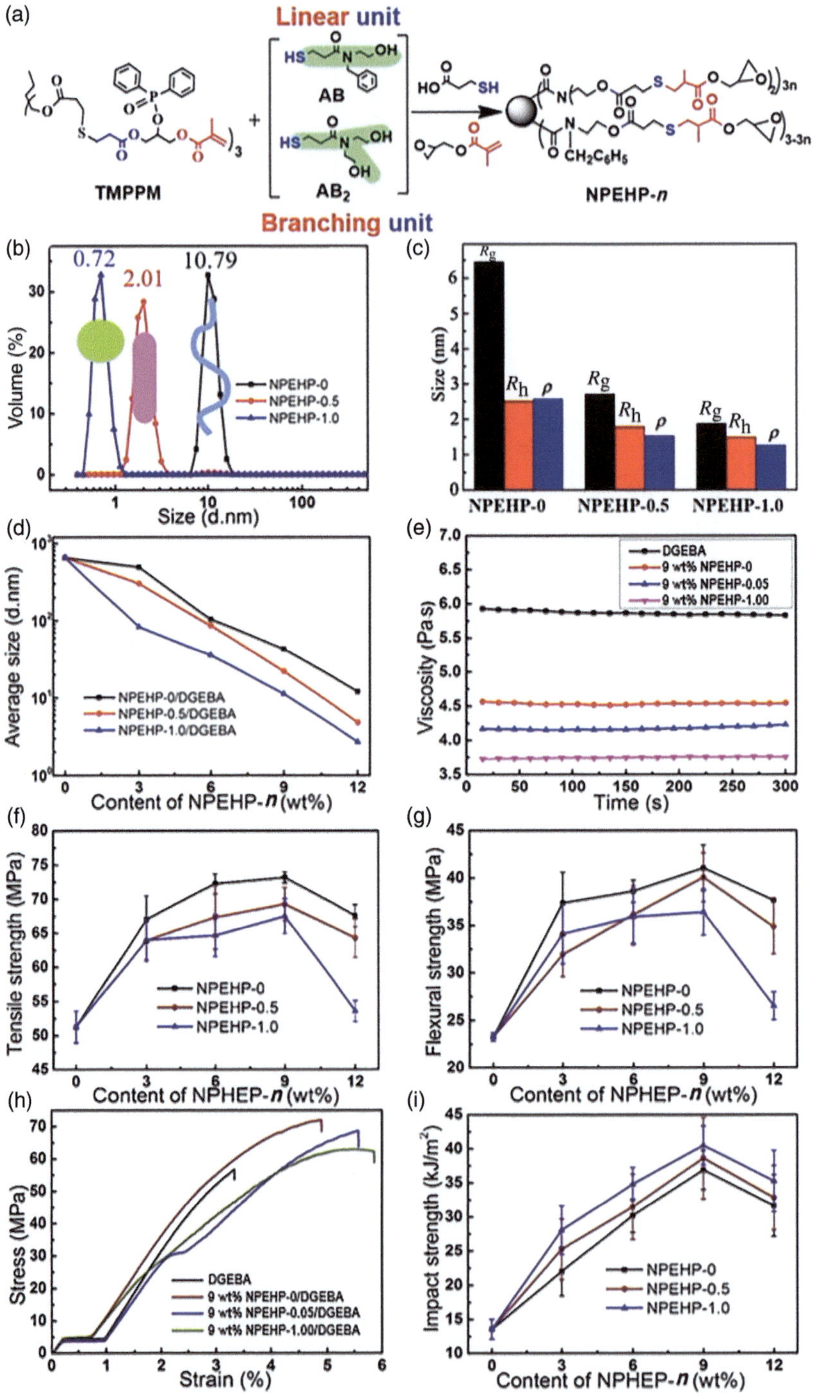

Figure 5.10 Controlling principle and properties of branching degree and topological structure of HBPs. (a) The synthesis scheme of different degrees NPEHP-*n*. (b) The particle size of NPEHP-*n*. (c) R_g, R_h, and ρ of NPEHP-*n*. (d) The particle size of NPHEP-*n*/DGEBA composites. (e) Rheological behavior of NPHEP-*n*/DGEBA blends. (f) Tensile strength. (g) Flexural strength. (h) Tensile stress. (i) Impact strength.

of relatively large (nanometer-scale) free-volume holes in conventional epoxy resins, as well as the lack of precise data regarding the size and distribution of subnanoscale free-volume holes [92, 93].

To deepen the understanding of the molecular-scale relationship between free volume and mechanical properties, we designed a series of novel hyperbranched epoxies (Bn, $n = 6, 12, 24$) featuring imide–cyanuric acid structures, which were synthesized via highly efficient thiol–ene click chemistry. The Bn molecules were tailored to penetrate the DGEBA matrix, enhance molecular packing density, and generate subnanoscale free-volume holes, while retaining rigidity and thermal stability over a wide temperature range. The subnanoscale free-volume characteristics of Bn/DGEBA copolymers—including hole diameter, size, fraction, and distribution—were analyzed using the LTv9 and CONTIN programs. PALS measurements demonstrated that with increasing B12 content and Bn molecular weight, both τ_3 and V_f of the Bn/DGEBA copolymers decreased, indicating a reduction in free-volume voids. This observation was consistent with the decreasing R values. The ellipsoidal morphology of Bn facilitated its dispersion and penetration into DGEBA voids, thereby reducing hole sizes. Although discrete-spectrum analysis via LTv9 yielded average hole sizes and distributions, CONTIN provided more reliable information. This is because large holes exert a disproportionate impact on mechanical strength through stress concentration, independent of the average hole radius or total free volume (Figure 5.11).

At a B12 content of 12%, the copolymer displayed the narrowest free-volume distribution (minimum full width at half maximum, FWHM) and the smallest void diameter (~0.48 nm), which is smaller than that of pure DGEBA (~0.51 nm). This enhancement is ascribed to the unique topology of Bn, where the total free volume consists of excess free volume and vibrational free volume. The ellipsoidal structure of Bn promotes dispersion, increases molecular packing density, and enables compact arrangement of the cross-linked network. Consequently, intermolecular chain interactions and load transfer are significantly enhanced. These findings indicate that even a slight reduction in free volume can result in substantial improvements in mechanical performance. In the 12% B12/DGEBA system, the compact cross-linked network imparts high strength by restricting chain sliding and dispersing stress. Meanwhile, the ellipsoidal hyperbranched structure dissipates and absorbs impact energy more efficiently than the linear DGEBA backbone, thereby improving toughness. SEM analysis of impact-fractured surfaces validated this observation: the 12% Bn/DGEBA system exhibited filamentous features (indicative of superior toughness and ductility), in contrast to the smooth, brittle fracture morphology of pure DGEBA (Figure 5.12).

We also proposed a novel strategy to develop high-strength and tough epoxy supramolecular thermosets using vanillin-based hyperbranched epoxy resin (VanEHBP), which involves introducing intermolecular hydrogen bonds and forming intramolecular and intermolecular cavities [94]. These supramolecular structures endow the thermosets with excellent energy dissipation capacity, thereby achieving high toughness and strength. The enhanced mechanical performance of EN-VanEHBP can be attributed to the supramolecular networks, with three key

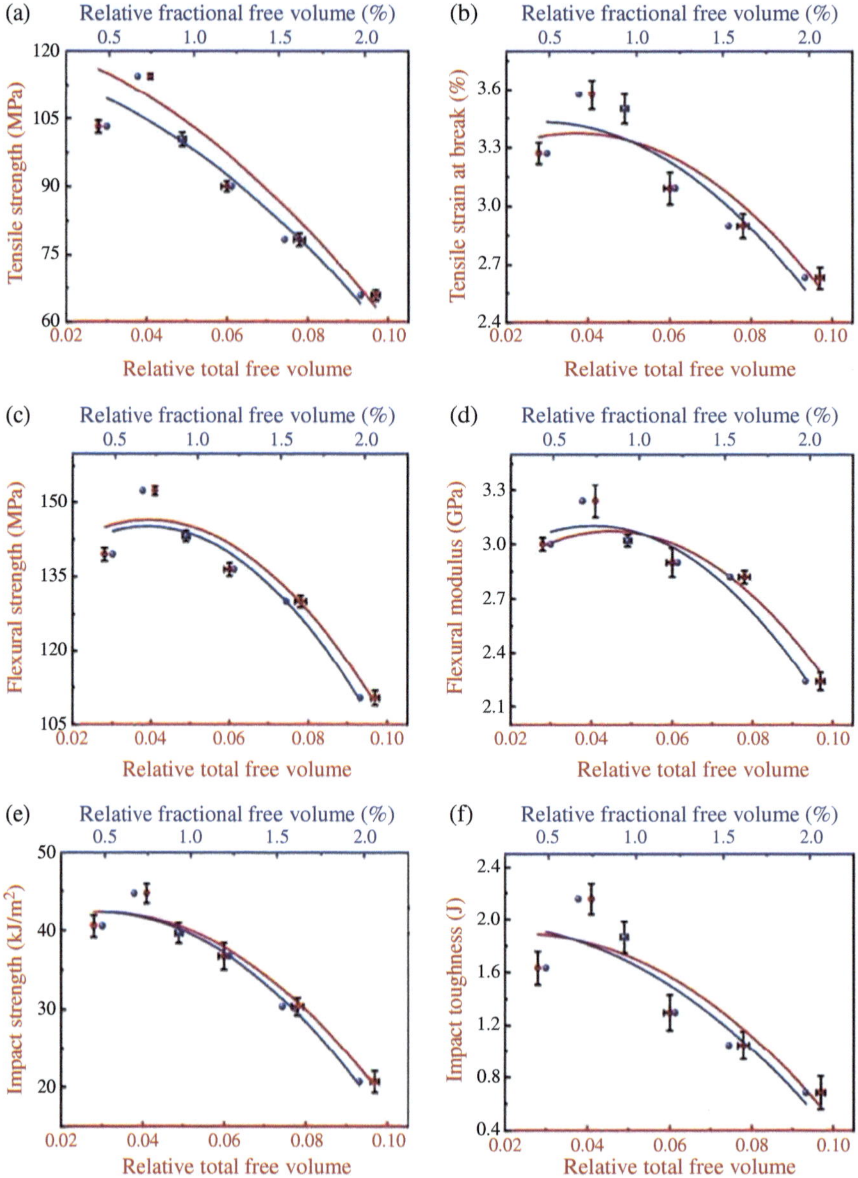

Figure 5.11 Correlation of mechanical properties and relative fractional free volume. (a) Tensile strength. (b) Tensile strain. (c) Flexural strength. (d) Flexural modulus. (e) Impact strength. (f) Impact toughness. The six data points from left to right were obtained in order from 15%, 12%, 9%, 6%, and 3% to 0% in the B12 content of the copolymer.

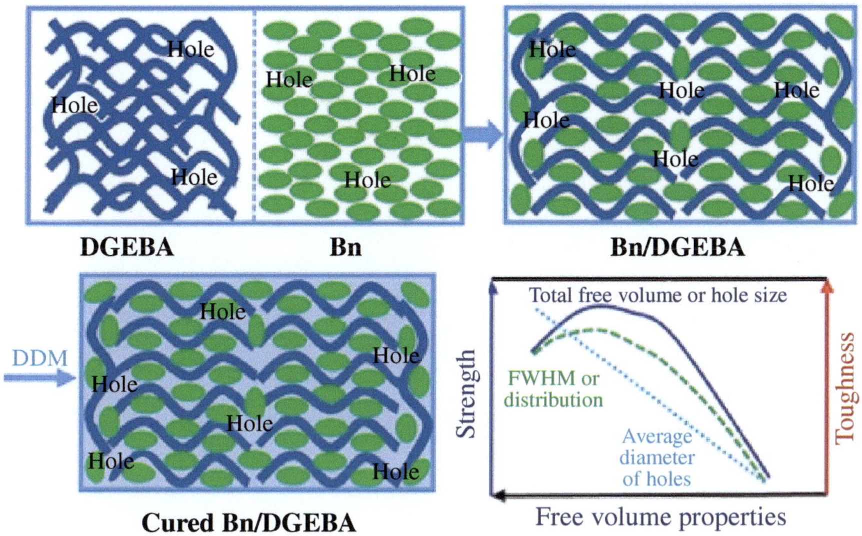

Figure 5.12 Process of performance improvement on Bn/DGEBA copolymer.

contributions: (i) hydrogen bonding interactions effectively stiffened and reinforced the materials. The reversible nature of hydrogen bonds enabled energy dissipation and stress redistribution prior to material failure, thus conferring EN-VanEHBP with high ductility and toughness. (ii) The deformability of intramolecular and intermolecular cavities provided an efficient pathway for energy dissipation, resulting in the rigid yet tough characteristics of EN-VanEHBP. Additionally, the local free volume associated with hyperbranched cross-links expanded the space available for kink motions, while the crankshaft structures in the molecular chains facilitated secondary relaxation. (iii) The introduction of soft segments reduced internal stress, and the arms of hyperbranched cross-links could readily redistribute forces, thereby alleviating stress concentration. VanEHBP enhanced the flexibility between cross-link points and increased the degree of conformational rearrangement during fracture. Consequently, the developed supramolecular epoxy thermosets exhibited outstanding mechanical performance (Figure 5.13).

Owing to its hyperbranched topological structure, hyperbranched epoxy resin (HER) has been employed to enhance the toughness and strength of epoxy resins. HER possesses cavities of varying sizes and multiple terminal epoxy groups; the latter can increase cross-linking density, thereby improving creep resistance and glass transition temperature (T_g). The cavities with diverse sizes and appropriate statistical spacing between dynamic covalent bonds may mitigate the adverse effects of rigid networks, which is favorable for the dynamic bond exchange reactions of covalent adaptable networks (CANs). Thus, incorporating HER into CANs can enhance the creep resistance and T_g of the latter without compromising reprocessability. Notably, we proposed an effective strategy to simultaneously improve the creep resistance, T_g, and stress relaxation rate of epoxy resins using vanillin-based hyperbranched epoxy resin (VEHBP) containing dual dynamic bonds [95]. The dual dynamic bonds in VEHBP reduced the activation energy for stress relaxation,

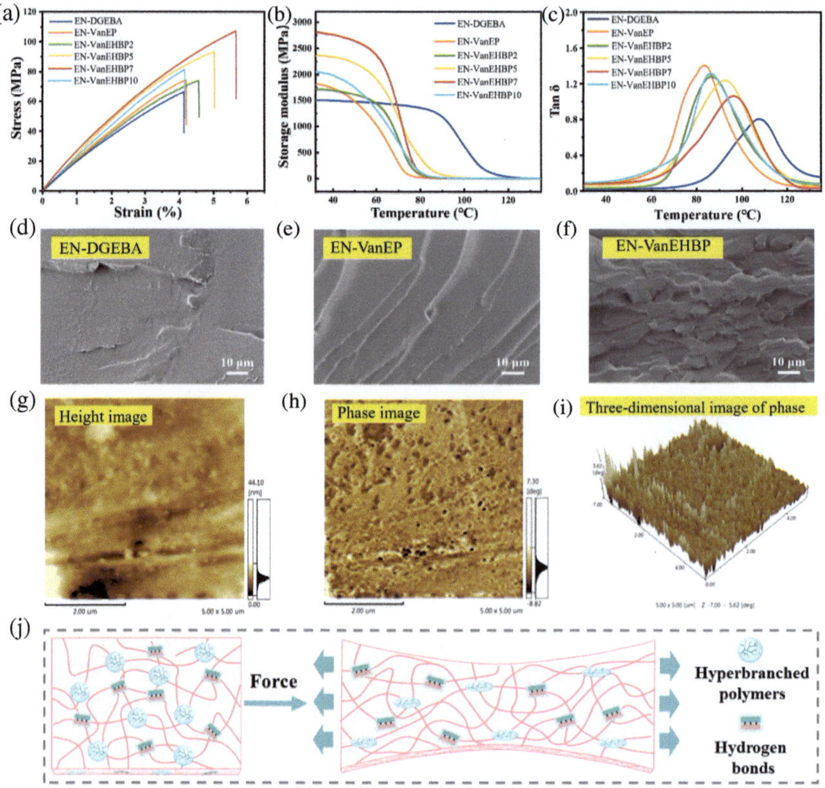

Figure 5.13 Mechanical properties and toughening mechanism of epoxy supramolecular thermosets. (a) Stress–strain curves of epoxy supramolecular thermosets at room temperature. (b) Storage modulus. (c) Tan curves of epoxy supramolecular thermosets. SEM images of (d) EN-DGEBA, (e) EN-VanEP, and (f) EN-VanEHBP7. (g) AFM images of EN-VanEHBP7 at room temperature. (h) Schematic illustration of the supramolecular networks of EN-VanEHBP upon deformation. *Source:* (d)–(g) [77] / Springer Nature / CC BY 4.0.

increased the mobility of dynamic bonds, and endowed the materials with excellent fast self-healing capabilities, rapid stress relaxation, remarkable reprocessability, and recyclability. It is observed that the fracture surfaces of pure epoxy resin and VEP-added epoxy resin were smooth and flat, a characteristic of brittle fracture. In contrast, the fracture surface of the 5% VEHBP/DGEBA/4-AFD composite was relatively rough, with a large number of "fibrils" emerging—indicating a significant effect of VEHBP on the crack propagation process. This morphological feature is attributed to the in-situ homogeneous reinforcement and toughening of HBPs. The terminal epoxy groups of VEHBP enhanced the interaction between the HBP and the epoxy matrix, enabling uniform dispersion of VEHBP within the cross-linked structure and increasing cross-linking density. Additionally, the intermolecular cavities in HER can regulate the epoxy resin structure and generate numerous intramolecular cavities. These cavities facilitate plastic deformation and shear yielding, inhibit crack propagation around VEHBP in the epoxy matrix, and thus transfer or absorb fracture energy to improve impact resistance (Figure 5.14).

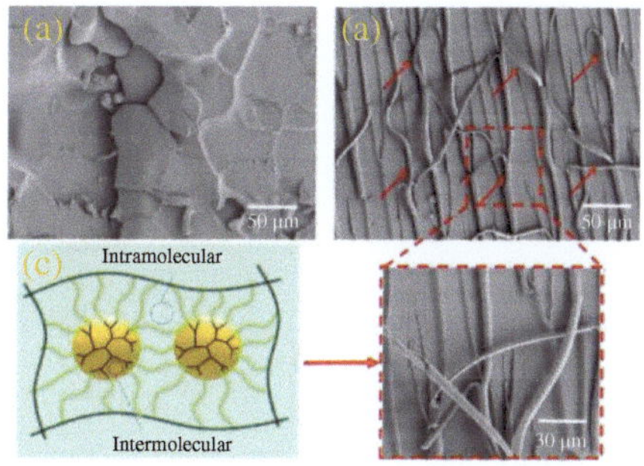

Figure 5.14 Scanning electron microscopy (SEM) images of (a) DGEBA/4-AFD and (b) 5% VEHBP/DGEBA/4-AFD. Mechanism of in-situ homogeneous reinforcement and toughening of VEHBP. *Source:* [95] / with permission of American Chemical Society.

Hot-melt adhesives (HEAs) are widely utilized in the automotive, electronic, and packaging industries, benefiting from their prominent advantages of being pollution-free, easy to peel off, and causing minimal damage to substrate surfaces. Nevertheless, the inadequate adhesion strength and poor reusability of traditional HEAs fabricated from DGEBA have severely constrained their practical applications. To address this challenge, we synthesized a hyperbranched epoxy resin (MHER) via an efficient microwave-assisted reaction and further constructed a high-strength reusable thermoplastic HEA by blending MHER with DGEBA and a polyether amine containing ureidopyrimidinone groups, without the involvement of any solvents [96]. This microwave-aided synthesis approach improved the preparation efficiency of MHER by approximately 6.3-fold and significantly simplified the cumbersome synthetic process. The incorporation of MHER effectively enhanced the mechanical properties of the prepared HEA, and the optimal performance was achieved when the MHER content in HEA was 8 wt% (designated as HEA-8). Compared with the MHER-free HEA-0, the adhesion strength, tensile strength, and tensile toughness of HEA-8 on stainless steel substrates were increased by 27.4%, 78.1%, and 475.0%, respectively. Specifically, HEA-8 exhibited an ultra-high adhesion strength of around 10.7 MPa and excellent reusability, retaining over 82% of its initial adhesion strength after five cycles of use. The superior reusability of HEA is attributed to the dynamic cleavage and reformation of noncovalent quadruple hydrogen bond interactions. It is anticipated that this work will provide a simple and environmentally friendly strategy for the fabrication of HEAs with excellent adhesion performance and reusability (Figure 5.15).

Shape-memory epoxy resins (SMERs) with excellent mechanical properties hold broad application potential; however, the design and synthesis of SMERs integrating high mechanical performance and short shape recovery time remain a significant challenge. Hyperbranched epoxy resin-based shape-memory materials provide a distinctive solution to this issue. For instance, a series of hyperbranched epoxy resins with varying molecular weights and structures (ETMP-n, ETHP-n, where n = 6, 9, 11) were synthesized via a simple and highly efficient thiol–maleimide click reaction [97]. The cured ETHP-n films exhibit shape fixity ratios and shape recovery ratios of up to 98.5% and 97.2%, respectively, even after seven shape-memory cycles. Specifically, the cured ETMP-9 and ETHP-9 films show rapid shape-memory responses, with shape recovery times as short as 450 and 430 ms, respectively—values significantly shorter than those of previously

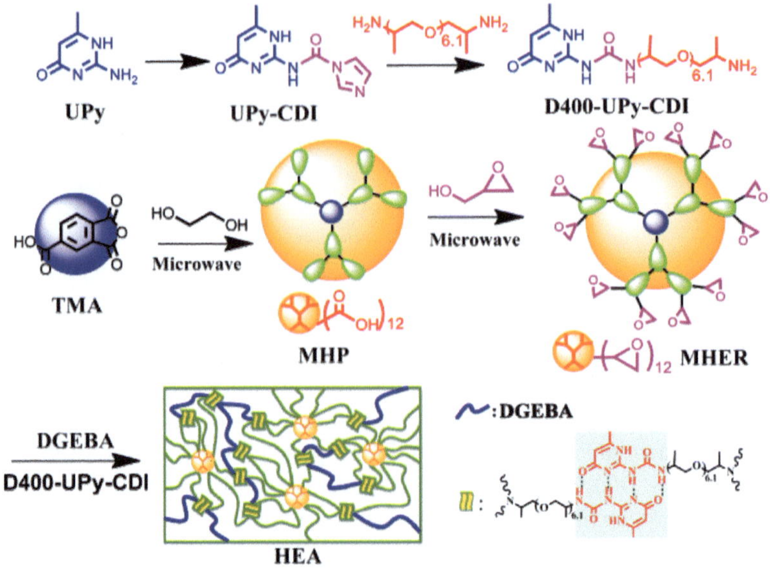

Figure 5.15 Preparation schemes of MHER, D400-UPy-CDI, and HEA.

reported SMERs. The maximum actuated stress of the cured films reaches 0.72 MPa, which is much higher than that of most mammalian skeletal muscles (0.35 MPa). Notably, the cured ETHP-9 films demonstrate superior shape-memory performances compared to the cured ETMP-9 films, which is attributed to their nitrogen heterocyclic rigid structures, appropriately high cross-linking density, and narrower molecular weight distribution. These outstanding shape-memory properties render the hyperbranched epoxy resins promising for practical applications in microelectronic devices and artificial muscles.

Previous studies have shown that epoxy resins modified with hyperbranched epoxy resins exhibit excellent performance. However, in the sustainable development of thermosetting materials, the efficient preparation of degradable hyperbranched epoxy resins and their composites with simultaneous high strength, elongation, modulus, and low-temperature resistance remains a major challenge. Herein, we developed self-curable hyperbranched epoxy resins (DSEHP-n, where n = 3, 5, 7, 11) for the modification of DGEBA. DSEHP-n not only simultaneously enhanced the tensile strength, flexural strength, impact strength, modulus, and elongation by 64.8%, 38.1%, 93.7%, 70.4%, and 50%, respectively, but also significantly improved the low-temperature resistance and degradation efficiency [98, 99]. The enhancement of mechanical properties is attributed to the synergistic effect of cross-linking density, free volume, intermolecular cavities, hyperbranched topological structure, and good compatibility. This originates from an in-situ reinforcement and toughening mechanism, which was verified by dynamic light scattering (DLS), SEM, DMA, and free-volume analysis. Detailed analysis of the degradation products via gas chromatography–mass spectrometry (GC–MS) spectra revealed that the degradation of the composites is caused by the cleavage of C—N and ester bonds (Figure 5.16).

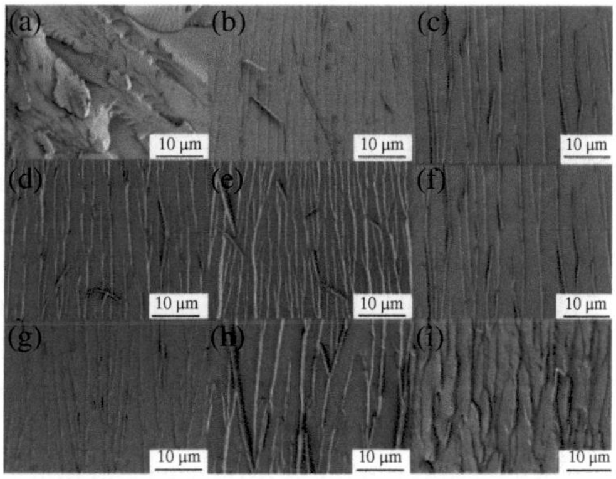

Figure 5.16 SEM micrographs of impact fracture surface of blend materials. (a) DGEBA, (b) 3 wt% DSEHP-5/DGEBA, (c) 6 wt% DSEHP-5/DGEBA, (d) 9 wt% DSEHP-5/DGEBA, (e) 12 wt% DSEHP-5/DGEBA, (f) 15 wt% DSEHP-5/DGEBA, (g) 12 wt% DSEHP-3/DGEBA, (h) 12 wt% DSEHP-7/DGEBA, and (i) 12 wt% DSEHP-11/DGEBA. *Source:* [99] / with permission of Elsevier.

5.4 Topological Phase Mechanism

IPNs have garnered considerable attention owing to their unique structure, which integrates two or more discrete cross-linked polymer networks. These networks are permanently entangled with one another yet lack any covalent linkages between their respective chains. Endowed with such a distinctive network architecture, IPNs typically display synergistic improvements in both thermal and mechanical performance, which far outperform those of their individual constituent polymers. This superior functionality is largely ascribed to the intrinsic characteristics of epoxy resins.

A triazine-containing HBPU was selected as the matrix material to prepare IPNs with varied HBPU/epoxy (EP) ratios [100]. Grafting-induced IPN formation promoted more robust cross-linking and tighter network interweaving, thereby endowing the resultant materials with enhanced thermal stability, mechanical performance, and shape-memory capability. In comparison with pristine HBPU, the incorporation of EP into the HBPU/EP IPNs led to a dramatic increase in tensile strength—from 12.5 up to 57.7 MPa—and a remarkable boost in tensile modulus, which soared from 4.8 to 1522.1 MPa. Such prominent enhancements stem from two key mechanisms: first, the addition of high-modulus EP contributes directly to the reinforcement of mechanical strength and elastic modulus and second, the construction of the interpenetrating network creates extra cross-linking junctions between HBPU and EP chains, which effectively impedes the slippage of polymer segments under external loading. Nevertheless, with a further rise in HBPU content within the HBPU/EP IPNs, EBA increases correspondingly, whereas both tensile strength and modulus start to exhibit a downward trend.

A poly(trimellitic anhydride-triethylene glycol) ester epoxy resin, designated as HTTE, was synthesized and subsequently toughened via incorporation into a DGEBA/4,4′-DDM epoxy resin system [101]. DMA results verified the occurrence of phase separation within the HTTE/DGEBA-amine system, which further evolved into an IPN structure. Specifically, the ternary blends presented as homogeneous solutions prior to curing; phase separation then took place as the curing reaction proceeded. During the curing process of the DGEBA/HTTE mixture, the epoxy groups of DGEBA underwent rapid reaction to generate hydroxyl groups, which diminished the compatibility between DGEBA and HTTE and eventually induced the formation of a biphasic microstructure. Meanwhile, the hardener molecules distributed at the phase interfaces reacted with both DGEBA epoxy groups and HTTE components, establishing chemical bonds that connected the rigid HTTE particles to the hard DGEBA matrix, thus achieving efficient toughening of the resultant composite.

The effects of HTTE content and generation number on the performance of the cured resin systems were investigated in detail. Compared with the unmodified DGEBA/4,4′-DDM system, the impact strength of HTTE/DGEBA blends was enhanced by 2–7 times. SEM observations of the fracture surfaces revealed the presence of cavitation in the central area and fibrous yielding at the edge, indicating that particle cavitation, shear yield deformation, and in-situ toughening constituted the primary toughening mechanisms. DMA characterizations further confirmed that phase separation in the HTTE/DGEBA-amine systems gave rise to an IPN architecture. This IPN-structured resin retained favorable transparency and exhibited a higher modulus than the neat DGEBA epoxy resin. Its glass transition temperature (T_g) showed a slight reduction relative to the neat epoxy; however, T_g increased progressively with the elevation of HTTE generation number (from the first to the third generation) as well as the concentration of hard segments.

Thermogravimetric analysis (TGA) data demonstrated that increasing HTTE content led to a marginal decline in the thermal stability of the cured systems, whereas thermal stability was enhanced with the increase of HTTE generation number from the first to the third. Additionally, HTTE was found to facilitate char formation in the HTTE/DGEBA blends. The improvement in thermal performance associated with the increase of HTTE generation number was attributed to multiple factors: the elevated molar mass and enhanced intramolecular hydrogen bonding of higher-generation HTTE, the strengthened interfacial interactions within the HTTE/DGEBA IPNs, and the increased cross-linking density arising from the greater abundance of terminal hydroxyl and epoxide functional groups in higher-generation HTTE.

5.5 Challenges and Perspectives

The interface is a distinctive microstructured composite domain that acts as both a medium and a bridge between two components—the matrix and the reinforcement phase. It exerts a profound influence not only on the interfacial and macroscopic properties of the composite but also on the interfacial chemical composition and microstructure. Therefore, it is imperative to conduct in-depth investigations into a series of interfacial characteristics, including its geometric traits, chemical bonding configurations, structural features, chemical and structural defects, interfacial stability, interfacial reaction behaviors, and the governing factors of these reactions. Nevertheless, the interface typically resides at the micro- or nanoscale, and its intricate chemical composition and structural complexity render its characterization and analysis extremely arduous. Moreover, the availability of precise, accurate, and quantitative methodologies for assessing interfacial morphology, dimensional parameters, and chemical composition distribution remains rather limited.

5.5.1 Raman Spectroscopy

In recent years, Raman spectroscopy has emerged as a widely utilized technique for probing material structures. Its exceptional capability to discern subtle variations in molecular composition enables the construction of images that are strictly based on such minor molecular discrepancies. Moreover, this technique is largely unaffected by the physical attributes of the tested materials; for instance, surface roughness and other inhomogeneities exert negligible influence on the generation and quality of Raman images.

This chemically selective yet robust characteristic sets Raman spectroscopy apart from other conventional phase separation analysis techniques, such as atomic force microscopy (AFM), SEM, and X-ray diffraction (XRD). Additional merits that augment its appeal over alternative methods include the straightforward setup of Raman spectrometers and the relatively short time frame required for image acquisition—typically within several minutes, contingent on the desired image size and resolution. A further major advantage lies in its minimal sample preparation requirements: samples as thin as a single molecular layer can be analyzed, and there is technically no upper limit to sample thickness. It should be noted, however, that Raman scattering can only be excited within the depth range that the incident light is able to penetrate the sample.

For example, Raman spectroscopy has been successfully employed to visualize phase separation in imine-based CANs at the scale of several micrometers. This observation was independently corroborated by AFM, which confirmed the presence of phase-separated domains within the polymer matrix. Notably, phase separation was detected, even though these materials were synthesized from miscible monomers—precursors that would intuitively be expected to yield homogeneous systems. Our investigations revealed that phase separation not only altered the surface appearance of the materials but, more significantly, exerted a pronounced impact on their thermomechanical performance. Specifically, CANs with identical aliphatic/aromatic monomer ratios that exhibited phase separation displayed both a higher crossover temperature ($T_{(cross)}$, the temperature at which $\tan(\delta) = 1$ and the material transitions from a rubbery to a viscous state) and an elevated elastic modulus (G').

By tailoring the architectural design of CANs, we were able to either suppress or promote phase separation. We postulate that this phenomenon is driven by favorable π–π stacking interactions between the aromatic moieties of the monomers. Our work further underscores the pivotal role of phase separation in CANs—even those fabricated from miscible components—and provides a viable strategy for regulating the dynamic material properties of such networks.

To further investigate the potential steric effects of dianiline monomers on phase separation behavior, a series of commercially available dianiline derivatives

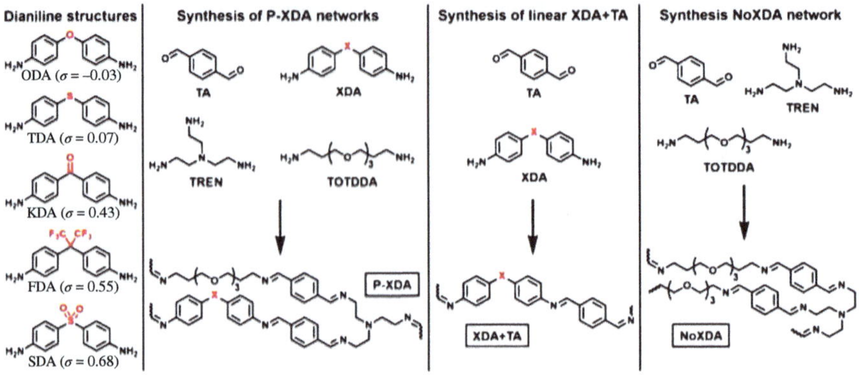

Figure 5.17 Overview of the building blocks for the synthesis of polyimines.

were selected for the synthesis of additional CAN specimens (Figure 5.17). First, 4,4′-methylenedianiline (MDA), a dianiline monomer with no steric hindrance, was chosen as the reference material. Based on the Hammett parameter of this monomer, if phase separation were governed by electronic effects, MDA-based CANs would be expected to exhibit phase separation behavior analogous to that of CANs derived from P-TDA and P-ODA. This hypothesis was subsequently verified by Raman imaging results (Figure 5.18). Next, a set of polymer films were prepared using dianiline monomers that possessed comparable electronic effects but featured increased steric bulk. Specifically, we selected dianiline monomers substituted with alkyl groups (methyl for Me-MDA and ethyl for Et-MDA) at both ortho positions adjacent to the amine groups on the aromatic rings. Additionally, a dianiline monomer bearing a bulky cyclohexyl substituent on the bridging sp^3-hybridized carbon atom (CyDA) was also incorporated into the study (Figure 5.17).

Raman imaging results of these specimens indicated an absence of phase separation in all cases, which strongly suggests that steric effects associated with π–π stacking indeed modulate the occurrence of phase separation. These observations are consistent with findings reported for ureidopyrimidone (UPy)-based supramolecular polymers, where microphase separation was found to diminish with increasing steric hindrance on the pyrimidone moieties. Previous studies have demonstrated that the branching nature of substituents inhibits molecular aggregation, as bulkier steric groups impede effective π–π stacking and thereby suppress phase separation. Analogously, the branched dianiline monomers employed in our work may inhibit the aggregation of XDA + TA domains, thereby preventing the occurrence of phase separation (Figure 5.18).

5.5.2 Positron Annihilation Lifetime Spectroscopy

PALS is a distinctive technique for quantifying the size, size distribution, and fraction of open-volume regions—including pores in porous materials and free-volume cavities in polymers. For polymeric systems, the annihilation lifetime of

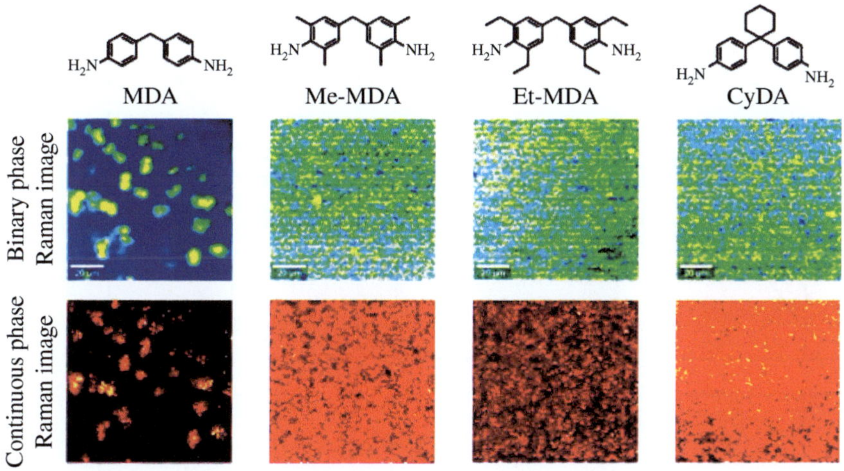

Figure 5.18 Structure and Raman images of dianilines with different substituents varying in steric hindrance. The Raman images show that phase separation was clearly observed for the nonhindered P-MDA, but not for the sterically hindered analogous materials. For all materials, the dianiline content was 20%. Binary and continuous color coding of the phases are included. *Source:* With permission of American Chemical Society. Schoustra, S. K. et al. / with permission of American Chemical Society.

ortho-positronium (o-Ps) is predominantly determined by the dimensions of free-volume holes and their surrounding chemical environment. The formation probability of o-Ps is dependent on both the volume of these free-volume cavities and the functional groups in the polymer backbone: highly electronegative or electropositive groups can hinder o-Ps formation by scavenging positrons or electrons. In the absence of such inhibitory effects, o-Ps formation is regulated by the size and density of free-volume holes.

In recent years, PALS has been widely employed to investigate the influence of free-volume holes on the static and dynamic properties of epoxy-based materials. This technique enables the acquisition of detailed information regarding the annihilation lifetimes and relative intensities of positrons and positronium. The free-volume size in polymers can be calculated from the o-Ps lifetime using the Tao–Eldrup model. In our work, temperature-dependent PALS measurements were performed to determine the free volume (V_f) and relative fractional free volume (f_r). Subsequently, the correlation between f_r and the mechanical as well as dynamic mechanical properties of the materials was established through PALS data analysis.

First, flame-retardant epoxy resins (FREPs) were fabricated using itaconic acid-derived hyperbranched epoxy resin (IA-EHBP) as the key precursor [102]. The temperature-dependent profiles of storage modulus (E'), loss modulus (E''), damping factor ($\tan\delta$), and free volume (V_f) for both neat epoxy (EP) and the as-prepared FREPs are presented in Figure 5.19. With rising temperature, the storage modulus (Trub(E')) of both EP and FREPs exhibits an increasing trend, which can be ascribed to the strengthened interfacial interactions between IA-EHBP and the epoxy matrix.

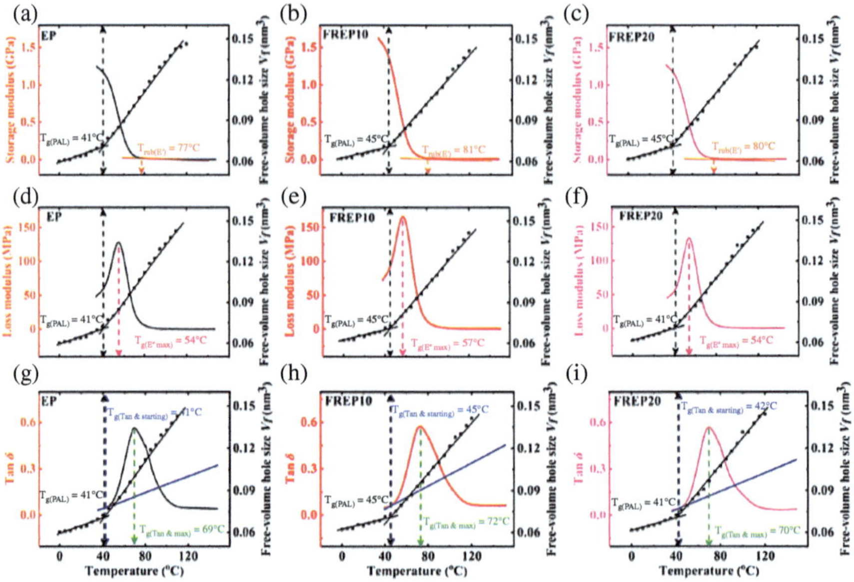

Figure 5.19 Temperature dependences of E', E'', tan δ, and V_f for EP, FREP10, and FREP20.

Below the glass transition temperature (T_g), the variation rate of V_f for both EP and FREPs remains relatively slow, accompanied by a gradual increase in loss modulus (E''). In the glassy state, the mobility of polymer chains is severely restricted, with molecular motions only confined to the alterations of bond lengths and bond angles along the chains. When the temperature surpasses T_g, the mobility of the originally frozen macromolecular chains is significantly enhanced, resulting in a rapid decline in E'' and a sharp elevation in V_f. Beyond the temperature corresponding to the maximum loss modulus (E''max), E'' decreases at an accelerated pace before stabilizing at a constant level.

In this work, the temperature at the onset of the tan δ peak was defined as T_g(tan δ starting), which shows a close correlation with T_g(PALS). Specifically, T_g(PALS) characterizes the initiation of main-chain motion, which is manifested by a rapid increase in the damping factor. For the tested system, the following relationships between the characteristic temperatures were summarized: T_g(PALS) is approximately 30–40 °C lower than Trub(E'); T_g(E'max) is roughly 10–15 °C higher than T_g(PALS); T_g(tan δ max) is about 30 °C higher than T_g(PALS); and T_g(tan δ starting) shows the closest correspondence to T_g(PALS). These findings confirm that PALS serves as a sensitive technique for probing the initial stage of polymer chain mobility, thus complementing the conventional DMA method (Figure 5.19).

Intramolecular hydrogen bonding within IA-EHBP constrains the mobility of polymer chains, which reduces the size of free-volume pores in the adjacent region and thereby lowers both the ortho-positronium lifetime (τ_3) and free volume (V_f). Specifically, the α-carbonyl and β-hydroxyl groups in IA-EHBP are spatially proximal, enabling the formation of intramolecular O—H···O=C

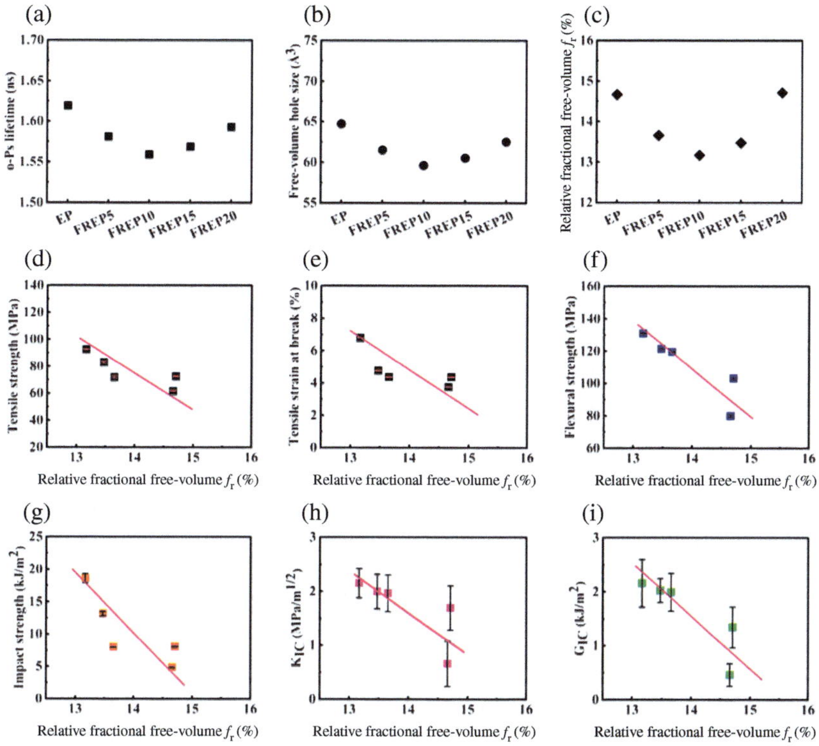

Figure 5.20 (a) o – Ps lifetime τ_3, (b) hole size V_f, and (c) relative fractional free-volume f_r of EP and FREPs. Variations in (d) tensile strength, (e) tensile strain at break, (f) flexural strength, (g) impact strength, (h) KIC, and (i) GIC as functions of relative fractional free-volume f_r.

hydrogen bonds. These intramolecular interactions not only restrict chain mobility but also enhance the positron trapping efficiency, inhibit positronium formation and ultimately reduce the observed positronium fraction (Pos fraction).

Furthermore, intramolecular hydrogen bonding contributes to an elevated cross-linking density (L_c) and a higher glass transition temperature (T_g) of the FREPs. Owing to the synergistic effects of free-volume reduction and intramolecular hydrogen bonding, the τ_3, V_f, and relative fractional free volume (f_r) values all reached their minimums when the IA-EHBP loading was 10 phr (Figure 5.20).

5.5.3 Confocal Microscopy Visualizes

The iterative and rational optimization of composite materials hinges on the appropriate selection of constituent components based on well-established principles, a process that necessitates a clear understanding of how individual ingredients modulate the overall fracture behavior of the composites. Polymer mechanochemistry, especially with the advancement of Optical Force Probes (OFPs), offers a direct means to visualize the response of polymer systems to applied mechanical forces.

Representative applications include the integration of spiropyran-based OFPs at the interface of SiO_2 NP fillers and PMMA (forming PMA–SiO_2 NP composites) or in SiO_2 NP-reinforced laminate composites, the incorporation of coumarin dimer OFPs into nanocomposites, and the design of SiO_2 NPs functionalized with dioxetane-based OFPs.

Currently, immobilized fluorescent proteins anchored at the fiber–epoxy interfaces are being utilized as effective indicators to monitor fiber breakage and interfacial debonding events. In addition, rhodamine spirolactam embedded in silk fiber–epoxy mixtures can be employed in tensile tests to visualize interfacial debonding, where localized damage zones are clearly identified via fluorescence lifetime imaging microscopy (FLIM) [103]. For instance, cross-linked rubbery poly(hexyl acrylate) particles modified with OFPs were dispersed in an epoxy matrix, and the fracture behavior of the composites as well as the interactions between particles and propagating cracks were observed in situ using confocal laser scanning microscopy (CLSM).

A variety of interaction modes and damage mechanisms were characterized both on and below the fracture surfaces, including particle deflection and crack pinning, crazing and microcracking within the epoxy matrix, and the deformation or rupture of rubber particles. CLSM further provided detailed insights into bond scission events occurring inside individual particles, revealing nonuniform stress distributions and localized particle tearing phenomena. These observations collectively demonstrate that OFP-based analysis is a powerful technique for elucidating the fracture behavior of polymer composites. In particular, it enables direct visualization of interfacial and interphase regions—an advantage that can be leveraged to guide future composite design through sophisticated synthetic strategies and the strategic placement of OFPs (Figure 5.21).

5.5.4 Nanoindentation

Nanomechanical testing has emerged as a widely recognized approach for investigating the mechanical behaviors of materials at the nanoscale. For composite materials in particular, this technique yields crucial insights into key properties including strength, hardness, durability, and structural integrity. These test results also provide valuable feedback for optimizing composite formulations and refining manufacturing processes.

Among various nanomechanical testing methods, nanoindentation is commonly employed to quantify material hardness, elastic modulus, and indentation depth at the nanometer level. For composite systems, nanoindentation enables the evaluation of the matrix's mechanical properties as well as the regulatory effects of functionalizing agents or dispersed phases. By performing nanoindentation across different regions of a composite, the material's uniformity and consistency can be assessed. Similarly, nanoscratch testing is used to evaluate nanoscale adhesion and wear resistance, offering insights into the mechanical strength of the matrix and its constituent phases.

A comprehensive understanding of multiphase materials requires clarifying not only the individual mechanical properties of each phase but also their interactions across nanoscale and microscale dimensions. Recent advancements in

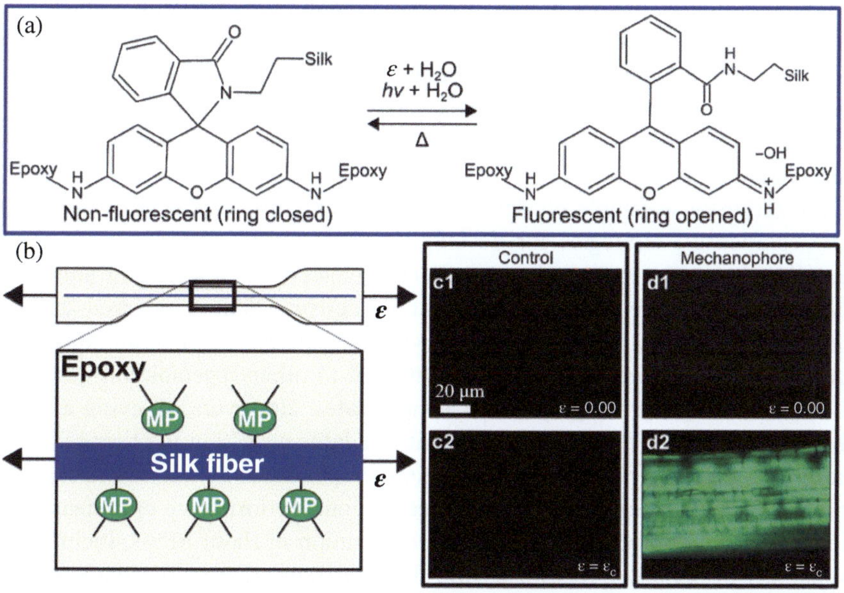

Figure 5.21 Imaging of stress transfer across a composite interface using mechanically responsive spirolactam (rhodamine 110, mechanophore). (a) Proposed activation mechanism for the ring opening of the spirolactam through applied energy and in the presence of water. (b) Schematic of a single fiber bundle composite tensile sample with MP covalently attached across the interface between the silk fiber bundle and epoxy matrix. (c1–c2, d1–d2) Two photon fluorescence lifetime microscopy (2P-FLIM) images of single fiber bundle silk epoxy composites (c1, d1) before and (c2, d2) after strain. Fluorescence images of the control (rhodamine-B functionalized silk fiber bundle in epoxy with nonreactive amine groups, CT), shows no activation within experimental parameters. (d1–d2) Significant activation of MP is observed in response to uniaxial mechanical strain. *Source:* Schoustra, S. K. et al. / with permission of John Wiley & Sons.

nanoindentation technology—such as eXpress Profiling Mode (XPM, also referred to as rapid indentation)—combined with high-resolution microscopy, have significantly shortened testing durations. XPM extends traditional single-indentation approaches by utilizing a piezoelectric scanner to traverse the sample surface, conducting rapid grid-based measurements to generate detailed hardness and modulus distribution maps over the designated test area. This method is particularly advantageous for multiphase polymeric materials, facilitating the efficient acquisition of spatially resolved mechanical property data. Overall, XPM test results indicate that rubber functionalization and the incorporation of self-healing agents do not substantially improve surface hardness or reduce the modulus.

5.5.5 Green Approaches for Epoxy Toughening

The global consumption of thermosetting resins, including epoxy resins, is on the rise. To address the inherent brittleness of epoxy resins and enhance their applicability as coatings and composite materials in structural engineering and the automotive industry, a variety of modification strategies have been developed. Meanwhile, stringent legislative frameworks and policies—such as the

Lead Market Initiative and the BioPreferred Program—are driving industries toward a sustainable economic model, leading to a growing market share of bio-based polymers. However, fully bio-based epoxy resins remain scarce in the commercial market. Consequently, there is an urgent need to adopt alternative approaches, such as substituting petroleum-derived additives and modifiers with bio-based equivalents.

The utilization of vegetable oil derivatives represents one of the most feasible and cost-competitive strategies for developing BBMs. This approach has been successfully adopted by automotive manufacturers such as Porsche and Subaru. Among various vegetable oil sources, soybean oil derivatives have emerged as the most extensively studied BBMs. Notably, soybean oil derivatives exhibit superior impact strength and fracture toughness compared to other vegetable oil-derived modifiers, rendering them highly effective candidates. However, achieving a balance between enhanced toughness and the unavoidable reduction in other critical mechanical properties—such as glass transition temperature (T_g), tensile strength, and modulus—remains an ongoing challenge. Beyond performance optimization, additional factors are pivotal for the commercialization of these BBMs, including the availability of bio-based feedstocks and the potential impacts of utilizing such feedstocks on the global food supply chain.

With the growing adoption of epoxy resins in structural applications, the demand for high cross-link density epoxy resins has surged. Given the exceptional efficacy of bio-based BCPs and HBPs in modifying high cross-link density epoxies, these materials hold significant promise as toughening agents for epoxy resins in structural scenarios. Incorporating bio-based HBPs thus emerges as an effective green strategy for epoxy toughening.

A comprehensive review of existing literature on the effects of HBPs on epoxy resins indicates that HBPs can enhance fracture toughness without compromising tensile strength or modulus. However, research focusing specifically on the application of bio-based HBPs for epoxy toughening remains relatively scarce. Commercially available HBPs and dendrimers are limited, with representative examples including DSM's Hybrane, Perstrop's Boltorn (a fatty acid-derived dendritic polymer), and Nanopartica's dendritic polyglycerol.

Glycerol stands out as a promising feedstock for HBP synthesis. As a low-cost co-product, glycerol can be utilized to produce cost-effective HBPs, effectively offsetting the high costs typically associated with HBP synthesis. Most current studies center on the synthesis of bio-based HBPs with oxirane functionality and the evaluation of their properties. Instead of using high-purity glycerol to synthesize hyperbranched epoxy resins as the matrix, an alternative approach involves adding small amounts of low-cost bio-based HBPs derived from crude glycerol into epoxy matrices, which is anticipated to improve toughness.

Considering the multiple purification steps and higher costs associated with pure glycerol, its use would further increase the production cost of already expensive HBPs, making commercialization unfeasible. As noted earlier, the effect of incorporating glycerol-derived HBPs into epoxy matrices remains unclear. It is, therefore, the responsibility of both academic researchers and industrial R&D teams to develop cost-competitive bio-based HBPs, systematically investigate their efficacy in epoxy toughening, and establish synthesis methods with potential for industrial scale-up.

The BCP industry still has considerable room for the development to achieve maturity, requiring collaborative efforts between academia and industry to synthesize cost-competitive BCPs. Currently, most commercially available BCPs are prohibitively expensive and are primarily utilized for toughening other polymer types such as poly(lactic acid) (PLA). Regulations aimed at limiting greenhouse gas emissions are likely to redirect R&D focus toward the use of bio-based monomers as building blocks for copolymers. Bio-based monomers including acrylated epoxidized soybean oil (AESO), lignin-derived monomers, and poly(propylene carbonate) (PPC) hold potential to replace conventional monomers in BCP synthesis. While research on the application of bio-based block polymers for epoxy toughening remains insufficient, studies on petroleum-based epoxies have confirmed the effectiveness of BCPs as toughening modifiers.

Although research on bio-based nanofillers is still an evolving field with many aspects of bio-based nanocomposites yet to be explored, the incorporation of nanofillers has demonstrated promising results in enhancing fracture toughness without compromising other mechanical properties. If commercialization barriers can be addressed, bio-based nanofillers are well-positioned to serve as effective modifiers for fabricating high-toughness epoxy composites.

In summary, the use of vegetable oil derivatives currently represents the most feasible approach, given the existing market conditions. However, due to their inherent negative impact on key mechanical properties such as strength, modulus, and glass transition temperature, alternative strategies involving bio-based HBPs, block polymers, and nanofillers are anticipated to become the future trend in epoxy toughening. Bio-based materials face numerous challenges in matching the physical, mechanical, and economic performance of their petroleum-based counterparts. Additionally, the cost-competitive scale-up of tough bio-based materials and large-scale production of renewable feedstock-derived materials remain ongoing hurdles. Despite significant advancements in related research, further efforts are needed to overcome these obstacles and advance the transition toward a sustainable thermoset economy.

References

1 Gu, H.B., Ma, C., Gu, J.W. et al. (2016). An overview of multifunctional epoxy nanocomposites. *J. Mater. Chem. C* 4: 5890–5906.

2 Vidil, T., Tournilhac, F., Musso, S. et al. (2016). Control of reactions and network structures of epoxy thermosets. *Prog. Polym. Sci.* 62: 126–179.

3 Xiang, Q. and Xiao, F.P. (2020). Applications of epoxy materials in pavement engineering. *Constr. Build. Mater.* 235: 117529.

4 Chen, Z.R., Liu, X.X., Chen, H.N. et al. (2024). Application of epoxy resin in cultural relics protection. *Chin. Chem. Lett.* 35: 109194.

5 Huo, S.Q., Zhou, Z.X., Jiang, J.W. et al. (2022). Flame-retardant, transparent, mechanically-strong and tough epoxy resin enabled by high-efficiency multifunctional boron-based polyphosphonamide. *Chem. Eng. J.* 427: 131578.

6 Mashouf Roudsari, G., Mohanty, A.K., and Misra, M. (2017). Green approaches to engineer tough biobased epoxies: a review. *ACS Sustain. Chem. Eng.* 5: 9528–9541.

7 Johnsen, B.B., Kinloch, A.J., Mohammed, R.D. et al. (2007). Toughening mechanisms of nanoparticle-modified epoxy polymers. *Polymer* 48: 530–541.

8 Wetzel, B., Rosso, P., Haupert, F., and Friedrich, K. (2006). Epoxy nanocomposites – fracture and toughening mechanisms. *Eng. Fract. Mech.* 73: 2375–2398.

9 Domun, N., Hadavinia, H., Zhang, T. et al. (2015). Improving the fracture toughness and the strength of epoxy using nanomaterials – a review of the current status. *Nanoscale* 7: 10294–10329.

10 Frounchi, M., Mehrabzadeh, M., and Parvary, M. (2000). Toughening epoxy resins with solid acrylonitrile-butadiene rubber. *Polym. Int.* 49: 163–169.

11 Celikbilek, C., Akovali, G., and Kaynak, C. (2004). Modification of epoxy by a liquid elastomer and solid rubber particles. *Polym. Bull.* 51: 429–435.

12 Chen, T.K. and Jan, Y.H. (1992). Fracture mechanism of toughened epoxy resin with bimodal rubber-particle size distribution. *J. Mater. Sci.* 27: 111–121.

13 Mimura, K., Ito, H., and Fujioka, H. (2001). Toughening of epoxy resin modified with in situ polymerized thermoplastic polymers. *Polymer* 42: 9223–9233.

14 Di Pasquale, G., Motto, O., Rocca, A. et al. (1997). New high-performance thermoplastic toughened epoxy thermosets. *Polymer* 38: 4345–4348.

15 Kang, S., Kim, D., Kim, T. et al. (2017). Toughening epoxy resins with core–shell (HTPB-PMMA/PMA) rubber particles. *J. Nanosci. Nanotechnol.* 17: 7429–7435.

16 Shi, W., Li, J., Wang, C.Y., and Ren, Q. (2025). Fabrication of polyurethane/polythiourethane core-shell microspheres via thiol-isocyanate click chemistry and their application in toughening epoxy resin. *Polym. Bull.* 82: 197–220.

17 Vita, F., Lucchetta, D.E., Castagna, R. et al. (2009). Effects of resin addition on holographic polymer dispersed liquid crystals. *J. Opt. A Pure Appl. Op.* 11: 024021.

18 Shen, W.B., Wang, L., Chen, G. et al. (2019). A facile route towards controllable electric-optical performance of polymer-dispersed liquid crystal via the implantation of liquid crystalline epoxy network in conventional resin. *Polymer* 167: 67–77.

19 Barcia, F.L., Amaral, T.P., and Soares, B.G. (2003). Synthesis and properties of epoxy resin modified with epoxy-terminated liquid polybutadiene. *Polymer* 44: 5811–5819.

20 Grubbs, R., Kübel, J., Broz, M., and Bates, F. (2000). Reactive block copolymers for modification of thermosetting epoxy. *Macromolecules* 33: 9522–9534.

21 Yao, J.W., Chang, H.C., Zhang, T., and Niu, Y.F. (2022). Synergistic toughening in the interleaved carbon fibre reinforced epoxy composites by thermoplastic resin and nanomaterials. *Polym. Test.* 115: 107769.

22 Kordi, A., Adibnazari, S., Imam, A., and Najafi, M. (2020). Effects of two- and three-dimensional graphene-based nanomaterials on the fatigue behavior of epoxy nanocomposites. *Mater. Today Commun.* 24: 101194.

23 Ratna, D. and Simon, G.P. (2002). Dendritic hyperbranched polymers and epoxy blends. *J. Polym. Mater.* 19: 349–355.

24 Boogh, L., Pettersson, B., and Månson, J.A.E. (1999). Dendritic hyperbranched polymers as tougheners for epoxy resins. *Polymer* 40: 2249–2261.

25 Zhang, J.H., Mi, X.Q., Chen, S.Y. et al. (2020). A bio-based hyperbranched flame retardant for epoxy resins. *Chem. Eng. J.* 381: 122719.

26 Memon, H., Liu, H.Y., Rashid, M.A. et al. (2020). Vanillin-based epoxy vitrimer with high performance and closed-loop recyclability. *Macromolecules* 53: 621–630.

27 Zeng, J., Schlup, J.R., and Fan, L.T. (2000). Synthesis and mechanical properties of leather-epoxy interpenetrating polymer networks. *J. Appl. Polym. Sci.* 78: 1224–1232.

28 Jia, L.Y., Zhang, C., Du, Z.J. et al. (2007). Preparation of interpenetrating polymer networks of epoxy/polydimethylsiloxane in a common solvent of the precursors. *Polym. J.* 39: 593–597.

29 Cascaval, C., Rosu, D., and Rosu, L. (2005). Advanced materials based on epoxy resins: 2. Interpenetrating polymer networks. *Mater. Plast.* 42: 41–46.

30 Jin, H., Zhang, Y.G., Wang, C.S. et al. (2014). Thermal, mechanical, and morphological properties of soybean oil-based polyurethane/epoxy resin interpenetrating polymer networks (IPNs). *J. Therm. Anal. Calorim.* 117: 773–781.

31 Morsch, S., Kefallinou, Z., Liu, Y.W. et al. (2018). Controlling the nanostructure of epoxy resins: reaction selectivity and stoichiometry. *Polymer* 143: 10–18.

32 Rodin, Y.P. and Molchanov, Y.M. (1988). Effect of conformational changes caused by a homogeneous static magnetic field on curing of epoxy resin. *Mech. Compos. Mater.* 24: 380–384.

33 Zhao, C.B., Huang, G.H., Zhang, H. et al. (2024). High-homogeneous recyclable self-cured epoxy resins based on imine. *Chem. Eng. J.* 501: 157047.

34 Foix, D., Yu, Y.F., Serra, A. et al. (2009). Study on the chemical modification of epoxy/anhydride thermosets using a hydroxyl terminated hyperbranched polymer. *Eur. Polym. J.* 45: 1454–1466.

35 Lagunas, C., Fernández-Francos, X., Ferrando, F. et al. (2014). New epoxy thermosets modified with amphiphilic multiarm star polymers as toughness enhancer. *React. Funct. Polym.* 83: 132–143.

36 Flores, M., Tomuta, A.M., Fernández-Francos, X. et al. (2013). A new two-stage curing system: thiol-ene/epoxy homopolymerization using an allyl terminated hyperbranched polyester as reactive modifier. *Polymer* 54: 5473–5481.

37 Foix, D., Serra, A., Amparore, L., and Sangermano, M. (2012). Impact resistance enhancement by adding epoxy ended hyperbranched polyester to DGEBA photocured thermosets. *Polymer* 53: 3084–3088.

38 Sun, Y.Y., Zhang, Z.Q., Moon, K.S., and Wong, C.P. (2004). Glass transition and relaxation behavior of epoxy nanocomposites. *J. Polym. Sci. Part B Polym. Phys.* 42: 3849–3858.

39 Zhang, J.H., Kong, Q.H., and Wang, D.Y. (2018). Simultaneously improving the fire safety and mechanical properties of epoxy resin with Fe-CNTs via large-scale preparation. *J. Mater. Chem. A* 6: 6376–6386.

40 Wan, Y.J., Tang, L.C., Gong, L.X. et al. (2014). Grafting of epoxy chains onto graphene oxide for epoxy composites with improved mechanical and thermal properties. *Carbon* 69: 467–480.

41 Naebe, M., Wang, J., Amini, A. et al. (2014). Mechanical property and structure of covalent functionalised graphene/epoxy nanocomposites. *Sci. Rep.* 4: 4375.

42 Qi, B., Zhang, Q.X., Bannister, M., and Mai, Y.W. (2006). Investigation of the mechanical properties of DGEBA-based epoxy resin with nanoclay additives. *Compos. Struct.* 75: 514–519.

43 Khanbabaei, G., Aalaie, J., Rahmatpour, A. et al. (2007). Preparation and properties of epoxy-clay nanocomposites. *J. Macromol. Sci. B* 46: 975–986.

44 Wang, F.Z., Drzal, L.T., Qin, Y., and Huang, Z.X. (2015). Mechanical properties and thermal conductivity of graphene nanoplatelet/epoxy composites. *J. Mater. Sci.* 50: 1082–1093.

45 Gojny, F.H., Wichmann, M.H.G., Köpke, U. et al. (2004). Carbon nanotube-reinforced epoxy-composites: enhanced stiffness and fracture toughness at low nanotube content. *Compos. Sci. Technol.* 64: 2363–2371.

46 Gojny, F.H., Wichmann, M.H.G., Fiedler, B. et al. (2006). Evaluation and identification of electrical and thermal conduction mechanisms in carbon nanotube/epoxy composites. *Polymer* 47: 2036–2045.

47 Zhou, K.Q., Liu, J.J., Shi, Y.Q. et al. (2015). MoS_2 nanolayers grown on carbon nanotubes: an advanced reinforcement for epoxy composites. *ACS Appl. Mater. Interfaces* 7: 6070–6081.

48 Zheng, Y.P., Zheng, Y., and Ning, R.C. (2003). Effects of nanoparticles SiO_2 on the performance of nanocomposites. *Mater. Lett.* 57: 2940–2944.

49 Merad, L., Benyoucef, B., Abadie, M.J.M., and Charles, J.P. (2014). Characterization and mechanical properties of epoxy resin reinforced with TiO_2 nanoparticles. *Exp. Tech.* 38: 59–66.

50 Yang, J.Y., Wang, H.X., Liu, X.H. et al. (2021). A nano-TiO_2/regenerated cellulose biohybrid enables simultaneously improved strength and toughness of solid epoxy resins. *Compos. Sci. Technol.* 212: 108884.

51 Guo, G.C., Wu, J.L., Zhang, J.W. et al. (2024). Functionalization of nano-objects in living anionic polymerization-induced self-assembly and their use for improving thermal properties of epoxy resins. *Polym. Chem.* 15: 4637–4649.

52 Prolongo, S.G., Campo, M., Gude, M.R. et al. (2009). Thermo-physical characterisation of epoxy resin reinforced by amino-functionalized carbon nanofibers. *Compos. Sci. Technol.* 69: 349–357.

53 Zheng, D.C., Tan, Q.Y., Zhu, G.Y. et al. (2025). Dense network formation in epoxy composites via hyperbranched polyester-mediated crosslinking: a route to enhanced toughness and corrosion resistance. *Prog. Org. Coat.* 204: 109270.

54 Wang, T.T., Yuan, L., Liang, G.Z., and Gu, A.J. (2022). Constructing nano SiO_2-anchored hyperbranched polymer hybrid architectures for thermosetting epoxy systems with outstanding mechanical property. *Polym. Compos.* 43: 5375–5389.

55 Ma, L.C., Song, G.J., Zhang, X.C. et al. (2021). Attaching SiO_2 nanoparticles to GO sheets via amino-terminated hyperbranched polymer for epoxy composites: extraordinary improvement in thermal and mechanical properties. *Eur. Polym. J.* 157: 110677.

56 Baruah, P. and Karak, N. (2016). Bio-based tough hyperbranched epoxy/graphene oxide nanocomposite with enhanced biodegradability attribute. *Polym. Degrad. Stab.* 129: 26–33.

57 Li, J.J., Zhu, W.J., Zhang, S.F. et al. (2019). Amine-terminated hyperbranched polyamide covalent functionalized graphene oxide-reinforced epoxy nanocomposites with enhanced toughness and mechanical properties. *Polym. Test.* 76: 232–244.

58 Liang, X., Li, X.J., Tang, Y. et al. (2022). Hyperbranched epoxy resin-grafted graphene oxide for efficient and all-purpose epoxy resin modification. *J. Colloid Interface Sci.* 611: 105–117.

59 Mi, X.Q., Wei, F., Zeng, L. et al. (2019). Incorporation of hyperbranched polyamide-functionalized graphene oxide into epoxy for improving interfacial and mechanical properties. *Polym. Int.* 68: 1492–1501.

60 Si, J.J., Shao, X.Y., Wei, W.Y. et al. (2024). Epoxy-ended hyperbranched polymer-grafted graphene oxide for cold-mixed epoxy asphalt modification. *Int. J. Pavement Eng.* 25: 2361344.

61 Li, H.F., Liu, C., Zhu, J.B. et al. (2024). Intrinsically reactive hyperbranched interface governs graphene oxide dispersion and crosslinking in epoxy for enhanced flame retardancy. *J. Colloid Interface Sci.* 672: 465–476.

62 Feng, Y.Z., He, C.G., Wen, Y.F. et al. (2018). Multi-functional interface tailoring for enhancing thermal conductivity, flame retardancy and dynamic mechanical property of epoxy/Al_2O_3 composites. *Compos. Sci. Technol.* 160: 42–49.

63 Chen, S.F., Chen, L., Wang, Y.M. et al. (2019). Load transfer of thiol-ended hyperbranched polymers to improve simultaneously strength and longation of CNTs/epoxy nanocomposites. *Eur. Polym. J.* 120: 109254.

64 Zhong, L.Y., Li, T.C., Zhang, J.H. et al. (2022). Simultaneously improving the flame retardancy and toughness of epoxy composites with hyperbranched phosphorus-containing polysiloxane functionalized halloysite nanotubes. *Eur. Polym. J.* 179: 111564.

65 Xiong, L., Lai, W.B., Pan, Z.Q., and Liu, F. (2022). A facile method for grafting hyperbranched poly(thiol ether-ester) onto MWCNTs via thiol-yne click chemistry and their dispersion and oxidation responsive properties. *Mater. Lett.* 311: 131579.

66 Li, S.P., Li, Y.B., Zhu, H.J. et al. (2018). Evaluation of glass-fiber grafted by epoxide-terminated hyperbranched polymer on the effect of mechanical characterization of epoxy composites. *Sci. Eng. Compos. Mater.* 25: 417–424.

67 Gryshchuk, O., Jost, N., and Karger-Kocsis, J. (2002). Toughening of vinylester–urethane hybrid resins by functional liquid nitrile rubbers and hyperbranched polymers. *Polymer* 43: 4763–4768.

68 Fan, P. and Lu, C.H. (2012). Grafting of hyperbranched poly (amidoamine) onto waste tire rubber powder and its potential application as the curing agent for epoxy resin. *Polym. Adv. Technol.* 23: 48–56.

69 Li, X.Y., Li, X., Guo, X. et al. (2024). Reactive hyperbranched core-shell architecture for developing high-performance bio-based adhesive. *Sustain. Mater. Technol.* 42: e01184.

70 Sajjad, M., Zhao, Z.F., Zhu, X.L. et al. (2021). Transition molecular structures between block copolymers and hyperbranched copolymers suitable for toughening epoxy thermosets. *Compos. Commun.* 25: 100762.

71 Saba, N., Jawaid, M., Alothman, O.Y. et al. (2016). Recent advances in epoxy resin, natural fiber-reinforced epoxy composites and their applications. *J. Reinf. Plast. Compos.* 35: 447–470.

72 Xiong, K., Xiong, L., Liu, F. et al. (2024). Simultaneous grafting of carbon nanotube and hyperbranched poly(thioether-yne) onto carbon fiber via thiol-alkyne click chemistry to improve interface properties of carbon fiber/epoxy composite. *Polym. Compos.* 45: 2646–2655.

73 Xiong, K., Liu, F., Chen, T., and Xiong, L. (2024). Facile and efficient grafting of graphene oxide onto carbon fiber with hyperbranched poly(thioether-yne) via click chemistry to improve interface and mechanical properties of composite. *Surf. Interfaces* 46: 104038.

74 Luo, Y.X., Shi, Z.C., Qiao, S.J. et al. (2025). Advances in nanomaterials as exceptional fillers to reinforce carbon fiber-reinforced polymers composites and their emerging applications. *Polym. Compos.* 46: 54–80.

75 Shi, L., Ma, L., Li, P. et al. (2019). The effect of self-synthesized hydroxyl-terminated hyperbranched polymer interface layer on the properties of carbon fiber reinforced epoxy composites. *Appl. Surf. Sci.* 479: 334–343.

76 Ma, L.C., Zhu, Y.Y., Wang, M.Z. et al. (2019). Enhancing interfacial strength of epoxy resin composites via evolving hyperbranched amino-terminated POSS on carbon fiber surface. *Compos. Sci. Technol.* 170: 148–156.

77 Zhang, D.H. and Jia, D.M. (2006). Synthesis of novel low-viscosity liquid epoxidized aromatic hyperbranched polymers. *Eur. Polym. J.* 42: 711–714.

78 Zhang, D.H. and Jia, D.M. (2006). Toughness and strength improvement of diglycidyl ether of bisphenol-A by low viscosity liquid hyperbranched epoxy resin. *J. Appl. Polym. Sci.* 101: 2504–2511.

79 Zhang, D.H. and Jia, D.M. (2006). Study on the performance of diglycidyl ether of bisphenol-a/hyperbranched aromatic polyester epoxy resin (HTME) system and their toughness mechanism. *Polym.-Plast. Technol. Eng.* 45: 1005–1011.

80 Zhang, D.H., Chen, Y.K., and Jia, D.M. (2009). Toughness and reinforcement of diglycidyl ether of bisphenol-A by hyperbranched poly(trimellitic anhydride-butanediol glycol) ester epoxy resin. *Polym. Compos.* 30: 918–925.

81 Zhang, D.H., Jia, D.M., and Chen, S.F. (2009). Synthesis and characterization of low viscosity aromatic hyperbranched poly(trimellitic anhydride ethylene glycol) ester epoxy resin. *Macromol. Chem. Phys.* 210: 1159–1166.

82 Zhang, D.H., Jia, D.M., and Zhou, Z.H. (2009). Synthesis and characterization of low viscosity aromatic hyperbranched polyester epoxy resin. *Macromol. Res.* 17: 289–295.

83 Wang, L.P., Liang, Y.Y., Yu, Q. et al. (2018). Synthesis of epoxy-ended hyperbranched polyesters with reinforcing and toughening function for diglycidyl ether of bisphenol-A. *Polym. Compos.* 39: E2046–E2055.
84 Chen, S.F., Zhang, D.H., Jiang, S.B., and Jia, D.M. (2012). Preparation of hyperbranched epoxy resin containing nitrogen heterocycle and its toughened and reinforced composites. *J. Appl. Polym. Sci.* 123: 3261–3269.
85 Zhang, D.H., Cheng, X.J., Zhang, A.Q., and Jia, D.M. (2011). Toughness and reinforcement of linear unsaturated polyester resins by unsaturated hyperbranched polymer and mechanism analysis. *Funct. Mater. Lett.* 4: 351–355.
86 Zhang, D.H., Wu, H.Y., Li, T.C. et al. (2011). Preparation of high-performance flame-retardant hybrid material by hyperbranched polyphosphate ester. *Polym. Compos.* 32: 36–43.
87 Yu, Q., Liang, Y.Y., Cheng, J. et al. (2017). Synthesis of a degradable high-performance epoxy-ended hyperbranched polyester. *ACS Omega* 2: 1350–1359.
88 Zhang, J.H., Liu, C.H., Cheng, J. et al. (2018). Simultaneous toughening and strengthening of diglycidyl ether of bisphenol-a using epoxy-ended hyperbranched polymers obtained from thiol-ene click reaction. *Polym. Eng. Sci.* 58: 1703–1709.
89 Wang, Y.M., Chen, S.F., Chen, X.C. et al. (2019). Controllability of epoxy equivalent weight and performance of hyperbranched epoxy resins. *Compos. Part B Eng.* 160: 615–625.
90 Wang, Y., Chen, S., Guo, W. et al. (2019). The precise effect of degree of branching of epoxy-ended hyperbranched polymers on intrinsic property and performance. *Prog. Org. Coat.* 127: 157–167.
91 Wu, J., Liang, N., Liu, X. et al. (2022). Controllability on topological structures and properties of hyperbranched epoxy resins. *Prog. Org. Coat.* 165: 106735.
92 Yu, C.L., Xu, Z.J., Ma, X. et al. (2021). Hyperbranched polymers containing epoxy and imide structure. *Prog. Org. Coat.* 151: 106031.
93 Liu, X., Wu, H.H., Xu, W. et al. (2024). Ultrastrong and high-tough thermoset epoxy resins from hyperbranched topological structure and subnanoscaled free volume. *Adv. Mater.* 36: e2308434.
94 Zhang, J.H., Jiang, C., Deng, G.Y. et al. (2024). Closed-loop recycling of tough epoxy supramolecular thermosets constructed with hyperbranched topological structure. *Nat. Commun.* 15: 4869.
95 Hao, Y.X., Zhong, L.Y., Li, T.C. et al. (2023). High-performance recyclable and malleable epoxy resin with vanillin-based hyperbranched epoxy resin containing dual dynamic bonds. *ACS Sustain. Chem. Eng.* 11: 11077–11087.
96 Han, X., Wu, Y., Han, H.R. et al. (2023). Super-strong reusable hot melt adhesives prepared from hyperbranched epoxy resin. *Mater. Chem. Front.* 7: 2676–2682.
97 Lu, Y.F., Xu, H.F., Liang, N. et al. (2022). High mechanical strength of shape-memory hyperbranched epoxy resins. *ACS Appl. Polym. Mater.* 4: 5574–5582.

98 Ma, X., Guo, W.Q., Xu, Z.J. et al. (2020). Synthesis of degradable hyperbranched epoxy resins with high tensile, elongation, modulus and low-temperature resistance. *Compos. Part B Eng.* 192: 108005.

99 Ma, X., Liang, Y.Y., Xu, Z.J. et al. (2020). The versatility of hyperbranched epoxy resins containing hexahydro-*s*-triazine on diglycidyl ether of bisphenol-A composites. *Compos. Part B Eng.* 196: 108109.

100 Wang, C., Zhang, Y.M., Li, J.M. et al. (2020). Shape memory properties of interpenetrating polymer networks (IPNs) based on hyperbranched polyurethane (HBPU). *Eur. Polym. J.* 123: 109393.

101 Fu, J.F., Shi, L.Y., Yuan, S. et al. (2008). Morphology, toughness mechanism, and thermal properties of hyperbranched epoxy modified diglycidyl ether of bisphenol A (DGEBA) interpenetrating polymer networks. *Polym. Adv. Technol.* 19: 1597–1607.

102 Wei, F., Zhang, J.H., Wu, C. et al. (2023). Closed-loop recycling of tough and flame-retardant epoxy resins. *Macromolecules* 56: 5290–5305.

103 Woodcock, J.W., Beams, R., Davis, C.S. et al. (2017). Observation of interfacial damage in a silk-epoxy composite, using a simple mechanoresponsive fluorescent probe. *Adv. Mater. Interfaces* 4: 1601018.

6

Degradation and Recycling of Epoxy Resins and Their Composites

Minghui Cui, Mengqiu Quan, Genzheng Sha, and Jing Chen

Ningbo Institute of Materials Technology and Engineering, Chinese Academy of Sciences, Key Laboratory of Biobased Polymeric Materials of Zhejiang Province, 1219 Zhongguan West Road, Zhejiang 315201, China

6.1 Introduction

As an important thermosetting polymer, epoxy resin is widely used in coatings, adhesives, electronic packaging materials, and the aerospace and automotive industries due to its excellent mechanical properties, chemical stability, and adhesive properties. However, with the increasing use of epoxy resins and their composites, the disposal of waste epoxy resins has become a growing concern. Traditional treatment methods, such as landfill and incineration, cause not only waste resources but also secondary pollution to the environment. Therefore, research on the degradation and recycling technology of epoxy resin to realize its sustainable use has become an important topic in materials science and environmental protection.

In recent years, with the increasing awareness of environmental issues and the growing demand for resource recycling, the degradation and recycling technology of epoxy resin has made significant progress. Early research mainly focused on physical degradation methods, such as mechanical crushing and cutting, but these methods have low recycling efficiency and a large impact on material properties. With the progress of technology, chemical degradation, biodegradation and other more efficient and environmentally friendly recycling methods have gradually become a research hotspot. Especially, the emergence of hyperbranched epoxy resins (HERs), due to their unique molecular structure and excellent performance, provides new ideas for the degradation and recycling of epoxy resins.

HER is also widely used in coatings, adhesives, electronic packaging materials, and other fields because of its unique molecular structure and excellent properties. However, with the increase in its use, the disposal problem of WEP and its composite materials is becoming increasingly prominent. Traditional disposal methods, such as landfill and incineration, cause not only waste resources but also secondary pollution to the environment. Therefore, it is of great significance to study the degradation and recovery technology of HER for realizing the sustainable development of materials and environmental protection.

The research on HER's degradation and recovery technology started late, but remarkable progress has been made in recent years [1]. Early research mainly focused on physical degradation methods, such as adding WEP as a filler to other polymers through mechanical crushing, cutting, etc., to achieve simple recycling. However, this method has low recycling efficiency and greatly impacts material properties. With the increasing awareness of environmental protection and resource recycling, researchers have begun to explore more efficient and environmentally friendly methods, such as chemical degradation and biodegradation.

Bio-based HER has become a research hotspot to replace traditional petroleum-based resin because of its environment-friendly and renewable characteristics, and bio-based to promote degradation performance. Zhang and coworkers [2] systematically summarized the preparation method and recycling strategy of bio-based HER. For example, a new biodegradable and recyclable thermosetting hyperbranched epoxy resin (EFTH-*n*) synthesized from bio-based 2,5-furandicarboxylic acid (FDCA) not only shows excellent performance but also improves impact strength, tensile strength, bending strength, energy storage modulus, and elongation, providing a new idea for the realization of green preparation and recycling of epoxy resin. The introduction of bio-based hyperbranched structure can not only balance the strength and toughness of the material but also achieve more efficient recycling through its biodegradable chemical structure. For example, by introducing biodegradable covalent bonds, epoxy resins can be decomposed into small molecular monomers or oligomers under mild conditions [3–6]. Bio-based HER is not only conducive to improving the thermal and mechanical properties of epoxy resin but also helps to realize its green preparation and recycling and achieve the goal of "carbon peak and carbon neutrality."

At the same time, HER can effectively strengthen and toughen epoxy resin due to its good compatibility, low viscosity, and other advantages and is widely used in coatings, adhesives, semiconductor packaging materials, high-performance composite materials, and other fields. The degradation mechanism of HER composites was investigated through relevant experiments and analytical tests. The results showed that a hyperbranched structure could significantly improve the degradation performance of epoxy resin, providing a theoretical basis for enhancing the degradation performance of composite materials and the possibility of recycling.

6.2 Degradation and Recycling of Common Epoxy Resins

6.2.1 Method of Degradation

6.2.1.1 Thermal Degradation

Thermal degradation is one of the most common forms of degradation for epoxy resins, typically occurring when the resin is exposed to high temperatures. As a thermosetting polymer, epoxy resins have a highly cross-linked structure, which makes them more resistant to thermal degradation compared to thermoplastic plastics. However, at sufficiently high temperatures, these cross-linked structures may break down, leading to the degradation of the resin and the release of smaller molecules [7–9].

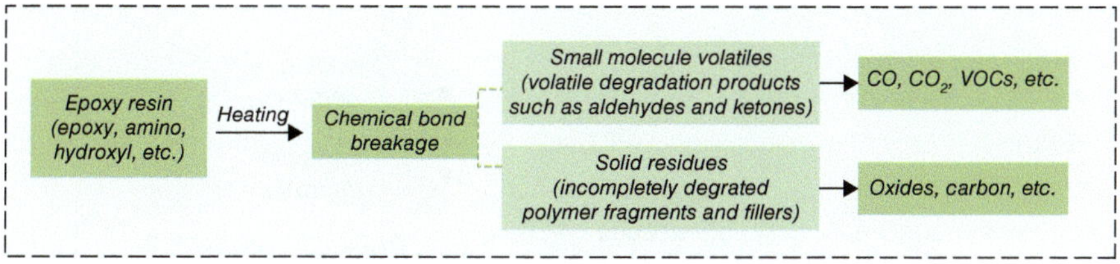

Figure 6.1 Thermal degradation of conventional epoxy resins.

The thermal degradation process begins with the breaking of weaker bonds in the polymer backbone, such as the ether and ester bonds in the structure of epoxy resins. The initial bond cleavage generates free radicals, which further facilitate the degradation process [9, 10]. As the temperature rises, the chemical structure of the resin starts to break down, ultimately releasing volatile degradation products [11–13].

Thermal degradation of epoxy resins is typically divided into two stages (Figure 6.1): bond cleavage and decomposition into smaller molecules and volatiles. The degradation rate increases with temperature, and the resin begins to degrade when it exceeds its T_g, which varies depending on the specific type of epoxy resin and curing agent used. The high thermal stability of epoxy resins is one of the reasons they are favored for high-temperature applications. Still, prolonged exposure to elevated temperatures can damage their mechanical properties and performance [9, 14].

In the study by Zhou et al. on the thermal degradation and decomposition mechanisms of carbon/epoxy composites under simulated lightning strike conditions, it was found that the thermal damage of the material can be categorized into three types: fiber bundle fracture, resin sublimation, and discoloration. The former is caused by the shockwave, while the latter two are mainly influenced by Joule heating. The degradation of epoxy resin begins with a dehydration reaction, accompanied by color changes and the generation of carbonyl compounds, and the release of volatile substances such as 2-propenal, acetylene, and aromatic ethers. Thermogravimetric analysis (TGA) indicates that the material's thermal decomposition follows a single-step reaction mechanism, with an activation energy of approximately 61.2–64.4 kJ/mol [10].

6.2.1.2 Chemical Degradation

Chemical degradation refers to the process in which epoxy resins decompose upon contact with reactive chemicals. The chemical environment plays a crucial role in this process, as certain chemicals can attack the polymer backbone or the cross-linking sites within the resin matrix, leading to material breakdown and a decrease in strength. The most common forms of chemical degradation in epoxy resins include hydrolysis, solvent interactions, and acid–base reactions (Figure 6.2).

Chemical degradation often occurs through solvent interactions, where solvents such as acetone, methanol, and ethanol penetrate the epoxy resin matrix, causing swelling, softening, or even dissolution of the material. These solvents can disrupt the cross-linked network structure, leading to degradation of the resin and a consequent loss of mechanical strength and stability [15–24].

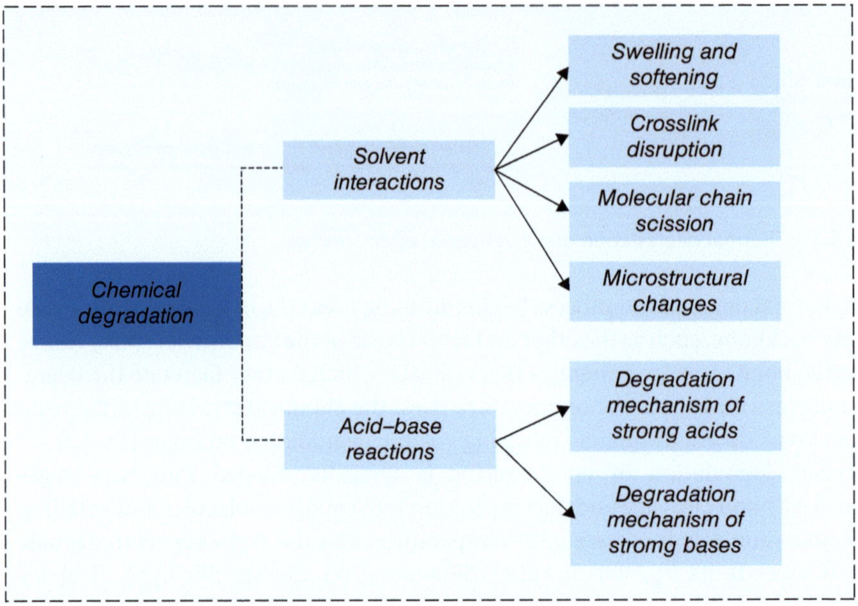

Figure 6.2 Chemical degradation of conventional epoxy resins.

Epoxy resins can degrade when exposed to certain solvents, as these solvents penetrate the resin network and cause swelling and softening. Such interactions may disrupt the cross-linked structure, break molecular chains, and alter the microstructure, ultimately reducing the material's mechanical strength and long-term stability [11, 25]. When epoxy resins are exposed to solvents, these interactions can further accelerate chemical degradation. Solvent molecules penetrate the polymer network, causing the resin to swell and soften while weakening the intermolecular forces that maintain structural integrity. As cross-links are disrupted and molecular chains become more mobile, the resin's microstructure gradually changes, leading to reduced hardness, rigidity, and overall mechanical stability [16, 26]. As swelling progresses, prolonged solvent exposure can also disrupt the cross-linked structure that underpins the stability of epoxy resins. Polar solvents such as acetone and methanol may form hydrogen bonds or solvation interactions with the cross-linking sites, weakening the network and making the material softer or even prone to disintegration. In stronger solvents like *N,N*-dimethylformamide (DMF) or dimethyl sulfoxide (DMSO), parts of the cross-linked structure can dissolve, leading to a notable loss in mechanical strength. As a result, the resin gradually loses its rigidity and exhibits a reduced glass transition temperature (T_g) [15].

With extended solvent exposure, the degradation of epoxy resins can progress to molecular chain scission. Certain solvents promote the cleavage of ether (—O—), ester (—COO—), or amine (—C—N—) bonds along the resin backbone, thereby reducing the molecular weight of the polymer. Such bond breakage may occur through solvent-induced hydrolysis or free radical reactions. As the polymer chains become shorter and less entangled, the resin's mechanical strength and toughness decline, often accompanied by the formation of low-molecular-weight degradation products [15, 27].

Solvent interactions can also induce microstructural changes in epoxy resins, such as the formation of microscopic pores that accelerate further degradation. When the solvent evaporates, surface cracks and increased internal porosity may develop, leading to a reduction in mechanical integrity. Overall, solvent exposure can cause epoxy resin degradation through a combination of swelling and softening, cross-link disruption, molecular chain scission, and microstructural damage. The extent of these effects depends on factors such as solvent polarity, solubility parameter, temperature, and exposure time. Building on these mechanisms, various solvent-based degradation strategies—such as solvent dissolution, solvent-assisted thermal degradation, solvent-catalyzed degradation, and solvent–ultrasonic degradation—have been explored for recycling. Future studies should aim to enhance degradation efficiency, develop environmentally friendly solvent systems, and minimize ecological impact to support the sustainable utilization of epoxy materials [18, 26, 28]. Xu et al. [17] developed an H_2O_2/DMF mixed system to achieve the oxidative degradation of epoxy resin. In this system, ·OH radicals attack the cross-linked structure, and DMF dissolves the fragments, forming a homogeneous reaction that accelerates the decomposition. The recovered fibers have smooth surfaces, a slight reduction in graphite content, and the tensile strength is maintained at 95%. This system enables efficient depolymerization and preservation of the fiber structure under mild conditions, making it suitable for the green recycling of thermosetting composites. Oliveux et al. [18] proposed a new strategy for the degradation of epoxy resin based on solvent-induced techniques. They used CO_2-expanded water at 300 °C and 82 bar to simulate supercritical fluid properties, combined with an 80% acetone–water mixed solvent to form a supercritical state. Within 30 minutes, the resin degradation rate reached 93%. CO_2 and phenolic by-products exhibited a synergistic catalytic effect, and acetone facilitated the dissolution of the products in the liquid phase, preventing fiber contamination. This system can be extended to the recycling of carbon fiber composites, offering an energy-efficient solution for the green degradation of thermosetting resins.

Acid–base reactions can further accelerate the chemical degradation of epoxy resins by attacking and breaking key chemical bonds within the polymer network. Exposure to strong acids or bases can cleave cross-links formed during curing, particularly those involving ether (—O—), ester (—COO—), and epoxy (—C—O—C—) groups. This bond cleavage weakens the cross-linked structure, resulting in a significant reduction in the resin's mechanical strength, thermal stability, and overall performance [27, 29–32].

Under strong acidic conditions (e.g., sulfuric, hydrochloric, or nitric acid), epoxy resins undergo a series of degradation reactions that progressively weaken their structure. If the resin is not fully cured, residual epoxy groups may participate in acid-catalyzed nucleophilic ring-opening reactions, forming alcohols or phenols that make the material softer and more brittle. Strong acids can also cleave ether bonds (—O—) within the polymer backbone, fragmenting the macromolecules into low-molecular-weight species and severely reducing mechanical strength. In anhydride-cured systems, ester bonds (—COO—) are particularly vulnerable to acid-induced hydrolysis, disrupting the cross-linked network and accelerating structural breakdown. For amine-cured resins, protonation of secondary or tertiary amine groups under acidic conditions further destabilizes the cross-linked framework, leading to decreased cross-link density and a notable loss in rigidity and toughness.

Ma et al. [25] used a nitric acid solution to selectively cleave the C—N cross-linking bonds of amine-cured epoxy resin. The resin decomposed into nitrated products with monomer/dimer structures (yield >60%). Zhang et al. [31] used dipropylene glycol dimethyl ether as a swelling agent combined with nitric acid oxidation to degrade amine-cured epoxy resin. The swelling pretreatment, due to the large molecular volume and strong permeability of DMM, loosened the dense resin structure, significantly improving the degradation efficiency. The resin was fully degraded within 1 hour at 100 °C. Xing et al. [32] developed a new method for recovering carbon fiber-reinforced epoxy resin (CFRP) based on acetic acid swelling. The research showed that acetic acid not only significantly swelled the cured epoxy resin through hydrogen bonding at high temperatures (140–220 °C) but also catalyzed mild degradation of the resin. This method achieves efficient, green recycling of thermosetting composites through the synergistic effect of physical swelling and chemical degradation.

In basic environments, epoxy resins cured with anhydrides are prone to ester bond saponification, which disrupts the cross-linked network and causes the material to become brittle or dissolve. Unreacted epoxy groups may also undergo base-catalyzed ring-opening reactions to form diols, resulting in a softer structure and a noticeable decline in mechanical strength [33]. If epoxy groups are not fully cured, basic conditions can promote ring-opening reactions that form diols, softening the resin. In amine-cured epoxy resins, bases can deprotonate amine groups, altering the cross-linked structure and reducing the overall degree of cross-linking, which decreases the material's mechanical strength. Ma et al. [25, 29] introduced phenyl phosphonate groups as dynamic degradation sites into the main chain of epoxy resin, successfully designing a novel epoxy resin that can be chemically degraded under mild conditions. The degradation mechanism is based on the cleavage of phosphonate ester bonds in an alkaline environment: in a methanol/NaOH system, the phosphonate ester bond undergoes cleavage via hydrolysis and ester exchange reactions, disrupting the cross-linked network of the epoxy curing agent. The study shows that the dual synergistic effect of the solvent (catalyzing the reaction and dissolving the products) is the key to degradation—methanol enhances the solubility and reactivity of the products, and the alkaline condition (OH^-) directly attacks the phosphorus center to induce chain cleavage, with both factors working together to achieve complete degradation within 40 minutes. This controllable degradability can be precisely regulated by adjusting the solvent ratio (methanol/alkaline solution) and NaOH concentration, providing an efficient, low-energy strategy for solving the recycling challenges of thermosetting resins while maintaining high glass transition temperatures and mechanical properties, showing its potential in the circular economy.

Ma and Webster [33] used natural carboxylic acids as cross-linking agents, combined with bio-based epoxy sucrose soy esters, to prepare high-performance fully bio-based degradable thermosetting materials through a water-assisted catalytic mechanism. During the cross-linking reaction, water facilitates the uniform dispersion of acids and catalyzes the esterification between epoxy groups and carboxylic acids, forming a stable cross-linked network. The resulting ester bonds, however, serve as vulnerable sites for degradation. Under strong alkaline conditions, such as 1 M NaOH at 90 °C, these ester linkages hydrolyze rapidly, leading to complete dissolution of the material within 13–235 minutes. Thermal degradation can also occur when secondary ester bonds break at 200–250 °C, generating carboxylic acids and hydroxyl compounds. In contrast, weak bases like ammonia water or sodium carbonate have only minor effects on epoxy resins; although prolonged exposure may cause surface changes such as slight whitening or superficial degradation, the overall structural integrity remains largely intact.

6.2.1.3 Biodegradation

Biodegradation refers to the process in which high-molecular-weight materials are broken down into smaller molecules through the action of microorganisms, enzymes, or biocatalysts, ultimately converting them into carbon dioxide, water, and biomass. However, due to the poor degradability of traditional petroleum-based epoxy resins, current research primarily focuses on enhancing their biodegradability, developing degradable epoxy resins, and optimizing degradation conditions.

Biodegradation Mechanism of Epoxy Resins The biodegradation of epoxy resins primarily involves microbial degradation, enzyme-catalyzed degradation, and degradation of bio-based epoxy resins (Figure 6.3).

(1) Microbial Degradation: Microorganisms (such as bacteria and fungi) can secrete extracellular enzymes that catalyze the breakdown of epoxy resins, degrading them into low-molecular-weight compounds. The degradation products mainly include small molecules such as alcohols, ketones, and phenolic compounds, which are eventually metabolized by microorganisms into CO_2, H_2O, and biomass. Microorganisms first secrete extracellular enzymes that bind to

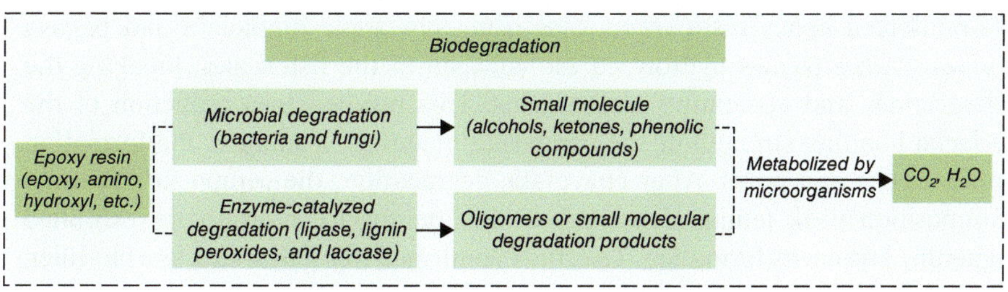

Figure 6.3 Biodegradation of common epoxy resins.

and penetrate the epoxy resin's surface. These enzymes then catalyze the cleavage of the cross-linked structure and the main chain, forming oligomers or monomers. These low-molecular-weight compounds then enter the microbial metabolic pathways and are ultimately converted into carbon dioxide, water, and biomass. Common fungi, such as white rot fungi and *Aspergillus*, possess strong lignin-degrading abilities and can break down the ether and amine bonds in epoxy resins. Common bacteria, such as *Pseudomonas* and *Bacillus*, secrete various hydrolytic enzymes that degrade the polymer structure [34–38]. De et al. [39] developed a biodegradable, high-toughness epoxy resin. The ester bonds in the resin provide water solubility, and testing with *Pseudomonas citronellolis* showed that after 60 days, the film's mass loss reached 30%. Microbial attack on the ester bonds leads to structural erosion, while traditional solid bisphenol A epoxy resins (SBEs) are nearly nondegradable due to the absence of ester bonds [33].

Pangallo et al. [36] investigated the microbial communities on the surface of epoxy resin statues exposed to outdoor conditions and their degradation mechanisms. Using both cultivation and noncultivation methods, they found that a variety of extremophilic microorganisms had colonized the resin surface, including bacteria such as *Bacillus* and *Staphylococcus*, which have lipase activity, as well as fungi like black yeast and *Aspergillus*. These microorganisms degrade the ester bonds in the resin by secreting lipases and esterases. Furthermore, photosynthetic microorganisms (such as green algae and *Cyanobacteria*) symbiotic with lichens exacerbated biofilm formation and physical–chemical erosion [30].

(2) Enzyme-Catalyzed Degradation: Specific enzymes (such as lipase, lignin peroxidase [LiP], and laccase) can catalyze the degradation of epoxy resins, breaking their cross-linked networks. These enzymes first bind to the epoxy resin matrix and catalyze oxidation or hydrolysis reactions. Then, the main chain breaks, forming oligomers or small molecular degradation products [34, 37]. These products are further biologically transformed and eventually degrade into CO_2 and H_2O. For example, laccase can catalyze the oxidative degradation of epoxy resins, LiP can break down phenyl ether bonds, causing the material to disintegrate, and lipase is effective at degrading ester bonds [40–43].

Faiza et al. [42] studied the enzymatic degradation mechanism of fish scale-reinforced epoxy resin composites using pineapple bromelain and papaya enzymes. Both enzymes hydrolyzed the collagen in the fish scales, breaking the peptide bonds and generating amino acids. This resulted in a reduction of the interfacial bonding strength of the composite materials, leading to the formation of microcracks and pores. After enzymatic degradation, the composite's thermal decomposition peak temperature shifted and the surface morphology exhibited roughening and cavity formation. The study confirmed that fish scales, as a bio-filler, not only regulate the mechanical properties of the material but also accelerate the environmental degradation of the polymer through enzymatic action.

Biodegradation Methods for Epoxy Resins

(1) Soil Degradation: In moist, high-temperature soil environments, microorganisms can slowly degrade epoxy resins. This method is suitable for environmental assessments of biodegradable epoxy resins.
(2) Composting Degradation: During the composting of organic waste, high temperatures and humidity facilitate the degradation of epoxy resins. This method is commonly used for testing the degradation of bio-based epoxy resins.
(3) Enzyme-catalyzed Degradation: Using biocatalysts such as laccase and lipase, the degradation of epoxy resins is accelerated. This method is suitable for the treatment of discarded epoxy materials.

Although progress has been made in the biodegradation of epoxy resins, several challenges remain. First, traditional epoxy resins have a low degradation rate, requiring modification or the introduction of bio-based components. Second, the efficiency of microbial and enzyme degradation is influenced by environmental conditions, so optimization of degradation conditions is necessary. Additionally, the ecological safety of degradation products remains to be studied, and the impact of residual materials after degradation on the environment needs to be assessed. Future research should focus on developing highly efficient degradation microbial and enzyme-catalyzed systems to enhance the biodegradability of epoxy resins, optimizing the structural design of bio-based epoxy resins to improve their degradability, and establishing industrial-scale biodegradation processes for epoxy resins to promote large-scale applications.

6.2.2 Recycling Method

6.2.2.1 Recycling of Epoxy Resins

Epoxy resins are commonly combined with fiber-reinforced materials (such as carbon fiber and glass fiber) to form composite materials, enhancing their structural stability and durability. However, it is this highly cross-linked chemical structure that makes epoxy resins difficult to remold or reshape through heating and melting, unlike thermoplastic plastics, posing challenges for recycling and reuse [44, 45].

Necessity of Epoxy Resin Recycling

(1) Environmental Protection Needs: With the increasing global plastic pollution problem, the management of waste epoxy resins and their composites has become a key issue in environmental protection. Epoxy resins, being non-degradable polymers, persist in the environment for long periods, potentially causing harm to ecosystems. For example, waste composite materials may accumulate in oceans and soils, leading to microplastic pollution. Traditional disposal methods, such as landfilling and incineration, not only consume large amounts of land but can also release toxic gases such as dioxins and chlorides, which threaten air quality and human health [12, 46, 47].
(2) Resource Recycling and Economic Value: Epoxy resins and their composite materials often contain high-value reinforcement fibers (such as carbon fiber and glass fiber) and other scarce resources, which can reduce dependence on

raw materials and lower production costs if effectively recycled. For instance, carbon fiber is a high-performance material with a production process that involves high-energy consumption and carbon emissions. Recycling carbon fiber from epoxy resin-based composites can reduce carbon emissions and lower the cost of new material production. Additionally, some of the recycled epoxy resins can be reused in low-cost industrial applications, such as coatings, fillers, or modified plastics [46, 48, 49].

(3) Regulatory and Policy Drivers: Governments and environmental agencies worldwide are increasingly strengthening regulations on waste recycling and resource reuse. For example, the European Union's "Circular Economy Action Plan" and China's "Dual Carbon" strategy have raised higher demands for the recycling of polymer materials. In industries such as automotive and aerospace, manufacturers are facing stricter environmental regulations, such as the European Union's "End-of-Life Vehicles (ELV) Directive" and the "Aircraft Environmental Management Regulation," pushing companies to adopt more sustainable materials management solutions. Therefore, epoxy resin recycling is not only necessary for companies to fulfill their social responsibility but also an inevitable trend in the future development of the industry [49].

Current Recycling Challenges and Opportunities Although epoxy resins hold great potential for recycling, their highly cross-linked network formed after curing makes degradation difficult, thus limiting the effectiveness of conventional mechanical recycling methods (e.g., crushing and grinding) in recovering high-quality raw materials. Meanwhile, existing chemical recycling approaches—such as pyrolysis, hydrothermal depolymerization, and catalytic cracking—still suffer from high costs, excessive energy consumption, and secondary pollution, hindering large-scale industrial adoption. Biodegradation technologies, on the other hand, remain at the experimental stage, and improving their degradation efficiency continues to be a key research challenge.

Recently, advances in materials science and engineering have led to the emergence of new recycling strategies. For instance, green solvent methods utilize environmentally benign solvents for selective degradation, enabling the separation of resins and reinforcing fibers; catalytic degradation technologies employ highly efficient catalysts to accelerate resin depolymerization and enhance resource recovery efficiency; and composite recycling strategies integrate multiple techniques to achieve more comprehensive material reuse. These technological innovations open new avenues for the sustainable recycling of epoxy resins and are expected to drive the development of a circular materials economy.

6.2.2.2 Recycling Methods for Epoxy Resins

Mechanical Recycling of Epoxy Resins Mechanical recycling refers to the process of using physical methods (such as crushing, grinding, and sieving) to process waste epoxy resins materials into reusable particles or powders, which can then be used to manufacture new products. This method does not alter the chemical structure of the epoxy resin and is primarily suited for lower-end recycling applications, such as

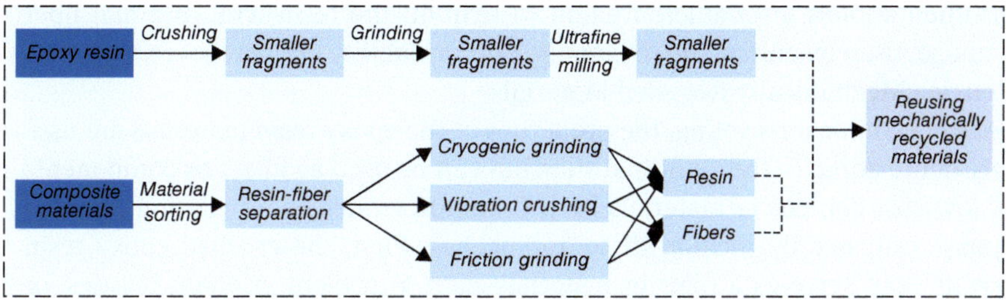

Figure 6.4 Mechanical recycling of common epoxy resins.

filler materials or modified composite materials. Since epoxy resins are thermosetting polymers, their molecular structure forms a three-dimensional cross-linked network after curing, making it impossible to remold or reshape them through heating like thermoplastics. Thus, mechanical recycling focuses on physical crushing, reuse as fillers, or modifying with other materials to enhance their performance (Figure 6.4) [50–53].

Main Methods of Mechanical Recycling

(1) Crushing and Grinding

Crushing and grindingare the fundamental steps in mechanical recycling, used to reduce WEP materials into smaller particles or powders for subsequent applications. These methods are applicable to larger-sized waste materials such as discarded glass fiber-reinforced epoxy resin (GFRP) and CFRP [50, 54].

Typical Crushing Methods

Coarse Crushing: Using shear, impact, or compression to break large epoxy resin materials into smaller fragments.

Fine Grinding: Further grinding of crushed epoxy resin into micron-sized particles to improve its performance as a filler in new materials.

Ultrafine Milling: Using high-energy ball milling or air milling to obtain finer epoxy resin powders for coatings, adhesives, or composite materials applications.

(2) Separation and Recycling of Composite Materials

Epoxy resin composite materials, such as GFRP and CFRP, are major targets for mechanical recycling [53, 54]. The recycling of composite materials generally involves three key stages. First, material sorting is carried out using spectral analysis or density-based separation to distinguish different types of composites, thereby improving the overall recycling efficiency. Next, resin–fiber separation, the core of mechanical recycling, focuses on recovering reinforcement fibers such as carbon or glass fibers from the epoxy resin matrix. Various physical methods have been developed for this purpose. For instance, cryogenic grinding cools waste composites to extremely low temperatures (often with liquid nitrogen), making them brittle and easier to crush, thus enabling efficient resin–fiber separation. Vibration crushing uses high-frequency oscillation to peel the interface between fibers and resin, improving separation efficiency, while friction grinding applies mechanical friction to strip off the resin with minimal fiber damage, thereby enhancing the quality and reusability of the recovered fibers.

(3) Reusing Mechanically Recycled Materials
After mechanical recycling, the properties of the epoxy resin materials are usually not restored to their original state but can be used as fillers or components of low-cost composite materials [54]. Common strategies for reusing recycled epoxy resin mainly include three approaches. First, the crushed epoxy resin powder can serve as a filler in materials such as cement, asphalt, plastics, or rubber to enhance mechanical strength and durability. When incorporated into construction materials like concrete, it can also improve compressive performance. Second, the recycled resin powder can be integrated into new resin systems to fabricate secondary composite materials, such as reinforced or engineering plastics, offering a cost-effective alternative for structural applications. The recovered carbon or glass fibers can likewise be reused in the production of lightweight components, including automotive parts and wind turbine blades. Finally, the recycled resin powder can be applied in the manufacture of low-end epoxy products, such as floor coatings, adhesives, and electrical insulation materials, where performance requirements are relatively low.

Kang et al. [50] used solid-state shear milling (S3M) technology to recycle WEP through mechanochemical effects. The S3M treatment pulverized the WEP into ultrafine powder of approximately 17 μm, which disrupted its cross-linking structure (the cross-linking density decreased from 3.43×10^{-3} to 2.69×10^{-3} mol/mL). During this process, C—N, C—C, and C—O—C bonds were broken, and a large number of hydroxyl and amino functional groups were generated on the powder's surface. These functional groups can react with the epoxy resin matrix and curing agents to form new cross-linked networks, significantly improving the interfacial bonding strength. The tensile strength, flexural strength, and impact strength of the regenerated composite (WEP/EP) increased by 62%, 41%, and 109%, respectively. This method is solvent-free, requires no high temperatures, is cost-effective, and efficient.

Wang et al. [51] also used S3M to recycle anhydride-cured epoxy resin. Through mechanochemical effects, the three-dimensional cross-linked network (C—C main chain and C—O bonds) of the epoxy resin was broken, producing reactive epoxy powder (REP) with active carbonyl (C=O) and hydroxyl (—OH) groups. After 20 milling cycles, the powder size was reduced to approximately 10 μm, and the surface radical concentration significantly increased. This powder can be used as an active filler to reenter the epoxy system, improving the tensile strength by 50% (to 51.6 MPa). It was successfully applied in hydrophobic coatings (contact angle of 144.1°) and oil–water separation (efficiency up to 99%), with stable performance after 10 cycles.

Advantages and Limitations of Mechanical Recycling

(1) Advantages of Mechanical Recycling
Simple Process, Low Cost: Compared to chemical recycling and pyrolysis, mechanical recycling does not involve complex chemical reactions, and the process is simpler with lower-energy consumption.

No Chemical Pollution: The entire process does not involve harmful chemical reagents, reducing environmental pollution risks.
Suitable for Large-scale Industrial Applications: It can serve as an economical and effective primary treatment method for composite material recycling, providing raw materials for subsequent chemical recycling or energy recovery.
Enhances Resource Recycling: Crushed epoxy resin powders or fibers can be used in a variety of industrial applications, improving the recycling rate of waste materials.

(2) Limitations of Mechanical Recycling
Despite its simplicity and low cost, mechanical recycling still faces several inherent limitations. First, because the cross-linked network of epoxy resins cannot be reversed through physical processes, the recycled materials often exhibit inferior mechanical and thermal properties, which significantly restricts their range of applications. Second, during the crushing and grinding stages, reinforcement fibers are prone to breakage, leading to a loss of mechanical strength and reduced reuse value. Third, the recycled epoxy resin powder is mainly utilized in filler or other low-performance products, making it difficult to regenerate high-quality resin materials. Finally, compared with chemical recycling, mechanical treatment may cause particle aggregation and surface degradation, thereby reducing dispersion uniformity and overall reuse efficiency.

Future Directions for Mechanical Recycling To improve the efficiency of mechanical recycling and the utilization of recycled materials, low-damage separation technologies can be developed, such as optimizing cryogenic grinding or using ultrasonic-assisted separation to reduce fiber damage and improve reuse rates. Additionally, combining chemical methods, such as solvent dissociation and catalytic degradation, with mechanical recycling can further recover useful components from epoxy resins. Research into surface modification technologies, such as plasma treatment or chemical grafting, can improve the dispersion and compatibility of recycled epoxy resin powders, making them suitable for more high-end applications.

Mechanical recycling is an important method for processing WEP, offering advantages such as simplicity, low cost, and no chemical pollution. However, due to the curing properties of epoxy resins, mechanical recycling cannot restore the materials to their original performance, and the recycled products are primarily used for low-end filler and modification applications. In the future, combining chemical recycling, optimizing separation technologies, and developing high-value reuse pathways will be key directions for improving the efficiency and economic value of mechanical recycling.

6.2.2.3 Chemical Recycling of Epoxy Resins

Chemical recycling is one of the key paths to achieving a closed-loop recycling process for polymer waste, particularly for thermosetting polymers such as epoxy resins and polyurethanes, which cannot be reprocessed by melting. Unlike traditional landfill, incineration, or physical crushing methods, chemical recycling uses

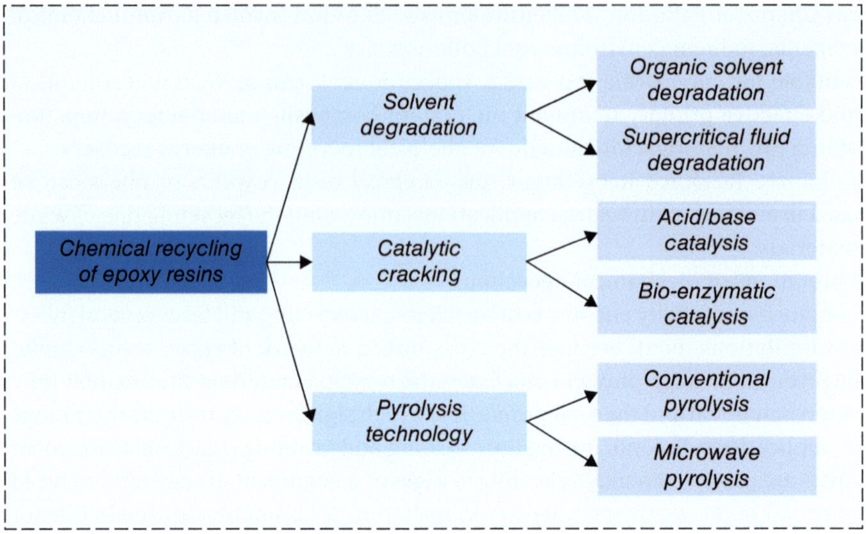

Figure 6.5 Chemical recycling of epoxy resins.

directed bond cleavage to transform waste polymers into monomers, oligomers, or platform chemicals. This not only reduces dependence on petrochemical resources but also lowers carbon emissions. Current mainstream technologies include solvent degradation, catalytic cracking, pyrolysis, and hydro/supercritical fluid depolymerization. The core goal is to efficiently recover resources through precise control of reaction conditions (temperature, pressure, catalyst) while avoiding the generation of harmful substances. With the progress of national "carbon-neutrality" strategies and circular economy regulations, this technology has become a research focus at the intersection of materials science and green chemistry (Figure 6.5).

Solvent Degradation Solvent degradation involves using specific solvent systems under high temperature or pressure to disrupt the three-dimensional cross-linked network of epoxy resins, enabling the separation of resin from reinforcement materials (such as carbon fiber or glass fiber).

Organic Solvent Degradation: Commonly used solvents include alcohols (e.g., ethylene glycol, propylene glycol) or amine-based solvents, with small amounts of acidic or basic catalysts to selectively break the ester or ether bonds in epoxy resins via nucleophilic substitution reactions. Reaction conditions (temperature, time) must be optimized based on the resin type. The degraded small molecular products can be repolymerized or used as chemical raw materials. However, the toxicity and high recycling costs of organic solvents limit large-scale application.

Hong et al. [45] developed a bio-based epoxy resin using vanillin and raspberry ketone. During recycling, this material undergoes acidic oxidation (using hydrochloric acid) to generate a secondary alcohol intermediate. This is then followed by triggering a retro-aldol reaction in an alkaline sodium hypochlorite solution, resulting in complete decomposition within 4 hours. The recycled carbon fiber (R-CF) retains 87% of its original strength, demonstrating both high practicality and potential for circular economy applications.

Reddy et al. [52] used hexylamine to selectively break the aliphatic ester bonds in the epoxy network. The degradation products consisted of small-molecule oligomers and amide compounds. Carbon fibers can be fully recovered through this process, and their surface morphology and mechanical properties remain consistent with the original fibers.

Supercritical Fluid Degradation: Supercritical carbon dioxide ($scCO_2$) or water (scH_2O) exhibits high diffusivity and solubility, allowing them to penetrate the resin–fiber interface. This interaction involves both physical swelling and chemical decomposition to separate fibers. Supercritical water's oxidative nature can further mineralize the resin to CO_2 and water, making it suitable for treating toxic resins (e.g., those containing halogenated flame retardants). However, the high corrosion resistance required for equipment is a challenge.

Kim et al. [55] developed a green method for recycling epoxy-based carbon fiber-reinforced plastics using supercritical water, without the need for catalysts or oxidants. By optimizing the treatment conditions (405 °C, 280 bar), the epoxy resin removal rate reached 99.5%. The R-CFs had a clean surface and retained high crystallinity, with a mechanical performance reduction of about 18–36%. The decomposition of the resin was achieved through the high diffusivity and solvent properties of supercritical water, avoiding the use of organic solvents or catalysts, making the process both environmentally friendly and efficient (Figure 6.6).

Catalytic Cracking Catalytic crackinginvolves using catalysts to lower the activation energy of epoxy resin degradation, improving the reaction efficiency and selectivity of the products. This process is typically divided into two methods: acid/base catalysis and bio-enzymatic catalysis.

Acid/Base Catalysis: Strong acids (e.g., H_2SO_4, H_3PO_4) or strong bases (e.g., KOH, NaOH) can break down the resin by protonating epoxy groups or hydrolyzing ester bonds. Acid catalysis tends to produce aromatic monomers (e.g., bisphenol A), while base catalysis may result in residual carbon. Supported heterogeneous catalysts (e.g., metal oxides/molecular sieves) can improve reaction controllability, but catalyst deactivation remains an ongoing issue. Chen et al. [47] used acidic catalytic conditions (such as phosphoric acid) to accelerate polymer decomposition and selectively break the chemical bonds in epoxy resin during the recycling process.

Bio-enzymatic Catalysis: Specific enzymes (e.g., esterases, laccases) degrade resins by hydrolyzing ester bonds or oxidizing phenolic structures. Enzyme catalysis is environmentally friendly, but challenges include the low permeability of enzymes to highly cross-linked resins and enzyme deactivation at high temperatures. Protein engineering or immobilization techniques are often used to optimize enzyme stability.

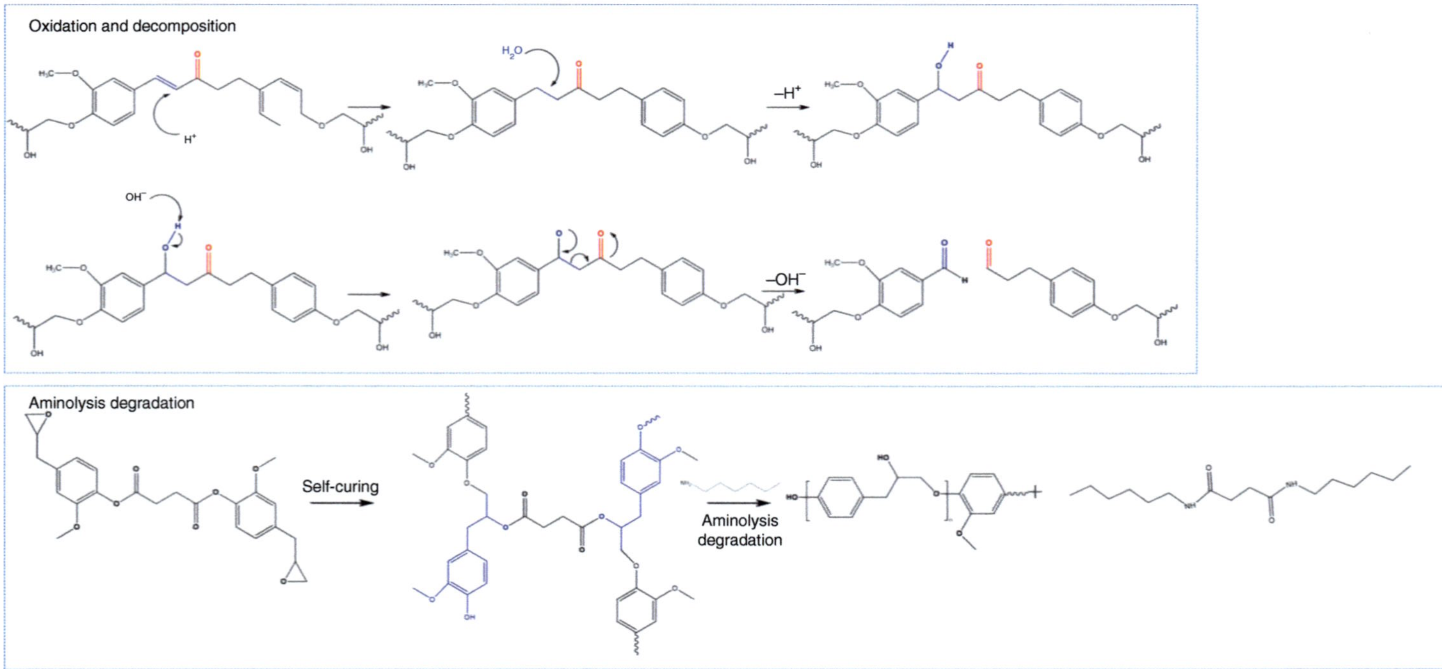

Figure 6.6 Conventional methodologies for solvent degradation. *Source:* [45] / with permission of Elsevier. [52] / with permission of American Chemical Society.

Pyrolysis Technology Pyrolysis involves the high-temperature decomposition of resin under anaerobic or limited oxygen conditions, primarily producing gaseous hydrocarbons (H_2, CH_4), liquid oils (aromatics, phenols), and solid residues, including carbon fibers.

Conventional Pyrolysis: Typically performed at temperatures ranging from 300 to 800 °C. At lower temperatures, resins decompose slowly into viscous liquid products, while at higher temperatures, they quickly vaporize. However, local overheating can cause fiber damage. A two-stage process (low-temperature predecomposition followed by high-temperature catalytic reforming) can be used to control the product distribution. Mezeix et al. [54] achieved material recycling by separating the epoxy matrix from the reinforcement materials under specific high-temperature (140 °C) chemical treatment, demonstrating the potential of recyclable bio-based epoxy in aerospace applications.

Microwave Pyrolysis: Uses electromagnetic waves to directly excite the vibration of resin molecules, enabling rapid and uniform internal heating. This method improves energy efficiency by 30–50% over conventional pyrolysis and avoids oxidation of the fiber surface, making it suitable for high-purity fiber recovery. Deng et al. [56] found that carbon fibers obtained by microwave pyrolysis had cleaner surfaces with less residual resin and that the chemical structure and graphitization degree of the fibers were not significantly affected. Microwave heating utilizes the rapid and uniform temperature rise within the material, avoiding the limitations of external heating in traditional pyrolysis, offering advantages in both efficiency and cleanliness. Hao et al. [57] used microwave pyrolysis technology to recycle carbon fiber/epoxy resin prepregs. Microwave pyrolysis effectively decomposes the epoxy resin, with the residual char decreasing as the temperature increases. The recovered carbon fibers have clean and smooth surfaces, with less than a 20% loss in tensile strength. The gaseous products are mainly CO, H_2, CO_2, and CH_4, while the liquid products contain phenols and aromatic compounds.

Challenges in Chemical Recycling The large-scale application of chemical recycling still faces multiple challenges. First, the complexity of polymer structures and interference from additives make depolymerization pathways difficult to standardize. There is a need to balance process efficiency, product purity, and energy consumption. Second, the high cross-linking density of thermosetting resins requires a reaction system that combines strong penetration ability with selective bond cleavage mechanisms. The development and regeneration cost of catalysts remain a bottleneck for industrial-scale applications. Additionally, the secondary utilization of recycled products (such as synthesizing new resins from regenerated monomers) needs to meet performance equivalency, and the technological validation chain is still incomplete. Future research should deeply integrate molecular dynamics simulations, high-throughput experiments, and artificial intelligence optimization to create a comprehensive solution for the entire life cycle, from waste design to high-value regeneration.

6.2.2.4 Biological Recycling Technologies and Applications

Biological recycling uses microorganisms or enzymes to degrade epoxy resins into reusable chemicals or environmentally friendly products. The core of this method lies in the controllability of the degradation reaction and the high-value conversion of the products. This section discusses engineering recycling strategies and practices, building on the biodegradation mechanisms discussed earlier.

Microbial-mediated Epoxy Resin Resource Recovery Microbial degradation is not only a breakdown process but also an important medium for carbon resource regeneration.

(1) Screening Functional Microbial Communities and Metabolic Regulation: Based on the degradation capabilities of white-rot fungi (e.g., *Phanerochaete chrysosporium*) and *Pseudomonas* bacteria, further efforts are focused on selecting strains that can tolerate high cross-linking densities or enhancing extracellular enzyme secretion (such as LiP and esterase) through genetic engineering to improve penetration efficiency into the epoxy resin network. The design of multispecies cooperative systems is also explored, for example, combining fungi (for large-molecule breakdown) with bacteria (for small-molecule metabolism) to optimize the directed generation of degradation products [39].
(2) Product Separation and Carbon Cycle Pathways: Degradation products (e.g., phenols, ketones) can be purified using membrane separation or adsorbent resins and then used as chemical raw materials (e.g., bisphenol A alternatives) or for synthesizing bio-based epoxy prepolymers. Microbial oxidation of CO_2 can be coupled with algae cultivation to convert it into biofuels or biodegradable plastics (e.g., polyhydroxyalkanoates [PHA]), creating a "resin waste-microbe-CO_2-bio-materials" closed loop [34].

Kadam et al. [58] conducted biodegradation tests using bacterial aggregates and found that the epoxy resin they developed achieved a degradation rate of 82% within 69 days and 95% within 259 days. The degradation mechanism relies on microorganisms breaking down the cross-linked structure of the epoxy resin through enzymatic action, with SEM observations showing the formation of surface cracks and pores. This indicates that the material can be fully mineralized through microbially mediated natural metabolic pathways (Figure 6.7).

Enzyme-catalyzed Recycling Technologies Enzymes in recycling systems have key advantages, including mild reaction conditions and high product specificity, but challenges remain with enzyme stability and reaction efficiency [42].

(1) Enzyme Immobilization and Reactor Design: Based on the catalytic mechanisms of laccase and LiP, efforts are directed toward developing covalent enzyme immobilization carriers (e.g., magnetic nanoparticles, porous resins) to enable the reuse of enzymes (with activity retention >90% after >20 cycles). Microfluidic enzyme reactors are being built to precisely control flow rates, temperature, and substrate contact areas, improving the spatial and temporal yield of epoxy resin degradation (e.g., monomer release per unit time).

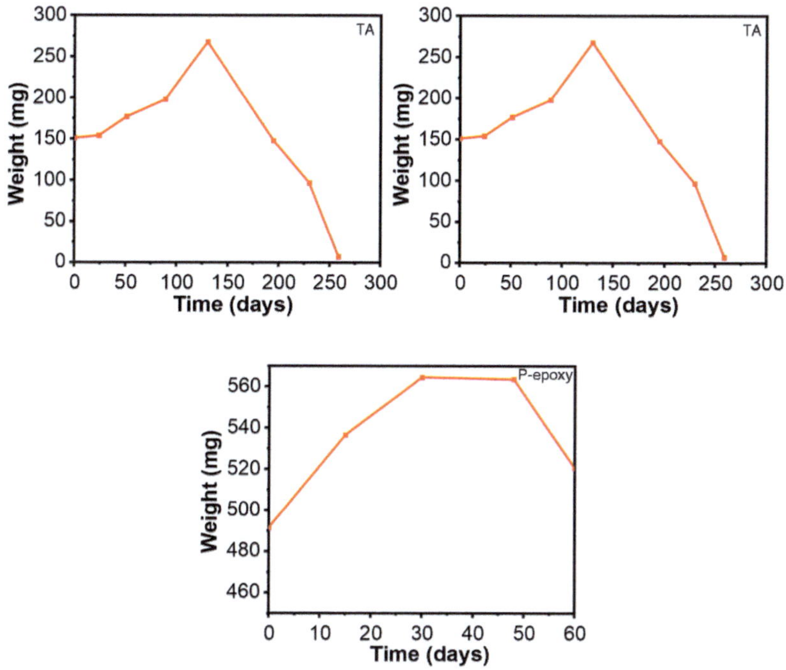

Figure 6.7 Mass loss percentages of various epoxy resins. *Source:* [58] / with permission of Elsevier.

(2) Multienzyme Cascades and Product Valorization: The design of laccase (for oxidative cleavage of ether bonds) and lipase (for hydrolyzing ester bonds) in a synergistic catalytic system can degrade epoxy resins into high-purity bisphenol A and polyol monomers, which can be used to regenerate epoxy resins or synthesize polyurethanes. Coenzyme regeneration systems (e.g., NADH cycle) can lower enzyme reaction costs and enhance industrial feasibility.

Challenges and Prospects of Biological Recycling Biological recycling faces two core challenges in scaling up:

(1) High Purification Costs for Complex Degradation Products: For example, after laccase-catalyzed decomposition of epoxy resins, the resulting phenolic derivatives are complex mixtures that require high-precision separation processes. Additionally, the cost of enzyme preparations needs optimization [59–62].
(2) Inefficiency of Traditional Batch Bioreactors: The handling efficiency of traditional batch bioreactors is insufficient, and continuous reaction systems must be developed to improve degradation throughput by employing dynamic feeding, real-time product separation, and other techniques.

Future breakthroughs should focus on interdisciplinary approaches. One direction is the use of synthetic biology to construct "whole-cell catalysts" (e.g., genetically modified *E. coli* performing both resin degradation and bio-based monomer synthesis), which would disrupt traditional multistep separation processes. Another approach is to accelerate the establishment of standards for regenerating materials, standardizing degradation product purity, toxicity, and reprocessing performance, ensuring a seamless transition from laboratory validation to industrial supply chain implementation.

6.2.2.5 Composite Recycling Technologies

Composite recycling technologies combine physical, chemical, and other multidimensional approaches to break through the efficiency bottleneck of single recycling methods and achieve both resource recovery and high-value conversion of epoxy resins. The core idea is to build a multistage interactive system that cooperatively addresses the degradation challenges of highly cross-linked epoxy resin structures while precisely controlling the product generation pathways and improving the market adaptability of regenerated materials. The following focuses on two major technological directions: physical–chemical synergistic recycling and pyrolysis–catalysis hybrid degradation.

Physical–Chemical Synergistic Recycling This method couples physical pretreatment with chemical selective degradation to overcome the inefficiencies of traditional single methods for highly cross-linked epoxy resins. The core logic is to weaken the resin network structure's density using physical methods (e.g., mechanical crushing, supercritical fluid swelling), thereby increasing the accessibility of chemical reagents to target functional groups (e.g., ester and ether bonds). This can be followed by ionic liquid-catalyzed hydrolysis or photothermal oxidation to achieve selective bond cleavage and monomer recovery. The technological framework includes physical enhancement mechanisms and chemical selective degradation.

Recent studies have explored both physical enhancement and chemical selective degradation strategies to improve the efficiency of epoxy resin recycling. In the physical domain, mechanical crushing and sorting techniques pulverize resin waste into micron-sized particles, which are then purified using cyclone or electrostatic separation to remove impurities such as residual glass fibers, thereby improving material uniformity. Moreover, supercritical fluid swelling using CO_2 effectively penetrates the resin matrix and weakens the cross-linked network, enhancing the efficiency of subsequent chemical degradation by 50–70%.

On the chemical side, ionic liquid-catalyzed hydrolysis has proven effective in selectively cleaving ester bonds within epoxy resins at moderate temperatures (120–150 °C), releasing bisphenol A monomers while minimizing side reactions. Additionally, photothermal synergistic oxidation, based on an Fe^{3+}/H_2O_2 photo-Fenton system, generates reactive oxygen species (·OH) under UV irradiation to target ether bonds, producing short-chain carboxylic acids as the primary degradation products.

Hughes et al. [48] combined cured epoxy resin with cellulose nanocrystals (CNCs) using a ball milling process. The abundant free hydroxyl groups on the surface of CNCs promoted the ester exchange reaction, transforming the permanent cross-linked network into a dynamically reconfigurable vitrimer network. The milled powder, after hot pressing, yielded high-performance nanocomposites.

This method requires no solvents or catalysts, with CNCs simultaneously serving as reinforcing fillers, significantly enhancing the mechanical properties of the material. This technique provides an environmentally friendly new strategy for the sustainable recycling of thermosetting polymers.

The main advantages of this approach include energy efficiency: physical pretreatment reduces the required chemical degradation temperature by 20–30%, with a combined energy consumption <1.8 kWh/kg. Additionally, product purity is improved: with staged processing, bisphenol A purity can reach over 95% (high performance liquid chromatography [HPLC] detection), making it suitable for repolymerization. However, challenges remain with impurities and high solvent recovery costs.

Pyrolysis–Catalysis Hybrid Degradation Technology The pyrolysis–catalysis hybrid process aims to convert epoxy resins into high-value carbon-based fuels or chemical raw materials [63]. The core concept is "thermal cracking of gas and catalytic reforming for upgrading." The graded pyrolysis strategy is as follows:

(1) Primary Rapid Pyrolysis: Resin is rapidly heated to 500–600 °C (heating rate >100 °C/s) in an oxygen-free atmosphere, breaking it into gaseous hydrocarbons and liquid tar, with a coke yield controlled to <10%.
(2) Secondary Catalytic Reforming: The pyrolysis gas is directed to a Ni–Co/Al_2O_3 bimetallic catalyst bed (temperature 400 °C, H_2 pressure 1.5 MPa), where hydrotreatment (HDO) converts oxygenated tars into alkanes (C_6—C_{12} hydrocarbon yield >75%).

The catalytic process is still limited by issues like halogen poisoning, catalyst deactivation, and uneven gas–solid mass transfer, hindering large-scale economic feasibility. Innovations in anti-poison catalysts (e.g., sulfur-modified metal sites) and smart reactor control systems will be key to breaking the technological bottleneck and achieving a "waste resin to low-carbon chemicals" closed-loop process.

6.2.2.6 Recycling and Reuse of Epoxy Resin

After efficient recycling of epoxy resin, the performance characteristics of the regenerated material differ from that of the original material. Therefore, reuse strategies must be adjusted according to the downstream application scenarios. Through material modification, blending, and functional design, it is possible to transition from low-value fillers to high-value products, establishing a closed-loop system of "recycling–regeneration–application." The current core pathways for reuse are as follows:

(1) Regenerated Epoxy Resin Matrix
   The recovered bisphenol A monomer can be repolymerized with epichlorohydrin to synthesize regenerated epoxy resin with a regeneration rate exceeding 80%. Incorporating nano-clay (montmorillonite) or carbon fiber (5–10 wt%) further enhances its mechanical performance, allowing the tensile strength to recover to 85–90% of that of the original resin. In addition, hydrocarbon oils

(mainly BTX aromatics) obtained from pyrolysis–catalysis processes can be blended with virgin epoxy resin at a 1:4 ratio to prepare thermosetting composite materials. The resulting materials exhibit a glass transition temperature (T_g) of 110–120 °C, which satisfies the performance requirements for electronic packaging substrates.

(2) Functional Fillers or Additives
Catalytically degraded carbon residues with a carbon content exceeding 85% can be compounded with magnesium hydroxide to prepare halogen-free flame-retardant synergists. When incorporated at 20 wt%, the resulting materials exhibit a Limiting Oxygen Index (LOI) above 32, making them suitable for polypropylene and other polymer applications. In addition, pyrolyzed carbon residues can be activated at high temperatures (around 800 °C, via KOH activation) to produce porous activated carbon with a specific surface area greater than 1200 m^2/g, which is suitable for use in supercapacitor electrodes and heavy-metal adsorbents.

(3) Energy and Chemical Raw Materials
Through catalytic hydrogenation, the degradation intermediates can be converted into C_8—C_{16} hydrocarbons, which are subsequently distilled and purified to produce diesel components with a cetane number above 45 and sulfur content below 50 ppm, meeting the EN 590 fuel standard. Meanwhile, bisphenol A recovered through hydrolysis can undergo hydroxyl protection and oxidation to yield terephthalic acid with a conversion efficiency exceeding 70%. This product serves as a key precursor for PET polyester synthesis, offering a sustainable pathway to reduce reliance on petroleum-derived raw materials.

However, the mechanical performance of regenerated epoxy resins currently declines significantly, with tensile strength reduced by 15–25% and fracture toughness lost by about 20–30%. The main reasons for this are the destruction of the cross-linked network and residual impurities. The thermal decomposition temperature (T_d) drops to 280–300 °C, which affects high-temperature applications. The chemical incompatibility leads to phase separation in composite materials, such as fluctuating dielectric constants in the blending of pyrolytic oil and resin, which exceeds industrial standards. Dynamic thermomechanical analysis (DMA) and molecular simulations need to be used to optimize interface design, with the application of nanomodification (e.g., SiO_2 interface coating) to compensate for performance deficiencies.

Additionally, the large-scale application of regenerated materials is limited by residual impurities (e.g., glass fiber fragments causing fracture risks) and insufficient cross-link density. Eddy current sorting and dynamic covalent networks (disulfide bonds) can improve purity to 98% and restore more than 85% of the mechanical strength. In terms of economic feasibility, the cost of regenerated resin is close to that of original resin. Therefore, continuous processes and catalyst recycling should reduce costs by 30%. Policy-driven mandatory blending can accelerate market acceptance, but a unified performance grading standard must be established.

6.3 Degradation and Recycling of HERs

6.3.1 Method of Degradation

6.3.1.1 Physical Degradation

HERscan be degraded by physical means, such as by thermal degradation. Under certain temperature conditions, the chemical bonds of the epoxy resin may break, thus achieving degradation.

In the research of polymer materials, epoxy resins are widely used in various fields due to their excellent properties. Zhang et al. [64] studied the thermal degradation kinetics of HER and bisphenol A diglycidyl ether (DGEBA) epoxy resin. The study found that the thermal degradation of these two resin systems can be divided into two stages: the low-temperature segment below 594 K and the high-temperature segment above 594 K. In the low-temperature stage, the thermal degradation activation energy of the hybrid resin system is similar to that of the pure DGEBA epoxy resin system, but in the high-temperature stage, the activation energy of the hybrid resin system is lower. With the increase in HER content, the thermal degradation activation energy of the hybrid resin system further decreased. In addition, when the molecular weight of HER increased or its content increased, because its cyclization degree was lower than that of DGEBA epoxy resin, the initial thermal degradation temperature of the hybrid resin decreased. The increase in hydroxyl content in the hybrid resin system will also affect its thermal degradation behavior. The thermal degradation reaction of DGEBA is divided into two steps, each with a different activation energy. The turning point is 594.2 K, at which time the dehydration of the hydroxyl group of epoxy resin occurs. At the high-temperature stage, the chemical bond breaking of the epoxy resin acts together with the intramolecular density of HER, rigid group density, epoxy equivalent mass, and cross-linking density of hybrid resin, resulting in significant changes in activation energy. Zhou et al. [65] studied the thermal degradation behavior and kinetics of hyperbranched polytrimeric anhydride glycol epoxy (HTME) and DGEBA. It was found that with the increase in HTME content, the thermal degradation activation energy of HTME/DGEBA hybrid resin decreased. Although the molecular weight or formation of HTME has little effect on its thermal degradation activation energy, the hydroxyl content of the hybrid resin system increases with the increase in the molecular weight or content of HTME, resulting in a decrease in its initial thermal degradation temperature (the temperature at 5% weight loss). The addition of HTME can effectively reduce the thermal degradation activation energy of hybrid resin. These studies not only enrich the theoretical system of thermal degradation kinetics of epoxy resin.

In recent years, with the continuous development of hyperbranched polymer technology, phosphorous hyperbranched flame retardants have gradually become the focus of research because of their excellent flame retardancy and good compatibility. Researchers have also analyzed the degradation performance of such materials. Battig et al. [66] investigated the pyrolytic mode of action and chemical decomposition mechanism of hyperbranched flame retardants in epoxies containing phosphorus. They found that the phosphate-based flame retardants decompose mainly through *cis*-elimination reactions while the amide-based flame retardants undergo hydrolysis reactions during heating. And both of these reactions can phosphorylate the flame retardants with epoxy resin substrates or other degradation products. In addition, the higher the content of epoxy group in the epoxy resin, the more secondary hydroxyl group is generated, which makes the phosphorylation reaction more likely to occur, thereby reducing the thermal decomposition temperature of the material. Wang et al. [67] used Michael addition polymerization to synthesize hyperbranched macromonomer containing phosphorus and mixed it with epoxy acrylate resin to prepare a flame-retardant coating. In the process of thermal degradation, the phosphorus-containing structure in the coating will preferentially degrade to form a flame-retardant layer. In addition, triacryloxyethyl phosphate plays a catalytic role in the thermal degradation, accelerating the degradation process of epoxy acrylate resin.

6.3.1.2 Chemical Solution Degradation

Chemical degradation involves the use of specific chemical solvents or reaction conditions to break chemical bonds in the epoxy resin. For example, by introducing degradable covalent bonds such as acetal bonds, disulfide bonds, Schiff base bonds, and ester bonds, the epoxy resin can be degraded into small molecular monomers or oligomers under mild conditions.

In the research of sustainable development of polymer materials, the degradation and recycling of epoxy resins have been the focus of scientists. The traditional thermosetting epoxy resin is difficult to achieve efficient degradation and recovery because of its irreversible cross-linked structure, which not only limits its application range but also causes a serious burden on the environment. In recent years, with the continuous progress of hyperbranched polymer technology, researchers began to explore the introduction of hyperbranched structure into epoxy resin to improve its degradation performance. Yu et al. [68] synthesized a carboxy-terminated epoxy hyperbranched polyester (DEHP) by the reaction of epichlorohydrin, citric acid, and maleic anhydride. This DEHP can effectively promote the degradation of its composite material formed with DGEBA. Specifically, with the increase in DEHP content and molecular weight, the degradation degree of the composite was also increased. For example, a composite containing 12% DEHP completely degrades in about 2 hours at 90 °C, while pure DGEBA degrades only 35%. The hydrolyzable ester bond in DEHP is more vulnerable than the ether bond in DGEBA, and with the increase in DEHP content and molecular weight, the number of ester bonds increases, further improving the degree of degradation. In the degradation process, the main product is formed due to the break of ester bond and ether bond and the partial break of C—N bond, in which citric acid is the product of the break of ester bond, and C—N bond remains intact. These results showed that the addition of DEHP significantly improved the degradation performance of the composites. Guo et al. [69] synthesized the hyperbranched polymer of glycol-isocyanate terminal through the "Click" reaction and further prepared the self-curing epoxy-terminal hyperbranched polymer, which possessed the characteristics of monomer degradation and regeneration. The study showed that the higher the concentration of HNO_3 solution, the greater the degradation degree of epoxy resin. When the concentration of HNO_3 reaches 1.0 mol/L, the epoxy resin can be completely degraded. When the DMF solution is added to blend, only 10 mL of 1.0 mol/L HNO_3/DMF solution can achieve complete degradation. Nitric acid provides free radicals during the degradation process, breaking the chemical bonds in the polymer, especially Se—C bonds, and oxidizing the $—NH_2$ group to $—NO_2$, which ultimately degrades the epoxy resin into a variety of small molecular structures. In the study of the degradability of polymer materials, hyperbranched polymers have gradually become the focus of research because of their unique molecular structure and excellent properties.

6.3.1.3 Synergistic Degradation

In the study of the degradation of polymer materials, the synergistic effect of temperature and chemical reagents is also a key factor to improve the degradation efficiency. The increase in temperature can significantly increase the speed of molecular movement, thus promoting the break of chemical bonds and the degradation reaction. In addition, the type and concentration of chemical reagents also have an important effect on the degree of degradation. Different chemical agents can effectively degrade different chemical bonds in epoxy resins. The synergistic effect of temperature and chemical reagents provides not only an effective technical means for the degradation of epoxy resin but also an important reference for the degradation of other polymer materials. Yu et al. [70] successfully achieved rapid and efficient degradation of cured thermosetting epoxies by synthesizing hyperbranched epoxies with degradation properties. They used 3-isocyanate-4-methyl epoxy-methyl phenyl carbamate, maleic anhydride, citric acid, and epichlorohydrin as raw materials to produce the resin. In a mixture of hydrogen peroxide (H_2O_2) and DMF at 90 °C, the resin can be completely degraded under atmospheric pressure in only 2 hours to produce raw citric acid. During the degradation process, H_2O_2 acts as a catalyst, and its increased content promotes the formation of free radicals, which attack the chemical bonds in the resin, leading to its decomposition. The degradation products can be completely dissolved in DMF solvent, and the degree of degradation increases with the increase in reaction time. High temperature helps to increase the degradation rate and thus shorten the reaction time. In addition, degradation is mainly achieved through the breaking of the C—O bond in the ester group and the C—N bond in the curing agent. The formation of the degradation product citric acid indicates the break of the ester bond, while other degradation products are formed due to the break of the C—N bond of other groups. This efficient degradation process provides a new idea for the recovery of epoxy resins. However, although this method performs well in degradation efficiency, its economy and environmental protection still need to be considered in practical applications.

In contrast, Ma et al. [71] further explored the degradation performance of DGEBA in different solvent environments by modifying it with a self-curing HER. The C—N and ester bonds in this composite break in H_2O_2 and HNO_3/DMF (7.5 g, 1.0 mol/L) solution. At 50 °C, the degradation rate of the material is 30.8%, and when the temperature is raised to 90 °C, it can be almost completely degraded after 6 hours. H_2O_2 can break down the structure of the cross-linked part, while DMF can dissolve the small molecules produced during the degradation process. This degradation process not only improves the recyclability of the material but also provides a new method for recycling the epoxy resin. However, this method has a high dependence on solvents during the degradation process and takes a longer time to achieve complete degradation. In addition to this, Ma et al. [72] did this by combining HERs containing 1,3,5-hexahydro-*s*-triazine with DGEBA. The higher the content and molecular weight of HER, the better the degradation performance of the material. At 95 °C, it only takes 2.5 hours to achieve complete degradation, greatly shortening the time. This hyperbranched epoxy contains an easily hydrolyzed ester group and an easily digestible triazine structure, while DGEBA contains ether bonds. Since the ester group and triazine structure are more easily degraded than the ether bond, when the hyperbranched epoxy content of 1,3,5-hexahydro-*s*-triazine in the composite increases, the degradation degree of the material also increases. By prolonging the degradation time and increasing the degradation temperature, the degradation efficiency of the composite can be further improved. During the degradation process, the formation of small molecules is mainly due to C—N bond break in the triazine ring, as well as the ester group break in HER and ether bond break in DGEBA. This degradation property opens up possibilities for material recycling and reuse, helping to drive the development of sustainable materials (Figure 6.8).

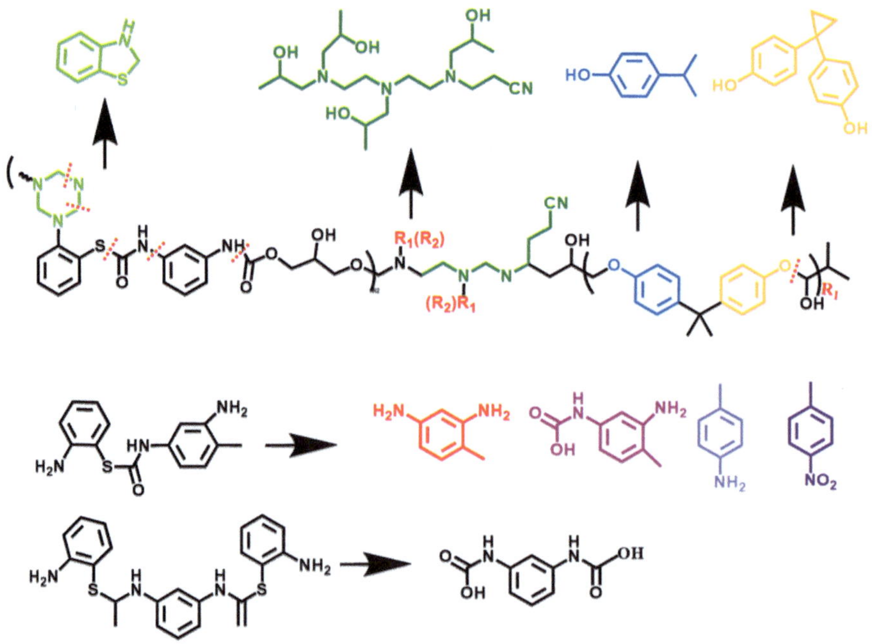

Figure 6.8 Degradation profile of hyperbranched epoxy resin. *Source:* [71] / with permission of Elsevier.

6.3.1.4 Biodegradation

Biodegradation refers to the use of microorganisms or their metabolites to break down epoxy resins. Biodegradation, as an environmentally friendly material treatment method, has been widely concerned in the field of epoxy resin in recent years. Bio-based HER has become a research hotspot due to its degradability and renewability. These materials can not only be used as matrix resins or tougheners but also biodegrade under certain conditions [34, 73–80]. Yuan et al. [81] prepared a polyarmic cashew phenol oligomer with cyclotriphosphonitrile as the structure center and epoxidized its unsaturated fat chain with *m*-chloroperoxy benzoic acid and acrylic group to synthesize a polyarmic bio-based acrylate. The environmental degradation performance was evaluated by soil burial test, and the results showed that the degradation rate of the sample increases with the increase in time. The addition of cashew phenol and polyhydroxyl alcohol can promote the degradability of the cured film, because the ester bond and fat chain in these components are the easily degradable parts of the cured film. The coating is capable of microbial degradation in the soil, which solves the problem that traditional coatings are not degradable. It provides a new idea for the application of bio-based epoxy resin. At the same time, Barua et al. [82] developed a hyperbranched epoxy adhesive for wound sealing in suture-free surgery, which is composed of triazinyl HER and polyamine hardener, and has good antibacterial properties. The results of in vitro degradation experiment in phosphate buffered saline (PBS) solution at 37 °C showed that the material had a degradation trend and that there was a possibility of biodegradation. The physiological fluids secreted at the wound site, such as water, enzymes, salt, and trypsin, provided the degradation conditions for the material.

Conventional hyperbranched epoxies have certain limitations in terms of biodegradability, which limits their application in some environmentally sensitive areas. Therefore, how to improve the biodegradability of HER by optimizing its chemical structure has become a key problem to be solved urgently in the field of materials science. Baruah et al. [83] prepared a tannin-based gallic acid-based hyperbranched epoxy thermosetting diethanolamide material using bio-based branched reactants. The study showed that the biodegradability of HER was enhanced by the introduction of phenol structure. *Pseudomonas aeruginosa*, a citrate-, oxidase-, and catalase-positive microorganism, was able to effectively degrade the phenol structure in the material, thereby improving its degradation performance. This strategy provides an important reference for designing HERs with specific degradation properties. At the same time, Bibekananda et al. [39] further expanded the application of bio-based hyperbranched bisphenol-based epoxies prepared by castor oil-based hyperbranched polyester polyols. The large amount of ester bond structure in this material makes it show good degradation performance under the action of hydrocarbon-degrading bacteria, and the degradation speed is accelerated with the increase in the content of castor oil-based polyester. Together, these two studies highlight the importance of optimizing the biodegradability of

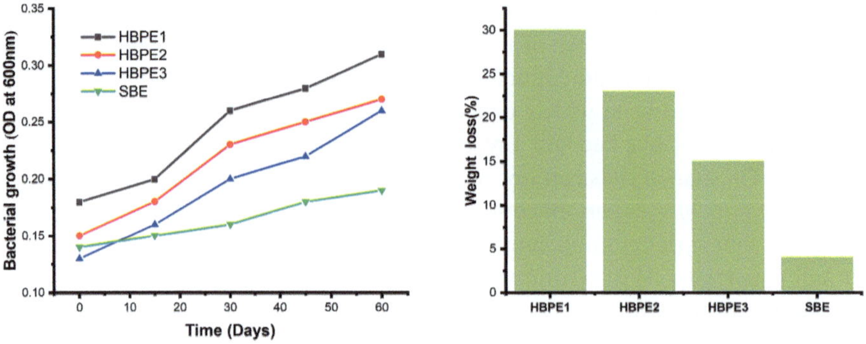

Figure 6.9 Biodegradation profile of epoxy resin. *Source:* [39] / with permission of American Chemical Society.

HERs by adjusting their chemical structure and provide new ideas for developing sustainable high-performance materials (Figure 6.9).

6.3.2 Recycling Method

6.3.2.1 Recovery by Physical Method

The degradation and recovery of epoxy resin can be promoted by controlling the temperature. For example, within 30 seconds at 120 °C, certain hyperbranched epoxies can be reprocessed quickly and efficiently. The original traditional physical method of recovering epoxies utilizes temperature control to facilitate their degradation and reprocessing, which is simple and easy to do, but usually only degradation recovery can be achieved [84–89]. Zhang et al. [90] developed an epoxy resin based on Itaconia-based hyperbranched polymer (IAHBP) and DGEBA, which has excellent overall properties and recyclability. This epoxy resin can be effectively degraded by a chemical recovery method using 10 wt% sodium hydroxide aqueous solution at 80 °C for 30 minutes without the use of organic solvents. The degradation process relies mainly on the hydrolysis of the ester group, and the recovery of the recovered monomer is as high as 95%. The recovered IAHBP (r-IAHBP) can be rereacted with DGEBA to prepare a new epoxy resin with similar mechanical and kinetic properties to the original epoxy resin. The study demonstrates a green and convenient epoxy recovery strategy, providing a new method for producing epoxies with excellent mechanical properties and environmentally friendly reprocessability. This method not only realizes the closed-loop recycling of epoxy resin but also maintains the mechanical and thermal properties of the material, opening up a new way for the sustainable development of epoxy vitreous body.

6.3.2.2 Recycling by Chemical Method

The use of specific solvents can promote the degradation and recycling of epoxy resin. Through the chemical degradation method, solvents can be used to assist the decomposition of epoxy resin, so as to achieve recovery. Traditional recycling methods usually involve high-temperature decomposition, which is less efficient and may produce secondary pollution, and has certain limitations. Therefore, for the purpose of environmental protection and resource recycling, the material structure is degraded by chemical solvent, and the raw material small molecules in the degraded components are recovered to become the research and development direction of HER materials. Wei et al. [44] developed an epoxy resin with flame-retardant properties, which is composed of HER derived from Itaconic acid and (1,3,5-hexahydro-*s*-triazine-1,3,5-triyl)benzyl mercaptan. The material can be chemically recovered under mild conditions to produce two monomers, 2-aminothiophene (ATP) and 9,10-dihydro-9-oxygen-10-phosphate phenanthrene 10-oxide with itaconic acid's product (DOPO-IA), with recovery rates of 86% and 75%, respectively. The recovery process was carried out in a tetrahydrofuran environment, using 1 mol/L of tetrahydrofuran phosphate solution as the degradation reactant and solvent, at 25 °C for 8 hours. Subsequently, the degradation solution was neutralized with 1 mol/L of NaOH and the organic layer was separated by column chromatography to obtain the recovered ATP (rATP). For the water layer, 1 mol/L HCl was used for dialysis and filtration after acidification, and the resulting filter residue was washed with de-ionized water and acetone, resulting in regenerated DOPO-IA (r-DOPO-IA). This study demonstrates the highly efficient closed-loop recyclability of this flame-retardant HER, providing a new avenue for the development of future high-performance chemically recyclable materials (Figure 6.10).

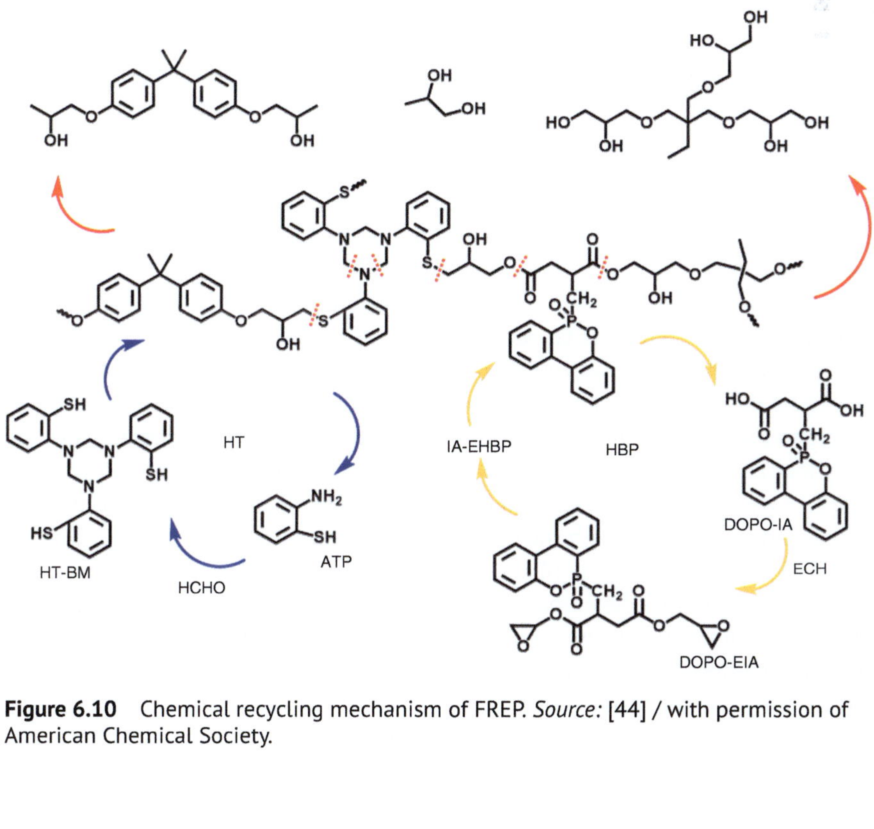

Figure 6.10 Chemical recycling mechanism of FREP. *Source:* [44] / with permission of American Chemical Society.

However, this method still relies on the use of organic solvents, which limits its environmental protection and economics to some extent. In the meantime, Chen et al. [47] have taken a different approach and developed a degradation recycling technique that does not use organic solvents. They prepared a biodegradable and recyclable thermosetting HER material using bio-based FDCA. The study showed that the hyperbranched topological components in the material were more easily hydrolyzed and degraded faster than the linear structural components. The degradation mainly occurs in the ether bond, some C—N bond, nitrile group, and ester bond, and these parts will be hydrolyzed, cracked, or fractured. In addition, the degradation process of the material does not use organic solvents, the conditions are green and mild, and some of the degradation products can be recycled, with a recovery rate of 56.8% showing its excellent environmental friendliness.

6.3.2.3 Multiple Conditions Collaborative Recovery

Traditional recycling methods face major challenges due to their irreversible cross-linked networks, especially under high-temperature and solvent conditions. Recent studies have shown that these thermosetting polymers can be effectively degraded through a synergistic effect of temperature with solvents such as acids. Xu et al. [91] designed a thermoset hyperbranched polymer containing a reversible hexahydro-*s*-triazine heterocyclic structure, aiming to achieve its degradation and recovery. Compared with concentrated acid, which is toxic and dangerous to some extent, phosphoric acid is safer and environmentally friendly. This material can be completely degraded in phosphoric acid solution in only 2 hours at 90 °C, and the recovery rate is as high as 85.1%. Phosphoric acid can break the C—N bond and ester bond in the structure of hexahydrotriazine, thereby releasing the original 2-aminobenzyl alcohol monomer. The recovered monomer can be reused to produce the hexahydrotriazine monomer, which realizes the closed-loop recovery of the material. The synergistic effect of temperature and solvent not only accelerates the breaking of C—N bonds and ester bonds, releasing recyclable monomers, but also provides an important solution for the sustainable development of materials.

6.3.2.4 Recovery of Composite Components

In the field of composite materials, another component of the composite can be recovered by means of a biodegradable and recyclable hyperbranched epoxy structure. Due to the excellent mechanical properties and lightweight characteristics of carbon fiber composites, the types and quantities of HERs and their composites are not few and are widely used in aerospace, automobile, sports equipment, and other industries. However, with the increase in its application, the problem of recycling and reuse of waste carbon fiber composite materials has become increasingly prominent. Xu et al. [92] developed a sulfhydryl hyperbranched polyimide synthesized from a monomer containing an imine structure and combined it with an HER to prepare a composite. The imide structure of this material enables it to be completely degraded under certain conditions, thus enabling the complete recycling of carbon fiber fabrics. The degradation process mainly relies on the attack of the ester bonds by hydroxyl radicals (·OH) and H^+ provided by H_2O_2. Studies have shown that when the optimal concentration of H_3PO_4 is 1.5 mol/L, the surface parts of epoxy resin and carbon fiber containing an imine skeleton structure can be almost completely degraded within 8 hours at 90 °C. The mechanical properties and surface morphology of the recovered carbon fiber were not much different from that of the original carbon fiber. This method not only realizes the efficient recovery of epoxy resin but also provides a new way for the sustainable utilization of carbon fiber composites.

These methods and mechanisms provide different ways of HERs as well as degradation and recycling of composites, helping to achieve recycling and environmentally friendly disposal of epoxy resins.

6.4 Economic Feasibility of Epoxy Resin Recycling

Epoxy resins are widely used in various industries due to their excellent mechanical and chemical properties. However, their nonbiodegradability and the challenges associated with recycling have raised concerns about their environmental impact. Traditional recycling methods for epoxy resins are often energy-intensive and involve the use of toxic chemicals, making them unsustainable [93, 94]. Recent advancements have focused on developing more efficient and eco-friendly recycling methods to address these issues [2, 95].

6.4.1 Cost-effectiveness of Recycling Methods

Several studies have explored the economic feasibility of different epoxy resin recycling methods. For instance, the use of hydrogen peroxide for recycling carbon fiber from epoxy matrix composites has shown favorable technoeconomic outcomes, with a net present value (NPV) of $15.56 million and a payback period of 10 years [96]. Another study demonstrated that recycling epoxy resins using a microwave-assisted green chemistry approach not only reduces energy consumption but also eliminates waste by-products, offering significant economic advantages [2]. These methods highlight the potential for cost-effective recycling processes that can be scaled up for industrial applications.

6.4.2 Environmental and Economic Benefits

The environmental benefits of epoxy resin recycling are significant. For example, recycling epoxy-based materials in asphalt pavement technology has shown a reduction in energy consumption and greenhouse gas emissions by approximately 44.8% and 39%, respectively, compared to traditional methods [97]. Additionally, the costs associated with recycled cold-mix asphalt (RCMA) are lower than those of traditional reclaimed hot-mix asphalt (RHMA), making it a cost-effective option [97]. These findings suggest that recycling epoxy resins can contribute to both economic savings and environmental sustainability.

6.4.3 Challenges and Opportunities

Despite the promising economic and environmental benefits, there are challenges associated with epoxy resin recycling. The permanent cross-linked structure of traditional thermosetting epoxy resins makes them difficult to degrade and recycle [94, 98]. However, the development of epoxy vitrimers with dynamic covalent bonds offers a potential solution. These materials can be reprocessed and recycled more efficiently, providing a sustainable alternative to traditional epoxy resins [94, 98]. The introduction of degradable or dynamically cross-linked structures into epoxy resins can enhance their recyclability while maintaining their original performance [99].

6.4.4 Future Prospects

The future of epoxy resin recycling lies in the development of novel materials and processes that balance economic feasibility with environmental sustainability. The use of bio-based HERs and the integration of dynamic covalent linkages in epoxy vitrimer networks are promising approaches [100, 101]. These innovations not only improve the recyclability of epoxy resins but also reduce the reliance on fossil-based raw materials, contributing to a circular economy. Continued research and development in this field are essential to overcome existing challenges and fully realize the economic and environmental benefits of epoxy resin recycling.

In conclusion, the economic feasibility of epoxy resin recycling is supported by various innovative methods that offer both cost savings and environmental benefits. While challenges remain, ongoing advancements in materials science and recycling technologies hold promise for a more sustainable future.

6.5 Achievements and Challenges

6.5.1 Achievements

In recent years, epoxy resin degradation and recovery technology has achieved some successful cases. For example, Zhang et al. [102] developed an epoxy supramolecular thermosetting material composed of vanillin-based hyperbranched epoxy resin (VanEHBP), which was constructed by introducing intermolecular hydrogen bonds and bidynamic covalent bonds, as well as forming intramolecular and intermolecular cavities. The material has rapid reprocessability and room-temperature closed-loop recyclability. At room temperature, the material can be effectively depolymerized by imide depolymerization, and its dynamic imide bond group covalent adaptive network can be degraded into oligomers containing aldehydes and amines under mildly acidic conditions. Specifically, the material is immersed in DMF and 0.1 M HCl aqueous solution (HCl aqueous solution/DMF = 1/10, v/v) for 180 minutes to obtain a clear orange degradation solution. After the solvent is removed, a regenerated oligomer can be obtained. The recovered material retains the structural integrity and mechanical properties of the original sample. This case reveals the hydrogen bond dissociation and reorganization process and movement mechanism of supramolecular epoxy resin, realizes the closed-loop chemical recovery of epoxy resin, and demonstrates the green and convenient epoxy resin recovery strategy.

6.5.2 Challenges

Although progress has been made, the epoxy resin degradation and recovery technology still faces some challenges. First of all, there are many kinds of degradation solvents, and some solvents have certain toxicity and danger, so it is necessary to find safer and environmentally friendly alternative solvents. Second, the degradation temperature is generally high, and the energy consumption is large, which is not conducive to industrial applications. In addition, the high-value recovery is difficult; how to efficiently transform the degradation products into high-value-added materials, to achieve the full utilization of resources, still needs further research.

6.5.3 Outlook

In the future, the development direction of epoxy resin degradation and recovery technology mainly includes the following aspects:

(1) Development of New Degradation Catalysts: Research and development of efficient, low-cost, environmentally friendly degradation catalysts, reduce degradation temperature and energy consumption, improve degradation efficiency and selectivity, to achieve rapid and complete degradation of epoxy resin.
(2) Optimize the Recycling Process: Improve the existing recycling process, improve the recovery rate and the quality of recycled products, reduce the generation of waste, and realize the high-value recovery of epoxy resin and the recycling of resources.
(3) Strengthen Basic Research: In-depth study of the degradation mechanism of HER and the chemical reaction in the recovery process to provide theoretical support for the development of more efficient degradation and recovery technology.
(4) Material Innovation: Developing epoxy vitrimers with dynamic covalent bonds and bio-based HERs to overcome recycling barriers;
(5) Promote Industry-University-Research Cooperation: Strengthen the cooperation between universities, scientific research institutions, and enterprises; promote the transformation and industrial application of research results; promote the rapid development of hyper-overrun epoxy resin degradation and recycling technology; and make greater contributions to achieving sustainable development and environmental protection.

References

1 Li, D., Zhou, S., Wu, Y. et al. (2025). Bifunctional hyperbranched modifier enabled interlocking interfaces for ultra-strong recyclable carbon fiber/bio-epoxy composites. *Chem. Eng. J.* 521: 166591.

2 Jiang, Y., Li, J., Li, D. et al. (2024). Bio-based hyperbranched epoxy resins: synthesis and recycling. *Chem. Soc. Rev.* 53 (2): 624–655.

3 Auvergne, R., Caillol, S., David, G. et al. (2014). Biobased thermosetting epoxy: present and future. *Chem. Rev.* 114 (2): 1082–1115.

4 Ma, S., Wei, J., Jia, Z. et al. (2019). Readily recyclable, high-performance thermosetting materials based on a lignin-derived spiro diacetal trigger. *J. Mater. Chem. A* 7 (3): 1233–1243.

5 Ruiz De Luzuriaga, A., Martin, R., Markaide, N. et al. (2016). Epoxy resin with exchangeable disulfide crosslinks to obtain reprocessable, repairable and recyclable fiber-reinforced thermoset composites. *Mater. Horiz.* 3 (3): 241–247.

6 Zhao, S. and Abu-Omar, M.M. (2018). Recyclable and malleable epoxy thermoset bearing aromatic imine bonds. *Macromolecules* 51 (23): 9816–9824.

7 Bao, X., Wu, F., and Wang, J. (2021). Thermal degradation behavior of epoxy resin containing modified carbon nanotubes. *Polymers* 13: 3332.

8 Qiu, C., Luo, J., Ling, Y. et al. (2022). Thermal degradation behavior and mechanism of organosilicon modified epoxy resin. *Macromol. Chem. Phys.* 223 (20): 2200164.

9 Levchik, S.V. and Weil, E.D. (2004). Thermal decomposition, combustion and flame-retardancy of epoxy resins—a review of the recent literature. *Polym. Int.* 53 (12): 1901–1929.

10 Zhou, G., Mikinka, E., Golding, J. et al. (2020). Investigation of thermal degradation and decomposition of both pristine and damaged carbon/epoxy samples with thermal history. *Compos. Part B* 201: 108382.

11 Feng, Y., Zhang, Z., Yue, D. et al. (2024). Recent progress in degradation and recycling of epoxy resin. *J. Mater. Res. Technol.* 32: 2891–2912.

12 Pimenta, S. and Pinho, S.T. (2011). Recycling carbon fibre reinforced polymers for structural applications: technology review and market outlook. *Waste Manag.* 31 (2): 378–392.

13 Nahil, M.A. and Williams, P.T. (2011). Recycling of carbon fibre reinforced polymeric waste for the production of activated carbon fibres. *J. Anal. Appl. Pyrolysis* 91 (1): 67–75.

14 Xing, W., Jie, G., Song, L. et al. (2011). Flame retardancy and thermal properties of epoxy acrylate resin/alpha-zirconium phosphate nanocomposites used for UV-curing flame retardant films. *Mater. Chem. Phys.* 125 (1): 196–201.

15 Kamimura, A., Akinari, Y., Watanabe, T. et al. (2010). Efficient chemical recycling of waste fiber-reinforced plastics: use of reduced amounts of dimethylaminopyridine and activated charcoal for the purification of recovered monomer. *J. Mater. Cycles Waste Manag.* 12 (2): 93–97.

16 Piñero-Hernanz, R., García-Serna, J., Dodds, C. et al. (2008). Chemical recycling of carbon fibre composites using alcohols under subcritical and supercritical conditions. *J. Supercrit. Fluids* 46 (1): 83–92.

17 Xu, P., Li, J., and Ding, J. (2013). Chemical recycling of carbon fibre/epoxy composites in a mixed solution of peroxide hydrogen and *N,N*-dimethylformamide. *Compos. Sci. Technol.* 82: 54–59.

18 Oliveux, G., Dandy, L.O., and Leeke, G.A. (2015). Degradation of a model epoxy resin by solvolysis routes. *Polym. Degrad. Stab.* 118: 96–103.

19 Li, J., Xu, P.-L., Zhu, Y.-K. et al. (2012). A promising strategy for chemical recycling of carbon fiber/thermoset composites: self-accelerating decomposition in a mild oxidative system. *Green Chem.* 14 (12): 3260–3263.

20 Huo, S., Song, P., Yu, B. et al. (2021). Phosphorus-containing flame retardant epoxy thermosets: recent advances and future perspectives. *Prog. Polym. Sci.* 114: 101366.

21 Nabipour, H., Niu, H., Wang, X. et al. (2021). Fully bio-based epoxy resin derived from vanillin with flame retardancy and degradability. *React. Funct. Polym.* 168: 105034.

22 Takahashi, A., Ohishi, T., Goseki, R., and Otsuka, H. (2016). Degradable epoxy resins prepared from diepoxide monomer with dynamic covalent disulfide linkage. *Polymer* 82: 319–326.

23 Chen, J.-S., Ober, C.K., and Poliks, M.D. (2002). Characterization of thermally reworkable thermosets: materials for environmentally friendly processing and reuse. *Polymer* 43 (1): 131–139.

24 Okajima, I. and Sako, T. (2019). Recycling fiber-reinforced plastic using supercritical acetone. *Polym. Degrad. Stab.* 163: 1–6.

25 Ma, C., Guo, Z., Fang, Z., and Li, J. (2022). Flame retardancy and chemical degradation of epoxy containing phenylphosphonate group under mild conditions. *Compos. Part B* 239: 109967.

26 Iwaya, T., Tokuno, S., Sasaki, M. et al. (2008). Recycling of fiber reinforced plastics using depolymerization by solvothermal reaction with catalyst. *J. Mater. Sci.* 43 (7): 2452–2456.

27 Liu, Y., Liu, J., Jiang, Z., and Tang, T. (2012). Chemical recycling of carbon fibre reinforced epoxy resin composites in subcritical water: synergistic effect of phenol and KOH on the decomposition efficiency. *Polym. Degrad. Stab.* 97 (3): 214–220.

28 Hyde, J.R., Lester, E., Kingman, S. et al. (2006). Supercritical propanol, a possible route to composite carbon fibre recovery: a viability study. *Compos. A: Appl. Sci. Manuf.* 37 (11): 2171–2175.

29 Baran, J. and Penczek, S. (1995). Hydrolysis of polyesters of phosphoric acid. 1. Kinetics and the pH profile. *Macromolecules* 28 (15): 5167–5176.

30 Dang, W., Kubouchi, M., Sembokuya, H., and Tsuda, K. (2005). Chemical recycling of glass fiber reinforced epoxy resin cured with amine using nitric acid. *Polymer* 46 (6): 1905–1912.

31 Zhang, X., Ma, C., Liu, P. et al. (2023). Recycling of ammonia-cured epoxy resin by oxidative degradation of nitric acid assisted by swelling agent. *Eur. Polym. J.* 186: 111823.

32 Xing, M., Li, Z., Zheng, G. et al. (2021). Recycling of carbon fiber-reinforced epoxy resin composite via a novel acetic acid swelling technology. *Compos. Part B* 224: 109230.

33 Ma, S. and Webster, D.C. (2015). Naturally occurring acids as cross-linkers to yield VOC-free, high-performance, fully bio-based, degradable thermosets. *Macromolecules* 48 (19): 7127–7137.

34 Eliaz, N., Ron, E.Z., Gozin, M. et al. (2018). Microbial degradation of epoxy. *Materials* 11: 2123.

35 Shah, A.A., Hasan, F., Hameed, A., and Ahmed, S. (2008). Biological degradation of plastics: a comprehensive review. *Biotechnol. Adv.* 26 (3): 246–265.

36 Pangallo, D., Bučková, M., Kraková, L. et al. (2015). Biodeterioration of epoxy resin: a microbial survey through culture-independent and culture-dependent approaches. *Environ. Microbiol.* 17 (2): 462–479.

37 Telegdi, J., Shaban, A., Trif, L., and El-Sherik, A.M. (2017). Chapter 8: Microbiologically influenced corrosion (MIC). In: *Trends in Oil and Gas Corrosion Research and Technologies* (ed. A.M. El-Sherik), 191–214. Boston: Woodhead Publishing.

38 Ma, J., Li, G., Hua, X. et al. (2022). Biodegradable epoxy resin from vanillin with excellent flame-retardant and outstanding mechanical properties. *Polym. Degrad. Stab.* 201: 109989.

39 De, B., Gupta, K., Mandal, M., and Karak, N. (2013). Biodegradable hyperbranched epoxy from castor oil-based hyperbranched polyester polyol. *ACS Sustain. Chem. Eng.* 2 (3): 445–453.

40 Liu, M., Baum, A., Odermatt, J. et al. (2017). Oxidation of lignin in hemp fibres by laccase: effects on mechanical properties of hemp fibres and unidirectional fibre/epoxy composites. *Compos. A: Appl. Sci. Manuf.* 95: 377–387.

41 Ochiai, B., Nakazawa, Y., Matsumura, Y., and Kawai, T. (2024). Hydrolytic degradation and biomineralization of amine-cured epoxy resin based on glycidate. *Polym. J.* 56 (4): 409–417.

42 Faiza, W., Firouzi, A., Islam, M.R. et al. (2021). Degradation analysis of epoxy resin composites reinforced with bioprotein: effects of hydrolysis using papain and bromelain. *Polym. Compos.* 42 (6): 2717–2727.

43 Sienkiewicz, A. and Czub, P. (2024). Studies on the enzymatic degradation process of epoxy-polyurethane compositions obtained with raw materials of natural origin. *Molecules* 29: 5667.

44 Wei, F., Zhang, J., Wu, C. et al. (2023). Closed-loop recycling of tough and flame-retardant epoxy resins. *Macromolecules* 56 (14): 5290–5305.

45 Hong, Y., Jeong, J., Oh, D. et al. (2023). On-demand and fast recyclable bio-epoxy. *J. Ind. Eng. Chem.* 117: 490–499.

46 Celiński, M. and Sankowska, M. (2016). Explosibility, flammability and the thermal decomposition of bisphenol A, the main component of epoxy resin. *J. Loss Prev. Process Ind.* 44: 125–131.

47 Chen, X., Chen, S., Xu, Z. et al. (2020). Degradable and recyclable bio-based thermoset epoxy resins. *Green Chem.* 22 (13): 4187–4198.

48 vom Saal Frederick, S. and Hughes, C. (2005). An extensive new literature concerning low-dose effects of bisphenol a shows the need for a new risk assessment. *Environ. Health Perspect.* 113 (8): 926–933.

49 Cao, X.-L., Corriveau, J., and Popovic, S. (2009). Levels of bisphenol A in canned soft drink products in Canadian markets. *J. Agric. Food Chem.* 57 (4): 1307–1311.

50 Kang, P., Yang, S., Bai, S., and Wang, Q. (2021). Novel application of mechanochemistry in waste epoxy recycling via solid-state shear milling. *ACS Sustain. Chem. Eng.* 9 (35): 11778–11789.

51 Wang, F., Wang, Q., and Yang, S. (2025). Mechanochemical recycling of acid anhydride-cured epoxy resin for functional applications. *ACS Sustain. Chem. Eng.* 13 (1): 647–657.

52 Reddy, K.S.K., Gao, W.-J., Chen, C.-H. et al. (2021). Degradation of thermal-mechanically stable epoxy thermosets, recycling of carbon fiber, and reapplication of the degraded products. *ACS Sustain. Chem. Eng.* 9 (15): 5304–5314.

53 Yue, L., Ke, K., Amirkhosravi, M. et al. (2021). Catalyst-free mechanochemical recycling of biobased epoxy with cellulose nanocrystals. *ACS Appl. Bio Mater.* 4 (5): 4176–4183.

54 Mezeix, L., Gupta, P., Bouvet, C., and Wongtimnoi, K. (2024). Mechanical characterization of recyclable and non-recyclable bio-epoxy resins for aerospace applications. *J. Compos. Sci.* 8: 191.

55 Kim, Y.N., Kim, Y.-O., Kim, S.Y. et al. (2019). Application of supercritical water for green recycling of epoxy-based carbon fiber reinforced plastic. *Compos. Sci. Technol.* 173: 66–72.

56 Deng, J., Xu, L., Zhang, L. et al. (2019). Recycling of carbon fibers from CFRP waste by microwave thermolysis. *Processes* 7: 207.

57 Hao, S., He, L., Liu, J. et al. (2021). Recovery of carbon fibre from waste prepreg via microwave pyrolysis. *Polymers* 13: 1231.

58 Kadam, A., Pawar, M., Yemul, O. et al. (2015). Biodegradable biobased epoxy resin from karanja oil. *Polymer* 72: 82–92.

59 Zhang, J., White, J.C., Lowry, G.V. et al. (2025). Advanced enzyme-assembled hydrogels for the remediation of contaminated water. *Nat. Commun.* 16 (1): 3050.

60 Kumar, V.V., Sathyaselvabala, V., Premkumar, M.P. et al. (2012). Biochemical characterization of three phase partitioned laccase and its application in decolorization and degradation of synthetic dyes. *J. Mol. Catal. B Enzym.* 74 (1–2): 63–72.

61 Antecka, A., Blatkiewicz, M., Boruta, T. et al. (2019). Comparison of downstream processing methods in purification of highly active laccase. *Bioprocess Biosyst. Eng.* 42 (10): 1635–1645.

62 Liu, L., Liu, X., Li, X. et al. (2024). An integrative chemical recycling approach for catalytic oxidation of epoxy resin and in situ separation of degraded products. *Angew. Chem. Int. Ed.* 63: e202405912.

63 Mahmoodi, R., Zabihi, O., Ahmadi, M. et al. (2024). Catalytic pyrolysis upcycling of waste thermosetting epoxy resin into fire-retardant additive. *J. Appl. Polym. Sci.* 141 (47): e56271.

64 Zhang, D., Jia, D., and Chen, S. (2009). Kinetics of curing and thermal degradation of hyperbranched epoxy (HTDE)/diglycidyl ether of bisphenol-A epoxy hybrid resin. *J. Therm. Anal. Calorim.* 98 (3): 819–824.

65 Zhou, J., Zhang, D., and Zhang, A. (2010). Effect of hyperbranched poly(trimellitic anhydride ethylene glycol) epoxy (HTME) on thermal degradation activation energies of HTME/diglycidyl ether of bisphenol-A epoxy hybrid resin by Kissinger and Flynn–Wall–Ozawa method. *Polym.-Plast. Technol. Eng.* 49 (2): 128–135.

66 Battig, A., Markwart, J.C., Wurm, F.R., and Schartel, B. (2019). Matrix matters: hyperbranched flame retardants in aliphatic and aromatic epoxy resins. *Polym. Degrad. Stab.* 170: 108986.

67 Wang, X., Xing, W., Song, L. et al. (2013). Flame retardancy and thermal properties of novel UV-curing epoxy acrylate coatings modified by phosphorus-containing hyperbranched macromonomer. *J. Polym. Res.* 20 (7): 165.

68 Yu, Q., Liang, Y., Cheng, J. et al. (2017). Synthesis of a degradable high-performance epoxy-ended hyperbranched polyester. *ACS Omega* 2 (4): 1350–1359.

69 Guo, W., Chen, S., Cheng, J. et al. (2019). Synthesis of renewable and self-curable thermosetting hyperbranched polymers by a click reaction. *Prog. Org. Coat.* 134: 189–196.

70 Yu, C., Xu, Z., Wang, Y. et al. (2018). Synthesis and degradation mechanism of self-cured hyperbranched epoxy resins from natural citric acid. *ACS Omega* 3 (7): 8141–8148.

71 Ma, X., Guo, W., Xu, Z. et al. (2020). Synthesis of degradable hyperbranched epoxy resins with high tensile, elongation, modulus and low-temperature resistance. *Compos. Part B* 192: 108005.

72 Ma, X., Liang, Y., Xu, Z. et al. (2020). The versatility of hyperbranched epoxy resins containing hexahydro-*s*-triazine on diglycidyl ether of bisphenol-A composites. *Compos. Part B* 196: 108109.

73 Ferreira-Filipe, D.A., Oliveira, L., Paço, A. et al. (2024). Biodegradation of e-waste microplastics by *Penicillium brevicompactum*. *Sci. Total Environ.* 935: 173334.

74 Han, J., Chen, Q., Feng, Y. et al. (2023). Seawater-degradable and antibacterial epoxy thermosets employing betaine ester linkages. *ACS Appl. Polymer Mat.* 5 (5): 3298–3305.

75 Haque, M.S. and Islam, M.A. (2024). Microbial degradation of the waste jute fiber and sawdust fiber-based epoxy composites in drainage system. *Heliyon* 10 (20): e39397.

76 Nuccetelli, P., Maisto, F., Kraková, L. et al. (2024). Evaluation of microbial degradation of thermoplastic and thermosetting polymers by environmental isolates. *Coatings* 14 (8): 982.

77 Jin, F.L., Li, X., and Park, S.J. (2015). Synthesis and application of epoxy resins: a review. *J. Ind. Eng. Chem.* 29: 1–11.

78 Leahy, J.G. and Colwell, R.R. (1990). Microbial degradation of hydrocarbons in the environment. *Microbiol. Rev.* 54 (3): 305–315.

79 Meier, M.A.R., Metzger, J.O., and Schubert, U.S. (2007). Plant oil renewable resources as green alternatives in polymer science. *Chem. Soc. Rev.* 36 (11): 1788–1802.

80 Varaprasad, K., Jayaramudu, T., Kanikireddy, V. et al. (2020). Alginate-based composite materials for wound dressing application: a mini review. *Carbohydr. Polym.* 236: 116025.

81 Yuan, Y., Chen, M., Zhou, Q., and Liu, R. (2018). Synthesis and properties of UV-curable cardanol-based acrylate oligomers with cyclotriphosphazene core. *J. Coat. Technol. Res.* 16 (1): 179–188.

82 Barua, S., Chattopadhyay, P., and Karak, N. (2015). *s*-Triazine-based biocompatible hyperbranched epoxy adhesive with antibacterial attributes for sutureless surgical sealing. *J. Mater. Chem. B* 3 (28): 5877–5885.

83 Baruah, P., Duarah, R., and Karak, N. (2016). Tannic acid-based tough hyperbranched epoxy thermoset as an advanced environmentally sustainable high-performing material. *Iran. Polym. J.* 25 (10): 849–861.

84 Anderson, B.J. (2013). Thermal stability and lifetime estimates of a high temperature epoxy by T_g reduction. *Polym. Degrad. Stab.* 98 (11): 2375–2382.

85 Chen, J.S., Ober, C.K., Poliks, M.D. et al. (2004). Controlled degradation of epoxy networks: analysis of crosslink density and glass transition temperature changes in thermally reworkable thermosets. *Polymer* 45 (6): 1939–1950.

86 Li, B., Zhu, G., Hao, Y., and Ren, T. (2022). An investigation on the performance of epoxy vitrimers based on disulfide bond. *J. Appl. Polym. Sci.* 139: 51589.

87 Shen, Y. and Wu, Y. (2025). Pyrolysis valorization of waste epoxy thermosets. *J. Anal. Appl. Pyrolysis* 186: 106908.

88 Tian, F., Wang, X.L., Yang, Y. et al. (2020). Energy-efficient conversion of amine-cured epoxy resins into functional chemicals based on swelling-induced nanopores. *ACS Sustain. Chem. Eng.* 8 (5): 2226–2235.

89 Yang, W., Xi, A., Yang, B., and Zeng, Y. (2022). Fabrication and properties of epoxy vitrimer based on multiply dynamic covalent bonds. *Gaodeng Xuexiao Huaxue Xuebao* 43 (11): 268–278.

90 Zhang, J., Gong, Z., Wu, C. et al. (2022). Itaconic acid-based hyperbranched polymer toughened epoxy resins with rapid stress relaxation, superb solvent resistance and closed-loop recyclability. *Green Chem.* 24 (18): 6900–6911.

91 Xu, Z., Liang, Y., Ma, X. et al. (2019). Recyclable thermoset hyperbranched polymers containing reversible hexahydro-*s*-triazine. *Nat. Sustainability* 3 (1): 29–34.

92 Xu, H., Hu, J., Liu, X. et al. (2023). Interface strengthening and high-value recycling of epoxy resin/carbon fiber fabric composites. *Chem. Eng. J.* 465: 142998.

93 Cafaro, F., Ferrari, F., Carallo, G.A. et al. (2025). A sustainable microwave-assisted process for chemical recycling and the reuse of epoxy resin matrices. *Polymers* 17 (7): 989.

94 Chen, Y., Pei, Z., Feng, D., and Yi, J. (2025). Environmental impact and comprehensive value assessment of epoxy-based binder for asphalt pavement recycling based on LCA and LCC. *Chem. Eng. J.* 511: 161973.

95 Klose, L., Meyer-Heydecke, N., Wongwattanarat, S. et al. (2023). Towards sustainable recycling of epoxy-based polymers: approaches and challenges of epoxy biodegradation. *Polymers* 15 (12): 2653.

96 Li, Y., Wu, Y., Li, K. et al. (2024). Recycling of epoxy resins with degradable structures or dynamic cross-linking networks: a review. *Ind. Eng. Chem. Res.* 63 (12): 5005–5027.

97 Liu, H., Wei, L., Sun, Z. et al. (2024). Effect of catalysts on the performance of vitrimers based on dynamic ester exchange. *Diangong Jishu Xuebao* 39 (16): 5134–5148.

98 Mabalane, P.N., Molnar, K., Mazibuko, P.T. et al. (2025). Recovery of carbon fibres from aged epoxy matrix composites using H_2O_2 as an oxidant: a thermodynamic and technoeconomic analysis. *J. Eng.* 2025 (1): 5807390.

99 Memon, H., Wei, Y., and Zhu, C. (2022). Recyclable and reformable epoxy resins based on dynamic covalent bonds – present, past, and future. *Polym. Test.* 105: 107420.

100 Zhang, S., He, E., Liang, H. et al. (2024). Chemical recycling of epoxy thermosets: from sources to wastes. *Actuators* 13 (11): 449.

101 Zhang, X., Shi, Y., Qu, C. et al. (2025). Mild and efficient degradation of amine-cured epoxy resins using mixed solvents. *J. Appl. Polym. Sci.* 142 (33): e57311.

102 Zhang, J., Jiang, C., Deng, G. et al. (2024). Closed-loop recycling of tough epoxy supramolecular thermosets constructed with hyperbranched topological structure. *Nat. Commun.* 15 (1): 4869.

7

Hyperbranched Epoxy Resins—Vitrimers

Xicheng Yuan and Zhihuan Weng

Dalian University of Technology, State Key Laboratory of Fine Chemicals, Frontiers Science Center for Smart Materials Oriented Chemical Engineering, No.2 Linggong Road, Ganjingzi District Dalian, 116024, P.R. China

7.1 Introduction

Accounting for approximately 70% of the market, epoxy resins (ERs) are the most widely used thermosetting resins globally [1]. They exhibit a combination of exceptional properties such as excellent dimensional and thermal stability, high mechanical strength and creep resistance, excellent electrical insulation and chemical resistance [2], making them suitable for a wide range of applications such as coatings [3], adhesives, insulating materials, electronic packaging, and composite matrices. Widely utilized across numerous industries, including automotive, aerospace, transportation, construction, household products, and electronics, EPs are frequently subjected to demanding service conditions. These involve exposure to high mechanical stresses or pressures, elevated temperatures, friction, and vibration [4]. Such exposures trigger a series of mechanical damages in epoxy products, such as creep, cracks, fractures, and punctures. These defects collectively compromise the product's service life and introduce significant operational risks [5]. A major obstacle with EPs, as with conventional thermosets, lies in monitoring and repairing microcracks. This difficulty stems fundamentally from their insoluble and infusible cross-linked networks [6, 7]. More critically, the irreversible nature of their covalent cross-linked networks fundamentally prevents EPs from being reprocessed, recycled, reshaped, remolded, and reconfigured after curing [8]. Consequently, most end-of-life epoxy products—whether aged, damaged, or simply discarded—are typically disposed of via incineration or landfilling. This practice results in significant environmental pollution and, concurrently, constitutes a considerable waste of valuable resources [9].

In response, the innovative concept of "vitrimers" emerged in 2011, offering a potential pathway to overcome the recyclability limitations of conventional thermosets [10]. A vitrimer is a class of polymeric materials featuring dynamic covalent networks, where bond exchange reactions enable the network topology to be reconfigured under specific conditions [11]. At their intended service temperatures, vitrimers possess the robust mechanical strength, chemical resistance, and dimensional stability characteristic of traditional thermosets. When heated above a threshold, they exhibit thermoplastic-like traits such as reparability and malleability due to activated bond exchange [12, 13].

Recent decades have seen a focus on introducing hyperbranched topology into EPs to create hyperbranched EPs (HERs). This architecture—characterized by a branched backbone, a functional periphery, and an intrinsically low viscosity—confers distinctive advantages to the resulting materials [14]. HERs' toughening effect is instrumental in the development of high-performance epoxy thermoset systems [15]. Unlike conventional epoxy toughening that relies on a dispersed second phase, such as rubbers, liquid crystal polymers, or nanomaterials, the incorporation of HERs creates a homogeneous toughening system within the epoxy matrix [16, 17]. A homogeneous toughening system is realized through the molecular-level dispersion of modifiers within the epoxy matrix. This facilitates the creation of interpenetrating networks, where molecular aggregates are seamlessly incorporated into the cross-linked architecture without triggering secondary phases or micro/nano-scale phase separation. Such a uniform integration is key to synergistically enhancing both the toughness and strength of the resin. Furthermore, HERs confer the additional benefit of markedly reducing matrix viscosity, thereby yielding low-viscosity epoxy formulations. This combination of enhanced mechanical properties and improved processability renders these systems highly promising for advanced composite applications [14, 18, 19]. The abundant functional groups within the glassy domains of hyperbranched epoxy vitrimers act as catalytic sites, significantly enhancing the efficiency of dynamic bond exchange reactions. Concurrently, the hyperbranched topology increases the free volume of the network and facilitates a higher effective cross-linking density. This unique architectural combination not only promotes dynamic exchange but also concurrently enhances the material's toughness [20].

7.2 Chemical Structures of Epoxy Resin Vitrimers

Over the past two decades, a major research focus has been the development of recyclable, reprocessable, and healable EPs, primarily through the strategic incorporation of reversible bonds into their cross-linked networks [11]. These dynamic bonds are primarily categorized into two classes: reversible non-covalent interactions [21] such as hydrogen bonds and host–guest interactions; and reversible covalent bonds [22], including but not limited to ester bonds, disulfide bonds, imine bonds, silyl ether bonds, and those formed via the Diels–Alder reaction. Due to their dependence on reversible physical bonds, epoxies with non-covalent cross-links are typically too fragile to endure applications involving significant mechanical stress [9]. As an alternative, a class of polymer networks constructed from reversible covalent bonds has been developed. These materials are termed "covalent adaptable networks" (CANs) [11, 23]. CANs are a class of materials that impart plasticity to thermosets through bond exchange. Based on their exchange mechanism, they are categorized into two distinct types: dissociative CANs and associative CANs [24, 25]. In dissociative CANs, network rearrangement occurs via a dissociative exchange mechanism, characterized by the reversible breakage of a chemical bond at one site followed by its reformation at another.

Vitrimers are dynamic networks whose properties are governed by temperature-dependent bond exchange. Below the topology freezing temperature (T_v), they behave as robust thermosets. Exceeding T_v rapidly activates exchange reactions, enabling the network to rearrange. This is key to their functionality: under stress, transesterification facilitates topology rearrangement and stress relaxation, granting the solid material a liquid-like, flowable behavior [26].

7.2.1 Dynamic Transesterification Reaction

The extensive research focus within the field of vitrimers has been on systems utilizing transesterification chemistry, surpassing the study of other dynamic networks and even traditional thermosets. This predominance is rooted in two inherent material advantages: the natural abundance of reactive epoxide groups in EPs and the widespread use of acids and anhydrides as their primary curing agents [11]. Scheme 7.1 illustrates the corresponding reaction mechanism. Thermally induced transesterification, facilitated by the high local concentration of ester and hydroxyl functionalities in the cross-linked network, is the fundamental mechanism responsible for the malleable behavior of these epoxy vitrimers. This network is readily constructed by melt polymerization of epoxy monomers/prepolymers with acids/anhydrides. A catalyst is commonly incorporated to facilitate the dynamic exchange. The curing reaction itself produces the essential dynamic sites (ester bonds and hydroxyls), enabling topology rearrangement upon heating [27].

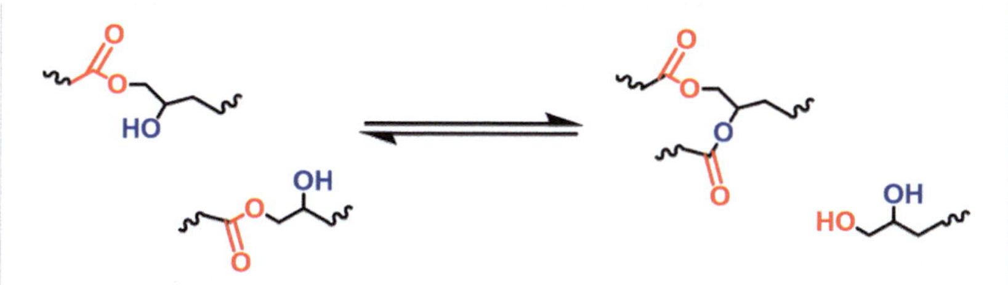

Scheme 7.1 Illustration of transesterification in hydroxyl–ester networks.

The most commonly employed small-molecule epoxy compounds or prepolymers are those of diglycidyl ether of bisphenol-A (DGEBA), including commercial grades such as E-51, E-44, DER 332, and DER 331. These materials form a critical and widely adopted component set for formulating both conventional and dynamic epoxy thermoset systems [28]. To ensure sufficient reactive sites for transesterification, curing agents that generate networks dense in ester bonds and hydroxyl groups are essential. For this reason, multi-carboxyl fatty acids and anhydrides are the favored choices, as they effectively build such cross-linked structures upon reaction with epoxies.

The fabrication of these epoxy vitrimers typically employs base catalysts to accelerate both the curing and transesterification processes. Common catalytic systems feature specific compounds like zinc acetylacetonate, zinc acetate dihydrate, 1,5,7-triazabicyclo[4.4.0]dec-5-ene (TBD), 1,2-dimethylimidazole (DMI), imidazoles, and triphenylphosphine. These represent the two dominant types: zinc(II)-based metal salts and a range of organocatalysts. However, while essential in most systems, these catalysts can introduce issues of potential toxicity and instability. These materials serve as fundamental and extensively utilized components in the development of both conventional and dynamic epoxy thermoset formulations [29].

The widespread application of traditional EPs is constrained by two major bottlenecks: high flammability and poor recyclability. As a promising solution, the bio-based, hexafunctional star-shaped epoxy monomer NEP, derived from a cyclotriphosphazene core, was synthesized by Wang et al. This was accomplished through a two-step procedure: nucleophilic substitution of hexachlorocyclotriphosphazene using eugenol and subsequent oxidation [30]. This work successfully developed an epoxy vitrimer (EV/NEP) that integrates intrinsic flame retardancy, recyclability, and shape memory function. With 9 mol% NEP, the material achieved an limiting oxygen index (LOI) of 26.2% and reduced peak heat release rate (PHRR) and total smoke release (TSR) by 24.9% and 35.4%, respectively, with its mechanisms analyzed. The dynamic network-enabled solvent- and heat-assisted recycling, as well as shape memory behavior. Combining these properties, EV/NEP emerges as a highly promising material for advanced applications in the aerospace and automotive sectors.

Over the past few decades, fiber-reinforced polymer composites (FRPC) have experienced tremendous growth. However, a significant end-of-life challenge persists: FRPC based on EPs is notoriously difficult to recycle due to the permanent, irreversible cross-linked network of the polymer matrix. This disposability issue is compounded by the high intrinsic value of the fibrous reinforcements. Consequently, developing effective recycling strategies for FRPC is not only an urgent environmental imperative but also offers considerable economic benefits by recovering valuable materials. Sangaletti et al. synthesized a recyclable and environmentally benign vitrimer utilizing epoxidized linseed oil as the bio-derived feedstock and a boronic ester as the dynamic cross-linker. This design leverages dynamic covalent chemistry to bridge sustainability with material reprocessability [31]. This vitrimer was successfully applied as a matrix to produce carbon fiber composites, achieving mechanical properties on par with traditional epoxy-based counterparts. Crucially, these composites address end-of-life challenges through a convenient recycling process. Based on the reversible hydrolysis of boronic ester linkages in a mild ethanol–water solution, both the valuable carbon fiber reinforcement and the reprocessable vitrimer matrix can be efficiently separated and recovered. This process establishes a practical pathway for achieving a closed-loop material cycle.

Zhang et al. successfully prepared a vitrimer material derived entirely from bio-based sources [32]. This research commenced with the ozone-mediated oxidation of Kraft lignin, yielding Oz-L enriched with carboxylic acid and phenolic moieties. This modified biopolymer was subsequently cross-linked with a sebacic acid-derived epoxy monomer using a zinc-based catalyst. The thermomechanical properties of the resulting materials—specifically the glass transition temperature (T_g), modulus, and tensile strength—were enhanced proportionally with increasing Oz-L content. The network's high density of ester and hydroxyl groups facilitated efficient zinc-catalyzed transesterification reactions (TERs). When heated above 160 °C, these TERs triggered rapid stress relaxation, imparting the material with excellent shape-changing, shape memory, self-healing, and malleable properties. Its application potential was demonstrated through the development of the Se-EP/Oz-L adhesive, which exhibited cohesive failure on aluminum substrates with a lap shear strength of 6.3 MPa. Remarkably, the adhesive could be rebonded at 190 °C after failure and could also be cleanly detached at elevated temperatures, both due to the reversible TER dynamics. This work establishes a straightforward method for fabricating fully bio-based vitrimers and underscores their promise as next-generation, recyclable adhesives.

While dynamic covalent chemistry presents a viable solution for recycling EPs and graphene composites, fundamental research challenges remain. A deeper understanding is required in several interconnected domains: the precise influence of curing agent architecture on network dynamics, the optimal pairing of dynamic covalent bond types with specific EP chemistries, and the underlying exchange mechanisms. Advances are contingent upon moving beyond empirical approaches to develop clear mechanistic models and robust design principles. The synergy of two bond exchange reactions is more conducive to improving the dynamic properties of materials. Starting from cardanol, Hu et al. designed a bio-based epoxy monomer, which was subsequently utilized to fabricate an epoxy vitrimer featuring a double dynamically cross-linked network [33]. This work charts a novel, sustainable path for advanced composites through bio-based epoxy vitrimers, which offer a compelling combination of straightforward processing, full life-cycle utility (including recycling/reuse), and excellent degradability. Their superiority stems from multi-stimulus responsiveness, multipath self-healing, and coupled mechanical/chemical recoverability. These properties are engineered by manipulating two key factors: the soft/hard segment architecture of the curing agent and the type of integrated dynamic covalent bonds. For instance, incorporating rigid furan rings yields vitrimers with enhanced mechanical strength (6.85 MPa) and thermal stability (T_g = 40.81 °C) over those with flexible chains. Moreover, a dual dynamic network design accelerates stress relaxation (achieving it in 5.6 minutes) and lowers the activation energy to 36.58 kJ/mol, outperforming single dynamic bond systems. The platform's functionality is further expandable, as demonstrated by creating conductive composites through graphene addition.

7.2.2 Dynamic Disulfide Exchange

A disulfide bond is a chemical bond with low bond energy, widely present in proteins. A common synthetic strategy for disulfide-based vitrimers involves the incorporation of small molecules bearing disulfide linkages into conventional curing systems, followed by cross-linking to form the dynamic network [34]. The specific reaction mechanism is shown in Scheme 7.2. Due to different reaction

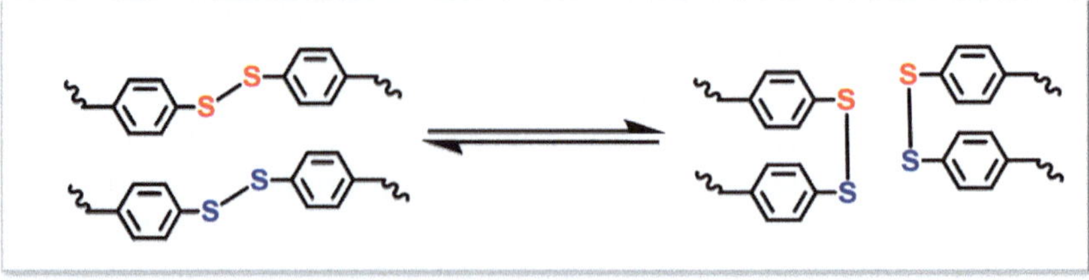

Scheme 7.2 Mechanism of disulfide bond exchange reaction.

mechanisms and substitution modes, the dynamic exchange reaction of disulfide bonds is a complex process. Disulfide bonds, characterized by their relatively low bond energy, serve as dynamic covalent cross-links capable of exchange via metathesis reactions at modest temperatures. This intrinsic feature allows them to impart reconfigurability to polymer networks under mild thermal conditions while retaining sufficient mechanical integrity. Notably, disulfide metathesis typically requires significantly lower activation temperatures compared to dynamic chemistries such as transesterification. Furthermore, the versatile reactivity of sulfur-based bonds enables their rearrangement under diverse stimuli, including heat, light (photolysis), or even ambient conditions, and can proceed with or without catalysts, offering broad adaptability in material design.

The relaxation process of disulfide bond exchange reaction is temperature-sensitive, exhibiting longer relaxation times at lower temperatures, while the relaxation time quickly shortens at higher temperatures. Its stimulus responsiveness (light, heat, or chemistry) can be applied in conductor materials, supercapacitors, and electronic skin [35].

In order to improve the degree of degradation and avoid deterioration of the physical properties of the material after multiple heat treatments, disulfide bonds are often mixed with other types of bonds. The mechanical and functional properties of the epoxy network are strategically tuned through specific dynamic bond combinations. Ductility and a broad glass transition zone, enabling multi-shape memory capability, are achieved by integrating disulfide bonds with heterogeneous silicon-based ether linkages. Shape stability is enhanced by incorporating a small amount of polyetheramine, which amplifies the modulus contrast between the glassy and rubbery states. Furthermore, the synergistic combination of disulfide and carbon–nitrogen bonds serves to elevate the material's heat resistance and overall mechanical performance [36].

Despite their significant advantages, vitrimers possess a critical limitation: their susceptibility to creep under service conditions. Creep refers to the time-dependent, progressive deformation of a material under sustained load and temperature. For vitrimers, while stress relaxation at elevated temperatures enables their malleability and reprocessability, the same exchange dynamics can lead to undesirable creep near their topology freezing transition temperature (T_v). In high-performance sectors like aerospace, where dimensional stability is paramount, this behavior severely restricts their application, presenting a major obstacle to replacing traditional thermosets. Addressing this challenge, Luzuriaga et al. made significant progress by developing a low-creep, aerospace-grade epoxy vitrimer. This material retains thermal and mechanical properties comparable to conventional aerospace epoxies while fully incorporating the unique capabilities of recyclability, repairability, and reprocessability [37]. To address the critical challenge of creep in high-performance applications, this work devised an aero-grade epoxy vitrimer by strategically embedding permanent cross-links within its dynamic network. This design successfully decouples creep resistance from reprocessability: the material retains excellent stress relaxation while its high-temperature dimensional stability is markedly improved. It achieves a glass transition temperature (T_g) of 175 °C and matches the mechanical and relaxation properties of classical aerospace-grade EPs, thereby meeting stringent aerospace specifications. The study thus outlines an effective and straightforward network design strategy for developing sustainable, high-temperature-resistant epoxy vitrimers suitable for aerospace components.

Schenk et al. established a systematic investigation framework for an epoxy vitrimer formulation. This framework is designed to build a comprehensive understanding by integrating the analysis of its fundamental reactivity, detailed curing kinetics, rheological behavior throughout processing, and final physicochemical properties [38]. This study successfully produced SR1710-V plates via resin transfer molding (RTM) that match the comprehensive physicochemical and mechanical profile of a benchmark thermoset. A key enabler for this was the optimization of the RTM process using a time–temperature–transformation (TTT) diagram, itself derived from curing kinetics modeled with the Kamal–Sourour equation. The vitrimer plates demonstrated superior performance in moisture resistance with only minor property loss in wet conditions, notwithstanding a distinctive dual-peak tan delta signature. Characteristic vitrimer behavior was confirmed through three-point bending relaxation tests, which showed linear viscoelasticity and an activation energy aligning with known disulfide-based systems. Furthermore, viable reprocessing parameters were identified, enabling mechanical grinding and thermoforming with substantial retention of initial mechanical properties. These results establish a foundation for extending this material platform to carbon fiber-reinforced composites, with future work aimed at assessing their reprocessing and repair potential.

Mounting ecological pressures associated with petroleum-derived polymers are driving a significant shift toward innovative, bio-based material platforms. This transition is particularly critical in the thermoset sector, which is heavily reliant on EPs, commanding approximately 70% of the market. Consequently, innovating sustainable and recyclable epoxy formulations has emerged as a central focus for materials research. Addressing this need, Yan and coworkers designed a novel epoxy vitrimer utilizing renewable feedstocks, achieving a significantly higher bio-based content compared to conventional epoxy formulations [39]. A fully bio-based vitrimer demonstrates exceptional property evolution upon recycling: after five reprocessing cycles, tensile strength surges from 1.87 to 18.47 MPa (a >900% increase), Young's modulus jumps from 2.98 to 268 MPa, and the glass transition temperature elevates from 7 to 25 °C. This self-strengthening material is synthesized from epoxidized starch amylopectin, garlic-derived diallyl disulfide, and pentaerythritol tetrakis (3-mercaptopropionate) (PETMP) thiol, forming a dynamic network. The key mechanism is a reprocessing-induced phase homogenization, where mechanical energy eliminates initial heterogeneity between components, resulting in a more dense and uniformly cross-linked network upon resetting. The study thereby presents an innovative strategy for creating recyclable thermosets whose performance improves with use.

The advancement of miniaturized, integrated, and high-density electronics creates a critical demand for thermal interface materials (TIMs) with superior heat dissipation capability. Concurrently, the growing issue of electronic waste underscores the urgent need for TIM recyclability. In response, Liu et al. developed a fully recyclable TIM by combining a synthesized epoxy vitrimer with boron nitride (BN) nanosheets. This composite achieves both high thermal conductivity and excellent conformability to rough surfaces [40]. In a practical thermal management test, an electronic device utilizing the developed composite achieved a 20 °C lower core temperature compared to conventional silicone-based solutions. This performance leap is rooted in a uniquely engineered microstructure: simple hot-pressing induces a highly oriented in-plane arrangement of BN fillers. At 40 wt% loading, this architecture yields a through-plane thermal conductivity of 3.85 W/m/K—surpassing the pristine polymer by 30 times and the non-oriented composite by a factor of 4.3. Beyond superior heat dissipation and mechanical compliance, the composite enables a closed-loop life cycle. Leveraging the multistage degradability of its vitrimer matrix, mild chemical recycling recovers 96.2% of the BN and 73.6–82.4% of the organic components. This work demonstrates an integrated strategy from microstructure design to application, culminating in a high-performance, fully recyclable TIM.

The development of high dielectric constant (high-k) polymer composites faces two principal challenges: the inherent compromise where elevated permittivity (ε_r) exacerbates dielectric loss ($\tan\delta$) and degrades breakdown strength (E_b), and the emerging need for intrinsic electrical damage repair to prolong component lifespan. While valued in electronics for their high energy density, processability, and light weight, overcoming these limitations is crucial for next-generation high-k PMCs. Targeting this dual objective, Zhang et al. reported a new epoxy-based composite vitrimer with high-k functionality [41]. This work employs a molecular engineering strategy to concurrently address electrical performance and multimode damage repair in a high-k composite. The strategy features a dual disulfide deployment: first, as dynamic cross-links within the epoxy network to grant intrinsic reparability and second, as surface grafts on nanoparticle fillers to enhance network dynamics and electrical properties. This engineered system delivers a low dielectric loss ($\tan\delta$ ~0.025 at 0.1 Hz, RT) and a high breakdown strength (E_b ~26.0 kV/mm). Repair is mode-specific: thermal healing at 180 °C for 10 minutes addresses mechanical scratches, while a thiol-mediated process at 80 °C for 60 minutes cleans electrical carbon tracks and restores >94% of the original insulation strength. The study outlines a definitive design blueprint for high-k composites integrating low loss, high E_b, and tailored reparability.

7.2.3 Dynamic Imine Bonds

The C=N bond in imines (Schiff bases) is a prominent dynamic covalent linkage, notable for its ability to undergo both associative and dissociative exchange pathways. Its reactivity is uniquely versatile, primarily manifesting in three forms: hydrolysis (cleavage by water), transamination (exchange with a primary amine), and imine metathesis (exchange with another imine). This multifaceted reversibility underpins its utility in constructing adaptive materials. Scheme 7.3 outlines the corresponding reaction mechanism. Based on this dynamic versatility, the imine bond has been extensively employed to fabricate catalyst-free polyimine vitrimers. These materials inherently possess a suite of desirable properties, including healability, recyclability, malleability, and degradability. A materials design paradigm that incorporates the beneficial traits of polyimine systems into epoxy architectures enables the creation of imine-based epoxy vitrimers. These engineered materials achieve high performance by unifying structural robustness with dynamic, responsive functionality [42].

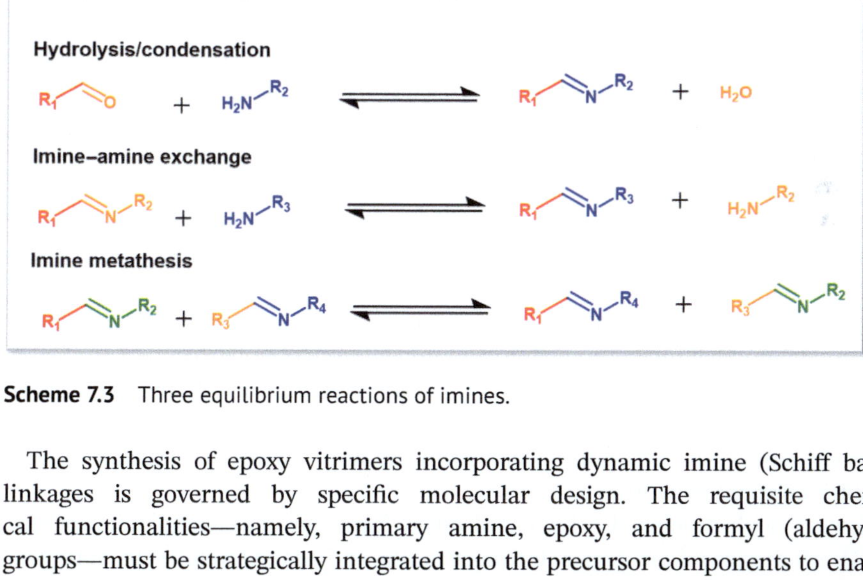

Scheme 7.3 Three equilibrium reactions of imines.

The synthesis of epoxy vitrimers incorporating dynamic imine (Schiff base) linkages is governed by specific molecular design. The requisite chemical functionalities—namely, primary amine, epoxy, and formyl (aldehyde) groups—must be strategically integrated into the precursor components to enable the concurrent formation of both the imine linkages and the epoxy cross-linked network. The imine bond can form a dynamic CAN under anhydrous conditions, which mainly undergoes three dynamic reactions: hydrolysis to compounds containing amino and carbonyl groups under acidic conditions, followed by the reformation of imine bonds under neutral conditions [43]. Two types of amines undergo exchange reaction to form imine bonds, and a double decomposition reaction can occur between the two imine bonds. The factors that affect reaction equilibrium include the type of solvent, acidity or alkalinity, temperature, etc. If under heating conditions, a repair rate greater than 100% can be achieved, and this reversible CAN can be recovered and reshaped at room temperature in the presence of water [44].

Employing a synergistic strategy that incorporates both imine and dynamic ester linkages within an epoxy network represents a promising design approach. Exemplifying this, Liu et al. synthesized a novel active ester hardener functionalized with imine bonds (TAI), which was then used to cure DGEBA EP, yielding a CAN material (DGEBA/TAI vitrimer) [45]. The synthesized epoxy vitrimer displays a well-rounded property profile: high thermal stability with an onset decomposition temperature of 364 °C and a char yield of 30% at 800 °C, low water uptake (0.25 wt%), and favorable dielectric characteristics (dielectric constant of 3.37, loss tangent of 0.013). A pivotal finding emerged from comparative scratch healing analysis: this rigid material lacks the intrinsic molecular-scale self-healing capability observed in softer vitrimers, clarifying a key mechanistic distinction. The observed scratch closure was attributed to macroscopic deformation recovery rather than molecular-scale bond reconnection. This finding suggests that the performance emphasis for such rigid vitrimers should pivot toward their recyclability and reprocessability. Accordingly, the resin was successfully recycled via hot-pressing and solvent methods, yielding recycled materials with well-retained tensile strengths (70 and 58 MPa, respectively, versus 69 MPa for the original). Beyond recycling, the dynamic network enables shape reconfiguration. To enhance the reliability of viscoelastic data, an alternative analytical approach was introduced. This method derives the relaxation time through stress relaxation curve fitting, offering a pathway to minimize experimental artifacts and independently verify data consistency.

Li and colleagues synthesized a bio-based epoxy vitrimer (epoxidized soybean oil [ESO]-VA) featuring dynamic Schiff base linkages. The material was derived from ESO, vanillin, and 4-aminophenol and exhibited excellent malleability and recyclability [46]. This work presents a sustainable life cycle for carbon fiber composites based on a Schiff base vitrimer matrix (ESO-VA). The composites themselves exhibit enhanced mechanical performance over the pure resin. Their end-of-life is addressed through designed circular pathways: chemical recycling via acid-triggered imine hydrolysis recovers valuable long CF fabric in pristine condition, while thermal recycling (e.g., compression molding) of shredded material produces new short-CF reinforced composites. Both pathways are intrinsically enabled by the dynamic covalent network.

Memon et al. developed a bio-based epoxy hardener containing imine bonds (ICH), synthesized from lignin-derived vanillin and methylcyclohexanediamine. This hardener is engineered to incorporate dynamic covalent bonds into epoxy networks, targeting enhanced recyclability of resulting composites. This work responds to an urgent demand in environmental economics for advanced recycling technologies, driven by the escalating waste volume of carbon fiber-reinforced polymer composites (CFRCs) [47]. This work presents a holistic life cycle management strategy for CFRCs, centered on an ICH-cured epoxy vitrimer. The vitrimer matrix itself exhibits a T_g ≥131 °C, tensile strength ≥82 MPa, and solvent resistance, while its dynamic imine bonds grant reprocessability (~90% strength retention post-reprocessing) and enable two distinct closed-loop chemical recycling paths, one achieving near-complete property recovery. In composite form, it matches conventional systems in flexural performance and allows repair (up to 92%

strength recovery). The strategy culminates in a full material cycle: amine-solvent degradation recovers pristine carbon fibers, and the dissolved matrix components are repolymerized into new resin. This integrated approach from durable material design to complete resource recovery offers a compelling, sustainable solution.

Developing epoxy vitrimers that integrate high performance with versatile recyclability is of paramount importance. To this end, Liu and coworkers synthesized two imine-containing hardeners, employing a comparative design strategy by using bio-based vanillin and petroleum-based *p*-hydroxybenzaldehyde as respective precursors [48]. The introduction of imine dynamic covalent bonds into the epoxy networks fundamentally dictates their multifunctional profile. These bonds not only grant the cured resins reprocessability and controlled degradability but also enable a practical closed-loop recycling where degradation products are directly repurposed into new resin. Consequently, the materials exhibit robust base performance—including a T_g surpassing 120 °C, tensile strength above 60 MPa, Young's modulus over 2500 MPa, and good solvent resistance—which is largely retained in both reprocessed and chemically recycled states. Crucially, the study verifies that the sustainable, vanillin-derived hardener yields EPs with thermal, mechanical, and recycling performance on par with those from petroleum-based chemistry.

7.2.4 Other Dynamic Cross-Linking Structures

The chemistry of siloxane equilibration, known for decades, experienced a significant revival in 2012 through the work of McCarthy [49]. Epoxy vitrimers based on siloxane equilibration can be readily prepared using common reagents. This chemistry endows the networks with efficient rearrangement capability, leading to excellent reprocessability. Pioneering this approach, Huang and co-workers demonstrated the fabrication of a stiff epoxy vitrimer for the first time. The curing reaction was achieved by combining diglycidyl ether of ethoxylated bisphenol-A with functional polysiloxane oligomers. In this system, aminopropyl side chains on the oligomers provided cross-linking sites, while terminal potassium silanolate groups catalyzed the subsequent siloxane exchange reactions [50]. This material exhibited a rapid stress relaxation rate alongside robust mechanical properties. In a related advancement, Lei and colleagues synthesized an epoxy thermoset featuring dynamic silyl ether (Si—O—Si) bonds. The resulting vitrimers, enabled by this exchangeable linkage, demonstrated exceptional performance, most notably in shape memory behavior and ease of reprocessing.

Silicon ether bond, as a new type of dynamic covalent bond used in material design, can cross-link styrene-based copolymers with dialkoxysilane through the hydroxyl groups of styrene monomers. Two types of dynamically cross-linked covalent bonds can be used to prepare a network: one introducing an amino group and the other not. The results show that introducing a neighboring amino group can accelerate the exchange rate of silicon ether groups by nearly three orders of magnitude. Combining this silicon ether bond into the silyl ether polymer network exhibits good ductility and reprocessing properties. In addition, uniform monodomain samples can also be prepared through reversible addition reactions of dynamic amino ester bonds and external stress relaxation methods. Due to the stepwise growth characteristics of hydroxyl and isocyanate groups, as well as the *trans–cis* photoisomerization properties of azobenzene, the network structure can exhibit reversible bending shape changes under visible light irradiation at 450 and 550 nm. Spontaneous thermal contraction/stretching deformation and reversible photo-induced bending/stretching behavior coupling in a system can be applied in emerging multifunctional devices [51].

7.3 Hyperbranched Epoxy Resin Vitrimer Matrices

Their widespread adoption across numerous industries is attributed to a combination of outstanding dimensional stability, robust mechanical properties, and strong resistance to both chemicals and creep. Their primary drawback (an inability to be recycled or reprocessed) is a direct consequence of their irreversibly cross-linked architecture. To address this, researchers have turned to dynamic covalent chemistry, which introduces bonds capable of exchanging within the network. When these cross-links efficiently rearrange between polymer chains, the material gains the ability to flow, enabling reprocessing while preserving its core characteristics.

In recent years, there has been growing research interest in bio-based epoxy vitrimers, primarily owing to the advantages of their plant-derived ingredients, which are environmentally benign, widely available, and renewable [52]. The synthesis of epoxy vitrimers from renewable sources often utilizes phenol monomers such as vanillin, guaiacol, and eugenol, which are derived from lignin, vegetable oils, or rosin. The key step involves converting their native functional groups (hydroxyl, aldehyde, methoxy, or C=C bonds) into epoxy groups, enabling their subsequent cross-linking into reprocessable vitrimer networks. A contemporary research strategy focuses on designing sustainable, high-performance epoxy materials through integrated molecular and topological engineering. This involves constructing the network from bio-based feedstocks and incorporating dynamic covalent bonds, thereby enabling essential life-cycle functions like self-healing and controlled degradation under environmentally benign conditions. Concurrently, integrating hyperbranched architectures into these dynamic networks addresses toughness limitations. The high density of functional groups' characteristic of hyperbranched structures further serves to accelerate the exchange kinetics of the dynamic bonds, enhancing the material's reprocessability [53, 54].

7.3.1 Hyperbranched Epoxy Vitrimers Based on Transesterification Reaction

The TER operates by rearranging network connectivity through the exchange of hydroxyl and ester functional groups. During the reaction, the total number of chemical bonds does not change, and the density of the material does not change either.

Due to its direct application in anhydride curing systems of EP materials and the easy formation of ester-based networks, the ester exchange reaction is currently the most studied reaction.

Zhong et al. achieved the production of catalyst-free, closed-loop recyclable epoxy vitrimers by synthesizing a key hyperbranched EP (ferulic acid-based hyperbranched epoxy resin [FEHBP]) derived from ferulic acid. Crucially, the resulting vitrimers retained their original thermal and mechanical properties [55]. The incorporation of hyperbranched FEHBP in the CA/FEP-FEHBP synthesis (Scheme 7.4) created epoxy vitrimers with a synergistic combination of robust mechanical properties and dynamic reparability. This synergy stemmed from FEHBP's dual role: its topological structure provided in situ reinforcement and toughening, while its hydroxyl groups catalyzed dynamic transesterification for facile reprocessing. These mechanisms endowed the material, particularly the 10 phr FEHBP formulation, with outstanding weldability, malleability, and a core set of high-performance metrics: a tensile strength of 126.4 MPa, a T_g of 94 °C, and a rapid stress relaxation time of 45 seconds at 140 °C, alongside a post-recycling strength retention above 88.3%. The closed-loop recyclability was further confirmed by reprocessing the degradation products under mild conditions into new vitrimers with properties mirroring the originals, underscoring the system's industrial potential.

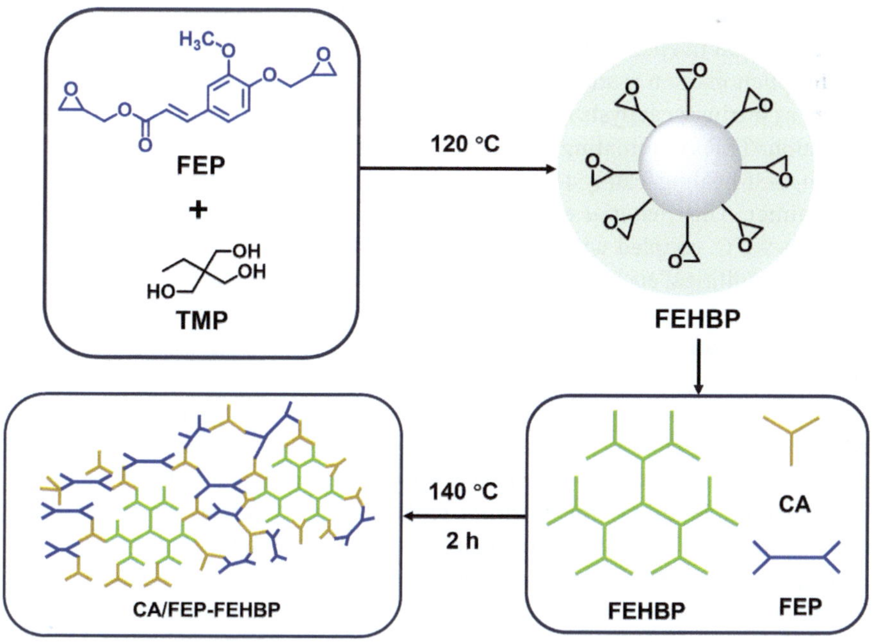

Scheme 7.4 Synthetic route of CA/FEP-FEHBP.

A catalyst-free epoxy vitrimer system was developed by Han et al., fabricated by the reaction of succinic anhydride (SA) with a pair of synthetically

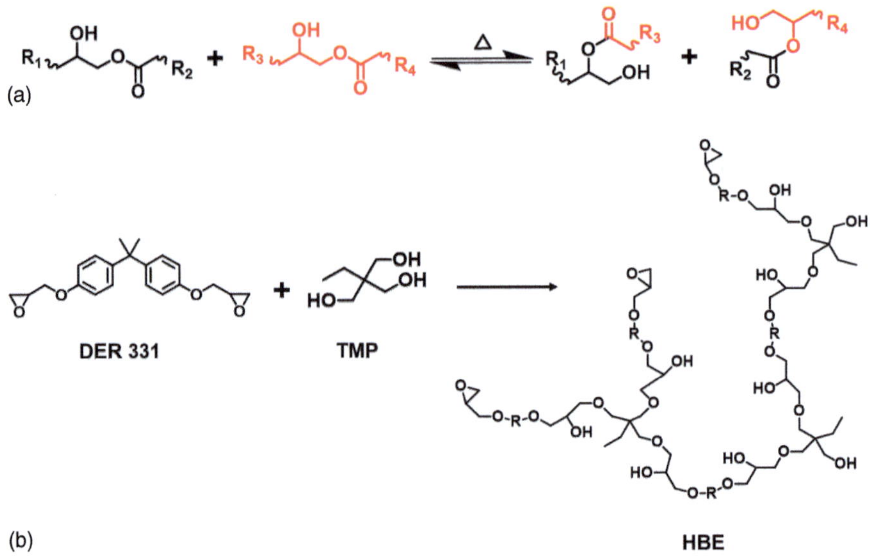

Scheme 7.5 (a) Schematic illustration of the dynamic transesterification in epoxy vitrimer and (b) synthetic route of hyperbranched epoxy (HBE).

tailored hyperbranched epoxy (HBE) resins [29]. Scheme 7.5a shows the dynamic transesterification in epoxy vitrimer, and Scheme 7.5b shows the synthetic route of HBE. This phenomenon is attributed to the abundant free hydroxyl groups in HBE, which act as intrinsic catalysts for both the initial curing and the subsequent transesterification (TER), eliminating the need for any external catalyst. The potential for developing self-healing, catalyst-free coatings is demonstrated, leveraging a unique epoxy vitrimer. This material exhibits a glass transition temperature (T_g) in the range of 70–96 °C, coupled with excellent thermostability and mechanical properties rivaling traditional epoxies. Its functionality stems from a temperature-sensitive dynamic network: upon heating above 120 °C, TERs are activated, enabling rapid stress relaxation within the cross-linked structure, which is the key to its reparability.

A catalyst-free and repairable epoxy coating system was constructed by Han et al. based on vitrimer chemistry. The matrix consisted of a blend of bisphenol-A epoxy (DER) and HBE, cured with SA [56]. This epoxy coating system delivers excellent hardness, strong adhesion, and intrinsic repair capability. These properties stem from its formulation: the abundant hydroxyls in HBE catalyze curing and bond exchange, while the DER epoxy provides low viscosity for processing and a high epoxy value for sufficient cross-linking. The healing process, triggered at 150 °C without applied pressure, enables surface scratches to be effectively remedied. This restoration allows the coating to fully regain its protective barrier function, thereby preventing corrosion of the underlying metal substrate when exposed to NaCl solution.

Zhang et al. prepared epoxy vitrimers by curing DGEBA with a synthesized itaconic acid-derived hyperbranched polymer (IAHBP) [57]. Scheme 7.6

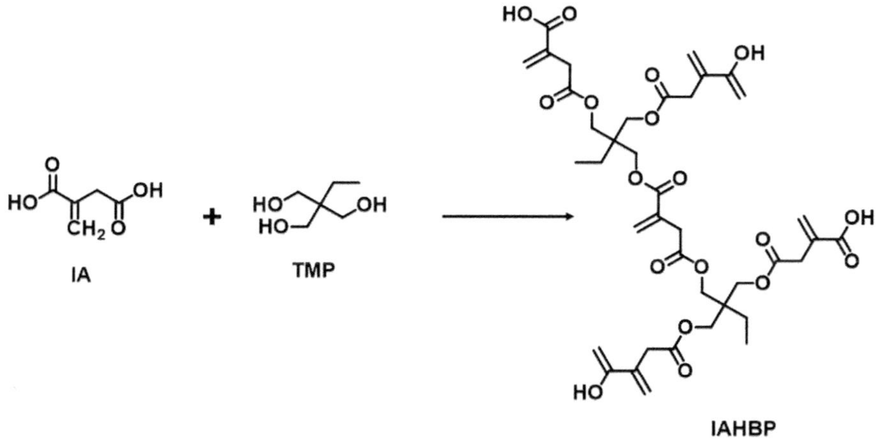

Scheme 7.6 Schematic preparation of IAHBP.

depicts the synthetic pathway employed. The DGEBA/IAHBP vitrimer exhibited tensile, flexural, and impact strengths of 91.0 MPa, 121.8 MPa, and 18.4 kJ/m [2], respectively. The hyperbranched topology of IAHBP promoted dynamic transesterification, endowing the material with outstanding malleability, shape reconfigurability, and reprocessability. It demonstrated exceptional shape memory performance over 10 cycles, with shape fixity (R_f) and recovery ratio (R_r) maintained at ~97% and ~99%, successfully executing grasping/loosening motions and recovering complex shapes (e.g., flower, 3D butterfly). The material exhibits integrated desirable properties: excellent solvent resistance stemming from its optimized free volume and cross-linking density, and efficient chemical recyclability via degradation in NaOH(aq.). The degradation products serve effectively as curing agents for fresh DGEBA, producing recycled vitrimers whose mechanical properties and T_g match the originals. This integrated performance and circular capability underpin a green and practical strategy for high-performance epoxy vitrimers.

Huang et al. successfully prepared epoxy vitrimer material based on ester exchange reaction using bisphenol F EP as the matrix, succinic acid as the curing agent, and *N*-methylimidazole as the catalyst, and modified the vitrimer material with hyperbranched polyesters (HBPs) Hyper C 102 [58]. The chemical structure and transesterification of epoxy vitrimers are shown in Scheme 7.7. The addition of HBPs reduces the cross-linking density of the material, which is due to the formation of voids in the internal microphase separation of the material. The presence of voids leads to a decrease in the cross-linking density and glass transition temperature of the material. As the content of HBPs continues to increase, microphase separation disappears and a continuous phase structure is formed, resulting in an increase in cross-linking density. The results of the relaxation experiment indicate that the addition of hyperbranched molecules prolongs the relaxation time of epoxy vitrimer materials. However, the relaxation time of each group of samples with different content of HBPs is within 30 minutes, and they can all relax to less than 10% of the initial modulus within 1 hour. The material demonstrates a synergistic

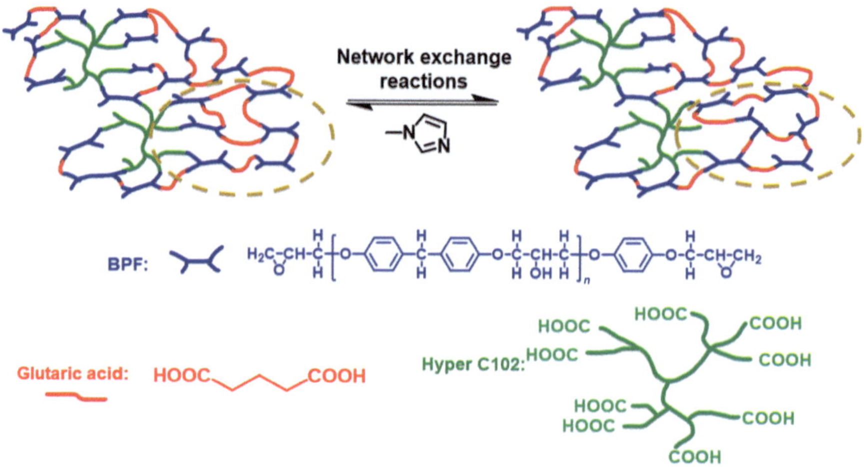

Scheme 7.7 Chemical structure and transesterification of epoxy vitrimer.

enhancement in properties. Final mechanical testing confirms that the addition of HBPs markedly improves mechanical performance, while the material maintains robust reprocessing and self-healing behavior. This advantageous combination is underpinned by a highly efficient internal TER, which is itself a consequence of the material's pronounced molecular chain mobility at high temperatures.

7.3.2 Hyperbranched Epoxy Vitrimers Based on Disulfide Exchange

Utilizing vanillin as a precursor, Hao et al. synthesized a hyperbranched EP (vanillin-based hyperbranched epoxy resin [VEHBP]) featuring disulfide and imine dynamic covalent bonds. This yielded a recyclable and malleable epoxy material possessing a high glass transition temperature, substantially improved creep resistance, and strong mechanical properties [17]. Scheme 7.8 shows the synthetic route of epoxy vitrimer. The distinctive hyperbranched topology and dual dynamic bonds of VEHBP impart a multifaceted enhancement when incorporated at 5% loading into an epoxy network. Structurally, this translates to remarkable mechanical gains: tensile strength (+34.1%), storage modulus (+19.7%), and most notably, tensile toughness (+173.3%). Thermally, it elevates the glass transition temperature to 175 °C and the creep temperature to 130 °C. Functionally, it confers excellent self-healing, reprocessability, and degradability. This clear structure–property relationship demonstrates a key advance in fabricating high-performance epoxy networks with adaptive covalent chemistry.

Zhang et al. synthesized hyperbranched polysiloxanes (HPBSi) functionalized with terminal methacrylate and epoxy groups. This design aimed to improve mechanical performance by creating a hierarchical network of permanent chemical cross-links within the material [59]. The specific chemical and dynamic structures are shown in Scheme 7.9. A series of room-temperature self-healing siloxane elastomers was constructed by cross-linking HPBSi with a linear dual-epoxy siloxane monomer, using an amino disulfide compound (AFD) as the dynamic linker.

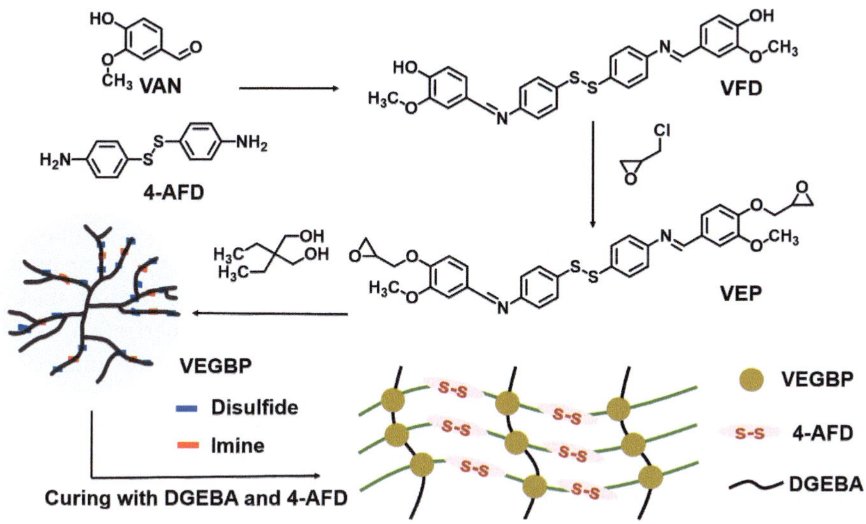

Scheme 7.8 Synthetic routes of VEHBP and dynamic covalent epoxy resin.

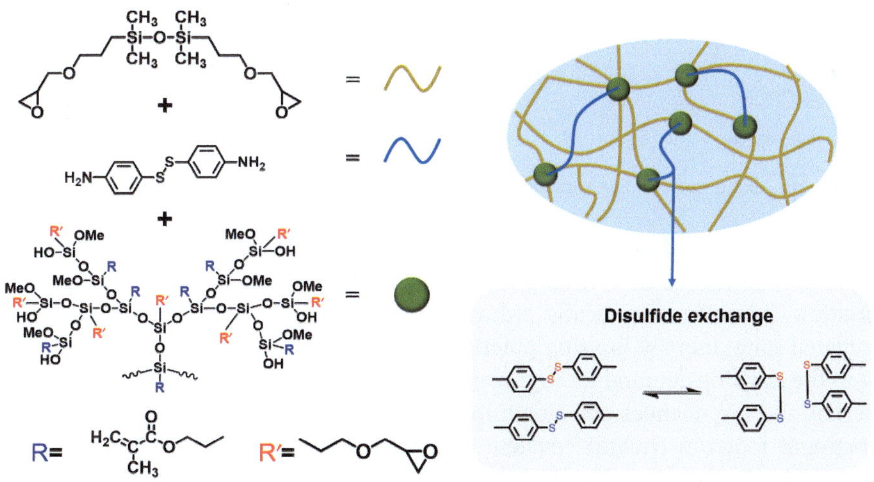

Scheme 7.9 Preparation of the siloxane elastomer.

The influence of HPBSi and AFD content was systematically investigated and found to be crucial in tuning the mechanical and self-healing properties. Specifically, the P1-5 elastomer with 5 wt% HPBSi achieved a tensile strength of 0.87 MPa, representing an approximately 410% enhancement over the HPBSi-free counterpart. Meanwhile, increasing the AFD content (as in P11-5) significantly accelerated the healing process, achieving the same efficiency in 12 hours compared to 48 hours required for P1-5. Ultimately, the application potential of this elastomer in flexible electronics was demonstrated by fabricating it into a functional strain sensor that successfully monitored a range of human motions, capitalizing on its inherent flexibility and self-healing capability.

7.3.3 Hyperbranched Epoxy Vitrimers Based on Imine Bonds

Zhang et al. fabricated a strong, tough epoxy supramolecular thermoset exhibiting both rapid reprocessability and the capability for closed-loop recycling at room temperature [60]. Scheme 7.10 shows the specific chemical structure and synthesis route. These thermosets were constructed using a vanillin-based hyperbranched EP, incorporating intermolecular hydrogen bonds, dual dynamic covalent bonds, and a unique cavity structure (both intramolecular and intermolecular). This supramolecular architecture endows the material with exceptional energy dissipation capabilities, resulting in high toughness and strength. A key feature of these thermosets is their capacity for efficient, room-temperature depolymerization, with recovered materials preserving the original structural and mechanical integrity. This property, alongside the ability to be rapidly and effectively reprocessed at 120 °C within 30 seconds, stems from the synergistic dynamic imine exchange and reversible non-covalent cross-links. Collectively, this strategy enables the design of tough, closed-loop recyclable epoxy thermosets viable for practical use.

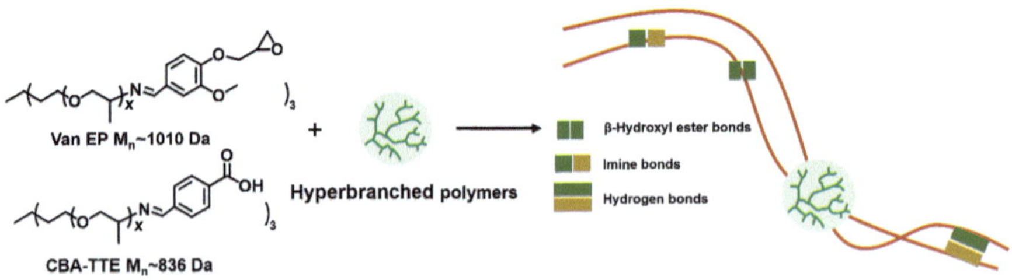

Scheme 7.10 Schematic of the synthetic route for epoxy supramolecular thermosets.

A major obstacle in fluorescent material science is the aggregation-caused quenching (ACQ) effect, prevalent in traditional systems built on rigid, planar π-conjugated aromatics. This effect drastically reduces fluorescence efficiency in the solid or aggregated state, thereby limiting practical applications. This stands in sharp contrast to the growing demand for high-performance fluorescent polymers in modern materials and life sciences. As shown in Scheme 7.11, Luo et al. developed a multifunctional material (hybrid covalent adaptive networks [HP/MB]) integrating mechanical robustness, self-healing, intrinsic fluorescence, and ion sensing [61]. Synthesized from hyperbranched polyamidoamine (HPAMAM) and MB via coupled dynamic reactions, the material achieves a tensile strength of 76.4 MPa and rapid thermal healing. Its fluorescence, originating from HPAMAM and correlating with cross-link density, enables real-time healing monitoring and, crucially, serves as a signal transducer via selective quenching by Fe^{3+} ions. This unique fusion of properties, stemming from its hybrid CAN design, highlights its potential as a platform for intelligent functional materials like self-reporting coatings and sensors.

7.4 Hyperbranched Epoxy Resin Vitrimer Composites

In most cases, pure polymers struggle to meet the ever-increasing performance demands across industries, and epoxy vitrimers are no exception to this limitation.

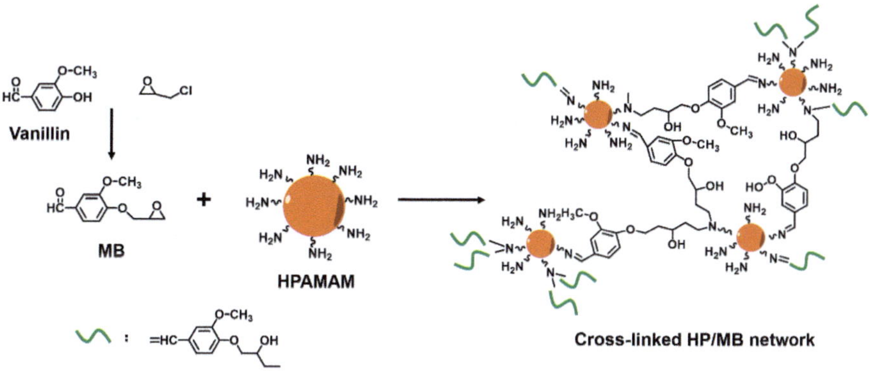

Scheme 7.11 Synthetic route and hypothesized network structure of the HP/MB material.

Consequently, the development of polymer composites—which synergistically combine a matrix with functional fillers—has become indispensable for achieving the required multifunctional property profiles [62]. Introducing nanoparticle fillers into vitrimer matrices is a widely adopted strategy to achieve multifunctionality. This strategy is directed toward the concurrent enhancement of multiple material properties, such as mechanical strength, thermal and electrical conduction capabilities, and stimulus responsiveness [63]. A fundamental design challenge for epoxy vitrimer composites lies in balancing property enhancement with the preservation of network dynamics. The incorporation of fillers, while improving mechanical or functional properties, inevitably restricts chain mobility and can inhibit the topological rearrangement underpinning vitrimer behavior. This leads to a practical trade-off: excessive filler content compromises reprocessability, while insufficient content yields marginal enhancement. This balance is further modulated by factors such as filler type, aspect ratio, loading, dispersion, and interfacial adhesion [64].

Designed to improve the local microenvironment of injured sites and facilitate tissue regeneration, biocompatible patches are indispensable noninvasive medical devices. Their efficacy in hemostasis, wound repair, and regeneration hinges on two key design criteria: mechanical compatibility with native soft tissues and tailorable biochemical properties. This is achieved through their capacity to serve as delivery platforms, encapsulating a range of active components including cells, extracellular vesicles, proteins, nucleic acids, and pharmaceuticals [65].

A novel synthetic route to HBPs with multiple epoxy chain ends has been developed recently. This method is based on the classical epoxy-amine curing chemistry but employs precise kinetic control during polymerization. The resulting multi-epoxy-functionalized HBPs are anticipated to be highly effective building blocks for constructing advanced cross-linked networks, capitalizing on the inherent advantages of the hyperbranched architecture [66]. This yields a cross-linked network, rapidly formed by reacting the resultant HBP with a thiol-functionalized polymer chain [67].

Zou et al. developed a hybrid biomedical patch system based on an epoxy network loaded with curcumin and Ti_3C_2Tx MXenes. The as-prepared composite patch exhibits a suite of advanced functionalities, including high stretchability, adhesion

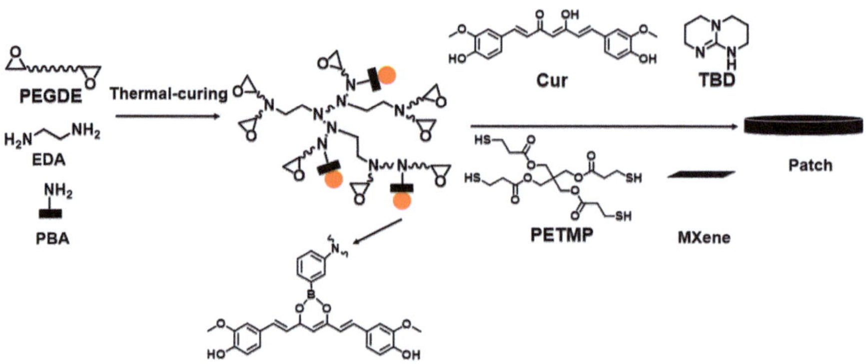

Scheme 7.12 Synthesis of EHBPE-PBA and the fabrication of biomedical patch.

to multiple substrates, self-healing capability, reprocessability, photothermal effect, and shape-morphing behavior [65]. Scheme 7.12 illustrates the precise chemical structure and the steps involved in its synthesis. The resulting biomedical patch concurrently offers robust functionalities critical for wound management: effective UV shielding, excellent reactive oxygen species (ROS) scavenging capacity, and high antibacterial efficacy against common pathogens such as *Staphylococcus aureus* and *Escherichia coli*. This comprehensive protective profile makes it particularly suitable for treating injuries to soft tissues, including skin wounds and various types of ulcerations.

The structural application of lightweight carbon fiber-reinforced polymer (CFRP) composites shows robust growth prospects in the wind and automotive sectors, where a steady compound annual growth rate (CAGR) of approximately 6% is projected. This expansion is primarily driven by their superior portfolio of properties, which includes excellent chemical and thermal resistance alongside outstanding mechanical performance [68]. Following a service life of 20–25 years, decommissioned wind turbine blades will generate a substantial stream of CFRP waste, projected to reach 38,000 tons annually after 2030. This looming waste crisis is exacerbated by the current low recycling rate, with only about 5% of CFRP being downcycled into lower-value secondary products. There is, therefore, an urgent and critical need to develop effective solutions to manage this waste and mitigate the associated environmental and resource challenges [69].

Ma et al. established a sustainable process to recycle both resin and fiber from CFRPs in a closed loop [70]. Scheme 7.13 shows the specific chemical structure. A novel composite (HT3/thiol-functionalized carbon fibers [TCF]) was fabricated using TCF and an epoxy-terminated degradable HBP (HT3). The composites exhibited robust acid resistance and mechanical strength, while the underlying network allowed controlled depolymerization with 89% recovery of chemical feedstock. These recovered chemicals were successfully utilized to resynthesize HT3 and its cross-linkable precursor. Furthermore, 100% of the carbon fibers were recovered without performance degradation and were directly reused to fabricate new composites. This work provides both a method for designing high-performance composites and a practical pathway for their high-efficiency, closed-loop recycling.

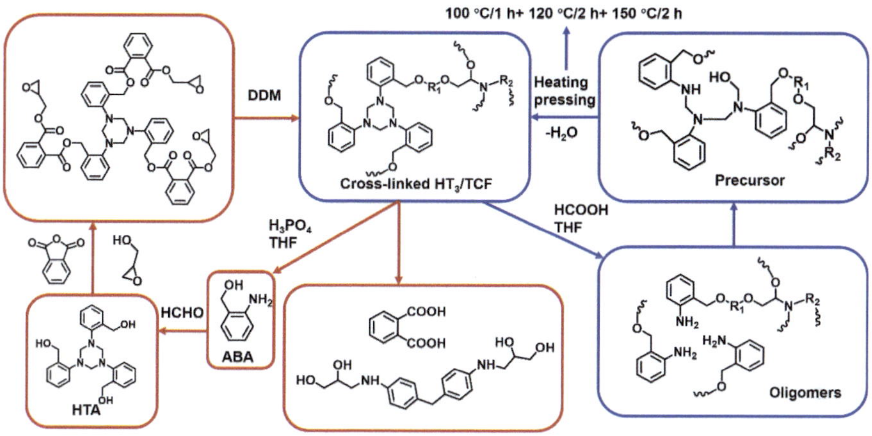

Scheme 7.13 Closed-loop process of composites.

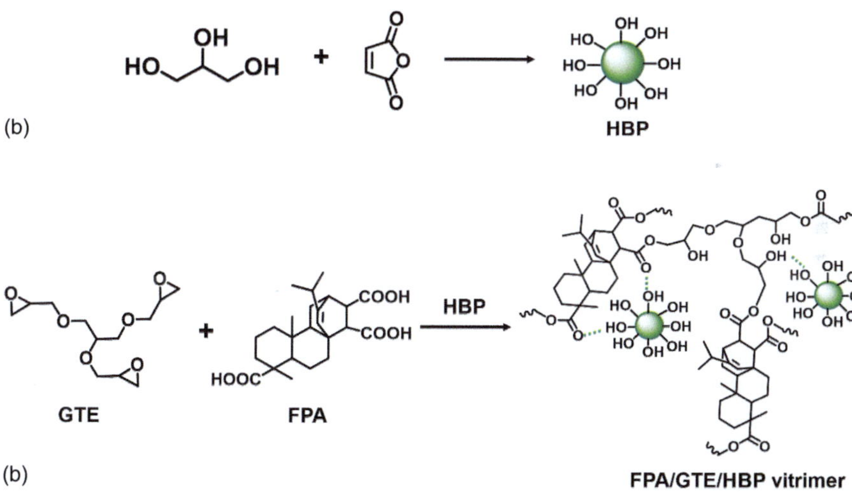

Scheme 7.14 Preparation of (a) hydroxy-terminated HBP and (b) FPA/GTE/HBP vitrimers.

Utilizing a blend of hydroxy-terminated hyperbranched polyesters (HBPs), bio-based fumaropimaric acid (from rosin), and glycerol triglycidyl ether (GTE), Xu et al. prepared a vitrimer matrix capable of closed-loop recycling [71]. The preparation route is as shown in Scheme 7.14. A fully recyclable CFRP strategy is demonstrated using a vitrimer matrix from FPA, GTE, and HBP. HBP provides catalytic hydroxyls for dynamic transesterification (enabling self-healing) and synergizes with rigid FPA for enhanced mechanics in both resin and composites (e.g., high interlayer shear strength). Crucially, HBP's hydroxyls also catalyze selective ethylene glycol degradation of the composite, enabling matrix closed-loop recycling while reclaiming undamaged carbon fibers (verified by SEM, Raman, tensile tests). This constitutes an eco-friendly design from the molecular level up.

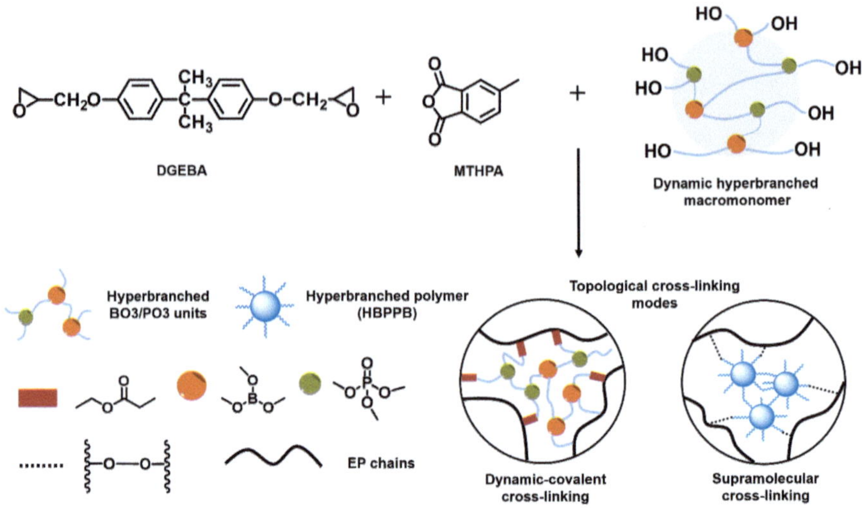

Scheme 7.15 The forming strategy of HDCNs.

Zhang et al. developed hyperbranched dynamic cross-linked networks (HDCNs) that integrate multiple dynamic bonds within a distinct topological cross-linking architecture [72]. Scheme 7.15 shows the forming strategy of HDCNs. A groundbreaking discovery reveals that HDCNs exhibit a unique ethanol-triggered reconfigurability, enabling them to function as a flexible elastomer at 25 °C, while attaining a supra-high flexural modulus of 5.45 GPa at −150 °C. This extraordinary behavior, mediated by ester exchange, stems from a uniquely designed architecture that permits solvent permeation for dynamic responsiveness while unifying high strength, transparency, and flame retardancy. Leveraging this architecture, conventional epoxy thermosets are transformed into degradable vitrimers via reversible dynamic bonds. The resulting material also supports mild solvent degradation and powder reprocessing at room temperature, facilitating composite recycling. This work unveils a material with unparalleled potential for applications spanning extreme thermal conditions.

7.5 Conclusions and Perspectives

Epoxy vitrimers have represented a leading edge of the dynamic and productive vitrimer research landscape over the past 10 years. Theoretically, epoxy vitrimers hold the ideal potential to replace conventional EPs across numerous applications. However, several critical challenges must be addressed to realize this vision. First, brittleness remains a common limitation for many epoxy vitrimers, akin to their traditional counterparts. This necessitates sophisticated molecular design, such as incorporating flexible segments or unique topological architectures, to enhance toughness. Second, the prevalent reliance on catalysts for curing and bond exchange raises concerns regarding cost, process complexity for mass production, and potential biocompatibility issues. Notably, we have demonstrated that transesterification-based reprocessing can be achieved in traditional epoxy networks without catalysts by leveraging the time–temperature equivalence principle, suggesting a promising hybrid pathway. Third, an inherent contradiction exists between the dynamic reversibility required for reprocessing and the long-term thermal stability needed for service. Engineering a sufficiently wide temperature window between these two states is a fundamental dilemma limiting practical deployment. Finally, beyond structural applications, significant opportunities lie in the biomedical field, where the unique dynamic properties of vitrimers could pave the way for next-generation diagnostic and therapeutic devices.

Introducing a hyperbranched topological structure into EP networks enables a homogeneous toughening mechanism, which simultaneously enhances both the toughness and strength of the material. This structural motif actively accelerates dynamic bond exchange reactions in the vitrimer network, thereby granting the epoxy vitrimers superior weldability, reparability, malleability, and reprocessability. While the incorporation of such dynamic bonds is essential for enabling chemical degradation and recycling of HERs, it often leads to an inevitable compromise in mechanical performance due to the relatively low bond energy of many dynamic linkages. Therefore, exploring novel dynamic bonds or chemical motifs—such as imide bonds or hexahydro-*s*-triazine structures—that combine high bond energy with efficient degradability is of great value. Future research should prioritize strategies that maintain or even enhance polymer performance through ingenious chemical design, while also critically evaluating the cost and synthesis efficiency of these new structures to ensure their practical applicability.

As conventional thermosets, cured EPs possess a highly cross-linked network that is inherently brittle. This brittleness stems from stress concentration at network defects and limited energy dissipation pathways. Introducing a hyperbranched topological structure into the EP network can effectively address this by enabling homogeneous toughening, which concurrently enhances both toughness and strength. To advance this strategy, research must focus on establishing a systematic relationship between topological structure and mechanical properties. Key scientific questions include: How does the hyperbranched architecture influence free volume distribution and, consequently, the macroscopic mechanical behavior? Investigating this requires the precondition of synthesizing HERs with precise, controlled structures. Furthermore, employing more advanced and precise analytical techniques is essential to accurately characterize the microstructures of both the HERs and their cured networks, thereby bridging the gap between molecular design and bulk performance.

Therefore, the rational design of novel hyperbranched epoxy vitrimers constitutes a pivotal frontier in sustainable polymer science, specifically aimed at reconciling the long-standing conflict between rapid dynamic exchange kinetics and robust thermal–mechanical stability. By leveraging the unique architectural features of hyperbranched topology—such as high-density, chain-end catalytic moieties, intrinsic free volume for stress dissipation, and nanoconfined dynamic networks—these engineered materials achieve programmable reconfiguration under mild stimuli while effectively resisting creep deformation during high-temperature service. This integrated capability fundamentally transcends the canonical "reprocessability-stability trade-off" that constrains conventional vitrimers. Consequently, hyperbranched epoxy vitrimer systems emerge as enabling platforms for next-generation technologies, including recyclable high-performance composites for aerospace, self-healing coatings for extreme environments, and materials conducive to closed-loop circular economies.

Acknowledgments

This work was supported by the Fundamental Research Funds for the Central Universities (DUT24ZD113) and National Natural Science Foundation of China (52073038 and 51873027).

References

1 Jin, F.-L., Li, X., and Park, S.-J. (2015). Synthesis and application of epoxy resins: a review. *J. Ind. Eng. Chem.* 29: 1–11.
2 Capricho, J.C., Fox, B., and Hameed, N. (2020). Multifunctionality in epoxy resins. *Polym. Rev.* 60 (1): 1–41.
3 Ou, B.L., Wang, Y.W., and Lu, Y. (2021). A review on fundamentals and strategy of epoxy-resin-based anticorrosive coating materials. *Polym. Plast. Technol. Mater.* 60 (6): 601–625.
4 Rahman, M.M. and Islam, M.A. (2022). Application of epoxy resins in building materials: progress and prospects. *Polym. Bull.* 79 (3): 1949–1975.
5 Cheng, L.L., Chen, R.Z., Yang, J.L. et al. (2025). Mechanisms, technical optimization, and perspectives in the recycling and reprocessing of waste wind turbine blades: a review. *Renew. Sustain. Energy Rev.* 218: 115834.
6 Fan, X.T., Zheng, J., Yeo, J.C.C. et al. (2024). Dynamic covalent bonds enabled carbon fiber reinforced polymers recyclability and material circularity. *Angew. Chem. Int. Ed.* 63 (40): e202408969.
7 Karger-Kocsis, J. and Kéki, S. (2018). Review of progress in shape memory epoxies and their composites. *Polymers* 10 (1): 34.
8 Li, W.D., Zhang, Y.C., Zhao, X. et al. (2023). Full-life-cycle eco-friendly polymeric insulating materials: research progress and future prospects. *J. Phys. D: Appl. Phys.* 56 (37): 373003.
9 Li, Y.Q., Wu, Y.J., Li, K.Y. et al. (2024). Recycling of epoxy resins with degradable structures or dynamic cross-linking networks: a review. *Ind. Eng. Chem. Res.* 63 (12): 5005–5027.
10 Montarnal, D., Capelot, M., Tournilhac, F. et al. (2011). Silica-like malleable materials from permanent organic networks. *Science* 334 (6058): 965–968.
11 Yang, Y., Xu, Y., Ji, Y. et al. (2021). Functional epoxy vitrimers and composites. *Prog. Mater. Sci.* 120: 100710.

12 Zhang, F.L., Zhang, L., Yaseen, M. et al. (2021). A review on the self-healing ability of epoxy polymers. *J. Appl. Polym. Sci.* 138 (16): 50260.

13 Zhang, S.H., He, E.J., Liang, H. et al. (2024). Chemical recycling of epoxy thermosets: from sources to wastes. *Actuators* 13 (11): 449.

14 Jiang, Y., Li, J., Li, D. et al. (2024). Bio-based hyperbranched epoxy resins: synthesis and recycling. *Chem. Soc. Rev.* 53 (2): 624–655.

15 Corcione, C.E. and Frigione, M. (2009). A novel procedure able to predict the rheological behavior of trifunctional epoxy resin/hyperbranched aliphatic polyester mixtures. *Polym. Test.* 28 (8): 830–835.

16 Dong, H.H. and Zhu, X.Y. (2017). Epoxy resins modified by hyperbranched polyurethanes with different composition. *Acta Polym. Sin.* (2): 342–350.

17 Hao, Y.X., Zhong, L.Y., Li, T.C. et al. (2023). High-performance recyclable and malleable epoxy resin with vanillin-based hyperbranched epoxy resin containing dual dynamic bonds. *ACS Sustain. Chem. Eng.* 11 (30): 11077–11087.

18 Kou, D.K., Feng, W.P., and Zhang, L. (2010). The preparation and properties of the E-51 epoxy resin modified by epoxy-terminated hyperbranched poly(amine-ester). *Proceedings of the 7th National Conference on Functional Materials and Applications*, Hunan, P.R. China.

19 Liang, X., Li, X.J., Tang, Y. et al. (2022). Hyperbranched epoxy resin-grafted graphene oxide for efficient and all-purpose epoxy resin modification. *J. Colloid Interface Sci.* 611: 105–117.

20 Lu, Y.F., Xu, H.F., Liang, N. et al. (2022). High mechanical strength of shape-memory hyperbranched epoxy resins. *ACS Appl. Polym. Mater.* 4 (8): 5574–5582.

21 Xie, H.F., Li, C.X., and Wang, Q.J. (2024). Thermosetting polymer modified asphalts: current status and challenges. *Polym. Rev.* 64 (2): 690–759.

22 Markouti, P., Tzouma, E., Paipetis, A.S. et al. (2025). Conventional thermoset composites and their sustainable alternatives with vitrimer matrix-waste management/recycling options with focus on carbon fiber reinforced epoxy resin composites. *Materials* 18 (2): 351.

23 Wang, B.B., Ma, S.Q., Yan, S.F. et al. (2019). Readily recyclable carbon fiber reinforced composites based on degradable thermosets: a review. *Green Chem.* 21 (21): 5781–5796.

24 Chen, C.W., Mao, H., Lee, H.K. et al. (2024). Upcycling recycled epoxy-based vitrimer for enhancing toughness of poly(ethylene terephthalate). *J. Polym. Environ.* 32 (2): 621–631.

25 Ding, X.M., Chen, L., Xu, Y.J. et al. (2023). Robust epoxy vitrimer with simultaneous ultrahigh impact property, fire safety, and multipath recyclability via rigid-flexible imine networks. *ACS Sustain. Chem. Eng.* 11 (39): 14445–14456.

26 Jouyandeh, M., Tikhani, F., Hampp, N. et al. (2020). Highly curable self-healing vitrimer-like cellulose-modified halloysite nanotube/epoxy nanocomposite coatings. *Chem. Eng. J.* 396: 125196.

27 Yang, H.K. and Wang, D. (2024). Comparing surface and bulk curing processes of an epoxy vitrimer. *ACS Appl. Mater. Interfaces* 16 (12): 15479–15486.

28 Wang, Y.J., Yan, X.L., Ai, T. et al. (2024). Preparation and characterization of novel epoxy vitrimer asphalt binders based on reversible covalent network. *Constr. Build. Mater.* 411: 134242.

29 Han, J., Liu, T., Hao, C. et al. (2018). A catalyst-free epoxy vitrimer system based on multifunctional hyperbranched polymer. *Macromolecules* 51 (17): 6789–6799.

30 Wang, D., Zhang, Z., Xu, X. et al. (2025). Cyclotriphosphazene based epoxy vitrimer with excellent recyclability and flame retardancy. *Polym. Degrad. Stab.* 238: 111346.

31 Sangaletti, D., Ceseracciu, L., Marini, L. et al. (2023). Biobased boronic ester vitrimer resin from epoxidized linseed oil for recyclable carbon fiber composites. *Resour. Conserv. Recycl.* 198: 107205.

32 Zhang, S., Liu, T., Hao, C. et al. (2018). Preparation of a lignin-based vitrimer material and its potential use for recoverable adhesives. *Green Chem.* 20 (13): 2995–3000.

33 Hu, Y., Tong, S., Hu, L. et al. (2023). Molecularly engineered cardanol derived epoxy vitrimers based on dynamic disulfide and dynamic ester exchanges with desirable dynamic response, degradability, and recyclability. *Chem. Eng. J.* 477: 147284.

34 Chai, S.B., Fang, Y.X., Chen, Z. et al. (2025). Constructing fully biobased epoxy vitrimer based on disulfide bonds with desired mechanical, recyclability, and degradability properties. *ACS Appl. Polym. Mater.* 7 (6): 3981–3990.

35 Luo, C.M., Liu, X.Y., Chen, Z.L. et al. (2024). Recyclable epoxy vitrimer with high strength and rapid stress relaxation under mild temperatures. *ACS Appl. Polym. Mater.* 6 (19): 12281–12289.

36 Ji, F.C., Liu, X.D., Sheng, D.K. et al. (2020). Epoxy-vitrimer composites based on exchangeable aromatic disulfide bonds: reprocessibility, adhesive, multi-shape memory effect. *Polymer* 197: 122514.

37 Ruiz de Luzuriaga, A., Markaide, N., Salaberria, A.M. et al. (2022). Aero grade epoxy vitrimer towards commercialization. *Polymers* 14 (15): 3180.

38 Schenk, V., de Calbiac, J., D'elia, R. et al. (2024). Epoxy vitrimer formulation for resin transfer molding: reactivity, process, and material characterization. *ACS Appl. Polym. Mater.* 6 (10): 6087–6095.

39 Tratnik, N., Tanguy, N.R., and Yan, N. (2023). Recyclable, self-strengthening starch-based epoxy vitrimer facilitated by exchangeable disulfide bonds. *Chem. Eng. J.* 451: 138610.

40 Liu, J., Feng, H., Dai, J. et al. (2023). A full-component recyclable epoxy/BN thermal interface material with anisotropy high thermal conductivity and interface adaptability. *Chem. Eng. J.* 469: 143963.

41 Zhang, Y.-C., Li, W.-D., Zhao, X. et al. (2023). Epoxy-based high-*k* composite vitrimer: with low dielectric loss, high breakdown strength and surface electrical damage repairability. *Chem. Eng. J.* 473: 145199.

42 de Calbiac, J., Olivier, P., Zbed, R.S. et al. (2025). Assessing the impact of the off-stoichiometric epoxy/amine ratio on the curing kinetics and properties of space-grade disulfide-based vitrimer resins. *ACS Appl. Polym. Mater.* 7 (8): 5271–5279.

43 Luo, C.M., Wang, W.C., Yang, W. et al. (2023). High-strength and multi-recyclable epoxy vitrimer containing dual-dynamic covalent bonds based on the disulfide and imine bond metathesis. *ACS Sustain. Chem. Eng.* 11 (39): 14591–14600.

44 Yan, T., Jiang, H., Pang, W.T. et al. (2024). Preparation and self-healing properties of epoxy vitrimer materials based on imine bonds. *J. Appl. Polym. Sci.* 141 (29): e55684.

45 Liu, J., Liu, X., Cui, X. et al. (2023). Imine-containing epoxy vitrimer cured by active ester: properties and theoretical analysis. *ACS Appl. Polym. Mater.* 5 (12): 10042–10052.

46 Liu, Y.-Y., He, J., Li, Y.-D. et al. (2020). Biobased epoxy vitrimer from epoxidized soybean oil for reprocessable and recyclable carbon fiber reinforced composite. *Compos. Commun.* 22: 100445.

47 Memon, H., Wei, Y., Zhang, L. et al. (2020). An imine-containing epoxy vitrimer with versatile recyclability and its application in fully recyclable carbon fiber reinforced composites. *Compos. Sci. Technol.* 199: 108314.

48 Memon, H., Liu, H., Rashid, M.A. et al. (2020). Vanillin-based epoxy vitrimer with high performance and closed-loop recyclability. *Macromolecules* 53 (2): 621–630.

49 Zheng, P.W. and Mccarthy, T.J. (2012). A surprise from 1954: siloxane equilibration is a simple, robust, and obvious polymer self-healing mechanism. *J. Am. Chem. Soc.* 134 (4): 2024–2027.

50 Wu, X., Yang, X., Yu, R. et al. (2018). A facile access to stiff epoxy vitrimers with excellent mechanical properties via siloxane equilibration. *J. Mater. Chem. A* 6 (22): 10184–10188.

51 Liu, Q.F., Jiang, L., Zhao, Y.Y. et al. (2019). Reprocessable and shape memory thermosetting epoxy resins based on silyl ether equilibration. *Macromol. Chem. Phys.* 220 (13): 1900149.

52 Zhang, J.H., Mi, X.Q., Chen, S.Y. et al. (2020). A bio-based hyperbranched flame retardant for epoxy resins. *Chem. Eng. J.* 381: 122719.

53 Varley, R.J. and Tian, W. (2004). Toughening of an epoxy anhydride resin system using an epoxidized hyperbranched polymer. *Polym. Int.* 53 (1): 69–77.

54 Wang, Y.M., Chen, S.F., Chen, X.C. et al. (2019). Controllability of epoxy equivalent weight and performance of hyperbranched epoxy resins. *Compos. B: Eng.* 160: 615–625.

55 Zhong, L., Hao, Y., Zhang, J. et al. (2022). Closed-loop recyclable fully bio-based epoxy vitrimers from ferulic acid-derived hyperbranched epoxy resin. *Macromolecules* 55 (2): 595–607.

56 Han, J., Liu, T., Zhang, S. et al. (2019). Hyperbranched polymer assisted curing and repairing of an epoxy coating. *Ind. Eng. Chem. Res.* 58 (16): 6466–6475.

57 Zhang, J., Gong, Z., Wu, C. et al. (2022). Itaconic acid-based hyperbranched polymer toughened epoxy resins with rapid stress relaxation, superb solvent resistance and closed-loop recyclability. *Green Chem.* 24 (18): 6900–6911.

58 Huang, X., Liu, H.C., Fan, Z. et al. (2019). Hyperbranched polymer toughened and reinforced self-healing epoxy vitrimer. *Acta Polym. Sin.* 50 (5): 535–542.

59 Zhang, T., Su, H., Shi, X. et al. (2022). Self-healing siloxane elastomers constructed by hierarchical covalent crosslinked networks and reversible dynamic bonds for flexible electronics. *J. Mater. Sci.* 57 (22): 10444–10456.

60 Zhang, J., Jiang, C., Deng, G. et al. (2024). Closed-loop recycling of tough epoxy supramolecular thermosets constructed with hyperbranched topological structure. *Nat. Commun.* 15 (1): 4869.

61 Luo, W., Yu, H., Liu, Z. et al. (2022). Construction and application of hybrid covalent adaptive network with non-conjugated fluorescence, self-healing and Fe^{3+} ion sensing. *J. Mater. Res. Technol.* 19: 1699–1710.

62 Chabert, E., Vial, J., Cauchois, J.P. et al. (2016). Multiple welding of long fiber epoxy vitrimer composites. *Soft Matter* 12 (21): 4838–4845.

63 Basu, A., Parasuram, S., Supriya, H. et al. (2025). Triple dynamic network-enabled vitrimer for repairable and recyclable carbon fiber-reinforced composites. *ACS Appl. Polym. Mater.* 7 (8): 4931–4943.

64 Li, M.Z., Yu, H.T., Liu, X.Y. et al. (2024). Dual rapid self-healing and easily recyclable carbon fiber reinforced vitrimer composites. *Chem. Eng. J.* 480: 148147.

65 Zou, Y., Jin, X., Zhang, X. et al. (2021). A multifunctional biomedical patch based on hyperbranched epoxy polymer and MXene. *Sci. China Technol. Sci.* 64 (12): 2744–2754.

66 Liu, H.W., Wilén, C.E., and Skrifvars, M. (2000). Reaction of epoxy resin and hyperbranched polyacids. *J. Polym. Sci., Part A: Polym. Chem.* 38 (24): 4457–4465.

67 Tang, X.D. and Tan, S.L. (2014). Study on the hyperbranched polyphosphate ester modified epoxy resin. *Proceedings of the 3rd International Conference on Material Engineering, Chemistry, Bioinformatics (MECB 2013)*, Hefei, P.R. China (26–27 October 2013).

68 Tian, Z.S., Wang, Y.Q., and Hou, X.L. (2022). Review of chemical recycling and reuse of carbon fiber reinforced epoxy resin composites. *New Carbon Mater.* 37 (6): 1021–1041.

69 Arano, F., Casado, U., Ferrero, I.Z. et al. (2024). Self-healing of microcracks and scratches in a carbon-fiber reinforced epoxy vitrimer by conventional or remote heating. *ACS Appl. Mater. Interfaces* 17 (9): 13170–13178.

70 Ma, X., Xu, H., Xu, Z. et al. (2021). Closed-loop recycling of both resin and fiber from high-performance thermoset epoxy/carbon fiber composites. *ACS Macro Lett.* 10 (9): 1113–1118.

71 Xu, Y., Zhang, H., Dai, S. et al. (2022). Hyperbranched polyester catalyzed self-healing bio-based vitrimer for closed-loop recyclable carbon fiber-reinforced polymers. *Compos. Sci. Technol.* 228: 109676.

72 Zhang, Y., Yan, H., Yu, R. et al. (2024). Hyperbranched dynamic crosslinking networks enable degradable, reconfigurable, and multifunctional epoxy vitrimer. *Adv. Sci. (Weinh)* 11 (2): e2306350.

8

Perspectives

Yu Jiang, Kai Cheng, Xiongjie Li, and Daohong Zhang

South-Central Minzu University, Hubei R&D Center of Hyperbranched Polymers Synthesis and Applications, Key Laboratory of Catalysis and Energy Materials Chemistry of Ministry of Education & Hubei Key Laboratory of Catalysis and Materials Science, No.182, Minzu Ave. Hongshan District, Wuhan 430074, P.R. China

Hyperbranched epoxy resins (HERs) have become a widely studied class of polymers, demonstrating immense potential to redefine the functionality of thermosetting materials. The unique architecture of HERs—their three-dimensional (3D) ellipsoidal topology, high density of terminal groups, intrinsically low melt and solution viscosity, and easily tunable molecular parameters—provides a powerful way to overcome the persistent limitations of conventional epoxy resins: inherent brittleness, high processing viscosity, and intractability post-curing. As detailed in the preceding chapters, a comprehensive foundation has been laid, including their historical development, synthesis pathways, characterization methods, diverse application landscapes, strengthening and toughening mechanisms, and burgeoning potential for degradation and recycling. However, despite remarkable progress over the past decades, a significant gap remains in translating fundamental research on HERs into widespread industrial application. To bridge this gap, this concluding chapter integrates the insights presented throughout the book to formulate a coherent and ambitious roadmap for HERs, focusing on strategic advances in synthetic innovation, precision structural control, multifunctional integration, sustainable development, and viable industrialization pathways.

8.1 Innovation in Synthesis and Precision Structural Control

The development of HERs begins with their synthesis, and future advancement demands a paradigm shift from established methods toward unprecedented precision and efficiency.

8.1.1 Novel Controlled Polymerizations and Click Chemistries

While current strategies like proton transfer polymerization, step-growth polycondensation of $A_2 + B_3$ monomers, and post-polymerization modifications (e.g., etherification, esterification, click reactions) are versatile, they often struggle with reproducibility and precise control over the complex interplay of molecular parameters. The future lies in developing intelligent synthesis platforms.

Developing novel synthetic approach with living or controlled polymerization of epoxy-containing monomers is a feasible direction. This requires the design of novel catalytic systems, potentially based on organocatalysts or tailored metal complexes that can suppress side reactions like etherification or uncontrolled branching, enabling precise control over molecular weight (MW) and chain architecture. The chloride ion-catalyzed proton transfer polymerization, which exhibits self-terminating behavior, is a precursor to this; future catalysts could offer external control over this "dormant" state, allowing for the synthesis of HERs with block-like or gradient structures.

The success of thiol–ene chemistry must be built upon. Future efforts should focus on the use of other highly efficient and orthogonal click reactions, such as copper-free strain-promoted azide–alkyne cycloaddition (SPAAC) or radical-resistant Diels–Alder reactions. These would allow for the modular, sequential construction of complex HER architectures under mild conditions, facilitating the precise integration of functional units to a hyperbranched core. Furthermore, multicomponent reactions and orthogonal synthetic strategies hold significant promise for constructing HERs with diverse architectures in a one-pot, efficient manner, minimizing purification steps and enhancing atom economy. Advanced oxidation technologies could also be explored to initiate or mediate specific polymerization pathways, potentially offering new routes for controlled HER synthesis under mild conditions.

8.1.2 AI-assisted Molecular Design and Big Data

The application of artificial intelligence (AI) and machine learning (ML) in organic synthesis is revolutionizing how chemists plan and execute molecular construction. These technologies are ideal candidates for the synthesis of HER with complex and multi-variable nature.

By aggregating vast datasets from the literature and experiments including encompassing monomer structures, catalyst types, reaction conditions, and the resulting polymer properties (DB, M_w, PDI, EEW, T_g, mechanical strength, etc.), ML algorithms can be trained to uncover hidden correlations. These models can predict the final properties of a HER from its synthetic recipe, drastically accelerating the design cycle.

Moreover, the introduction of AI allows the inverse design of HER's structure. A materials engineer can input a set of target properties required for a specific application, and the AI system would propose optimal synthetic pathways including polymerization strategy, monomer combinations, and reaction stoichiometries to achieve them.

8.1.3 Green and Energy-efficient Synthesis

Most HER syntheses are currently confined to the laboratory scale. Scaling up HER production necessitates a paradigm shift toward greener and more efficient synthetic processes to minimize environmental impact and economic costs. For example, microwave-assisted synthesis should become a standard technique, as it offers unparalleled advantages through molecular-level "internal heating." This method significantly reduces reaction times from hours or days to mere minutes, thereby significantly lowering energy consumption. Moreover, the rapid and uniform heating profile effectively suppresses side reactions, such as epoxy hydrolysis or premature gelation, leading to products with higher purity and more defined branching architectures.

Beyond microwave technology, several other green synthesis strategies hold great promise. For instance, the adoption of continuous-flow reactors is critical for overcoming the challenges of viscous intermediates encountered in traditional batch processes. The system's high surface-to-volume ratio provides exceptional heat and mass transfer control, preventing hotspots and ensuring consistent product quality with narrow MW distributions. Ultimately, all these technologies will form the cornerstone of intelligent, sustainable, and scalable production platforms for the next generation of HERs.

8.2 Structural Parameter Control and Design for Performance Optimization

The properties of HERs are intimately tied to their structural parameters. The future lies in moving from coarse tuning to fine, independent control of these parameters.

Accurately characterizing the complex structure of HERs is a prerequisite for precise control. Conventional 1D and 2D nuclear magnetic resonance (NMR) spectroscopy can be used to determine the degree of branching (DB) of HERs. Dynamic light scattering (DLS) and diffusion-ordered spectroscopy (DOSY) NMR can be applied to predict the MWs of HERs based on the linear correlation between the logarithm of diffusion coefficient (log D_{τ}) and the molecular weights (log M_{W}) and the external calibration curves. GPC with differential refractive index detection and ultraviolet (UV) detection followed by calibration with a standard polymer (polystyrene) is the most frequently used method to determine the MW and polydispersity index (PDI) of HERs. Matrix-assisted laser desorption/ionization time-of-flight mass spectrometry (MALDI-TOF MS) technique will provide absolute MWs and confirm architectural details. For the size determination of HERs, DLS and DOSY NMR can be utilized to study the diffusion behavior of HERs in solutions and the hydrodynamic radius can be determined through the Stokes–Einstein equation. Proper characterization methods should be taken to elucidate the structural parameters of HERs before any structure–property relationships can be established.

8.2.1 Precision Control of the Degree of Branching (DB)

DB is a critical parameter that influences the solution behavior, rheological properties, and, ultimately, mechanical performance of HERs. Most current methods only allow for "coarse-tuning" of DB. Future work should focus on designing novel AB_n or $A_2 + B_3$ monomer systems and leveraging reaction kinetics to achieve precise DB control. The use of core-dilution techniques or slow monomer addition can be optimized computationally to target specific DB values. This would enable the establishment of more definitive structure–property relationships and the tailoring of materials for specific applications.

Notably, current DB values are averages. Future research should develop methods to quantify the distribution of DB within a HER sample, as molecules with high and low DB in the same batch can lead to property averaging and defects in the cured network. Advanced chromatographic separation coupled with sophisticated detection could be a pathway to address this.

8.2.2 Precision Control of Molecular Weight Distributions

The PDI of HER is a major parameter for achieving uniform and predictable performance in high-reliability applications. A deeper mathematical and experimental understanding of the reaction kinetics in $A_2 + B_3$ systems could lead to protocols that naturally produce narrower distributions. Optimizing polymerization processes including slow monomer addition, segmented temperature control, and the use of chain transfer agents can enable effective control over PDI. Furthermore, the combination of size exclusion chromatography (SEC) with absolute detection methods like multi-angle light scattering (MALS) and MS will be essential for the precise analysis of different MW fractions.

8.2.3 Strategic Design of End-group Functionality and Topology

The type and number of end-groups directly determine the reactivity of HERs and the structure of the final cross-linked network. Current end-capping strategies rely heavily on a limited set of reagents like epichlorohydrin. Future research should expand the library of capping agents to include epoxy-functionalized compounds containing phosphorus, silicon, or fluorine, enabling the integration of flame retardancy, low dielectric constants, and hydrophobicity into the HER structure.

Moreover, multifunctional terminal groups may be introduced into the HERs. For instance, a single HER molecule could be designed with epoxy groups for cross-linking, tertiary amines for catalytic activity, long alkyl chains for hydrophobization, and clickable handles (e.g., azides) for post-modification. This requires sophisticated protection–deprotection strategies or sequential post-polymerization modifications.

Research will also move beyond simply "hyperbranched" to control the specific topology—whether it is a more globular, dendrimer-like structure, or a more elongated, "pear-shaped" one. As seen in Chapter 2, the shape factor ($\rho = R_g/R_h$)

significantly influences HER's properties. Synthesizing HERs with controlled spacer lengths between branch points is a powerful strategy to tailor backbone flexibility, free volume, and, ultimately, performance.

8.2.4 Establishing Topological Structure–mechanical Property Relationship

As the conventional thermosetting resins, the cured epoxy resins present a highly cross-linked structure with defects in the toughness. A homogeneous toughening effect could be realized by introducing the hyperbranched topological structure into the cross-linked network of epoxy resins, improving both the toughness and strength of epoxy resins. Future work should focus on the construction of a systematic topological structure–mechanical property relationship. How does the hyperbranched topology influence the free volumes and thus the mechanical properties of the resins? Structure-controlled synthesis of HERs is the precondition for this investigation. In addition, more precision analysis and testing technologies are expected to be used to characterize the microstructure of the HERs and the cured resins.

8.3 Functional Integration and Application Expansion

8.3.1 Multifunctional Integrated Design

Future HERs will evolve beyond improving a single property toward "multifunctional integration." Designing HER molecules that combine flame retardancy, self-healing, conductivity, and shape memory within a single structure will create materials capable of adapting to extreme environments. The challenge is to integrate multiple, sometimes contradictory, properties into a single material system without resorting to physical blending, which can lead to phase separation and compromised performance. This requires ingenious molecular design to overcome chemical reactivity and compatibility conflicts between different functional groups.

8.3.2 Dynamic Covalent Chemistry and Reconfigurable Networks

The incorporation of dynamic covalent bonds including esters, imines, disulfides, and hexahydro-*s*-triazines, has proven effective in endowing HERs with recyclability, malleability, and self-healing capabilities. Future research should explore novel dynamic bonding systems to create reversible cross-linked networks that operate under even milder conditions.

A critical challenge on the path to next-generation recyclable composites is overcoming the "reprocessability–stability trade-off" inherent to conventional vitrimers. The unique hyperbranched topology of HERs offers a distinct advantage: it can be strategically designed to catalyze dynamic exchange reactions (e.g., transesterification) by concentrating catalytic groups like hydroxyls within a constrained local environment. When integrated into epoxy networks, this topology can induce homogeneous toughening, simultaneously enhancing both toughness and strength.

The rational design of hyperbranched epoxy vitrimers thus represents a pivotal frontier, aiming to reconcile rapid bond exchange with robust thermal and mechanical stability. Furthermore, a promising direction lies in constructing hierarchical networks by combining dynamic covalent bonds with strong, reversible supramolecular interactions (e.g., UPy-based quadruple hydrogen bonding). Such materials could maintain high strength and toughness at service temperatures while becoming readily malleable and recyclable upon heating, achieving an unprecedented combination of properties.

8.3.3 Expansion of Application Scenarios

As the performance of HERs continues to improve, their application scope will expand into high-tech fields. For instance, leveraging click chemistry techniques could enable the synthesis of low-viscosity, halogen-free HERs, making them ideal candidates for advanced electronic packaging applications in semiconductors and chips, where processability and performance are paramount. Furthermore, the exceptional combination of high strength, high toughness, and lightweight inherent to HERs positions them as next-generation structural materials for demanding sectors such as aerospace, the emerging low-altitude economy, and intelligent robotics, enabling the development of lighter, more durable, and high-performance components. Collaboration with end users in co-developing application-specific resins is important in the future. This approach will enable a versatile product landscape capable of delivering both versatile, multi-use materials and highly customized solutions for specialized needs.

8.4 Sustainable Development and Circular Economy

8.4.1 Expansion of Bio-based Monomer Resources and Performance Enhancement

While progress has been made in bio-based HERs, efficient, economical, and large-scale production of bio-based monomers remains a challenge. The future development of bio-based HERs is strongly dependent on the development of biorefineries, which supply high-quality chemicals as the building blocks for HERs. Achieving efficient, economical, and large-scale synthesis of bio-based monomers from abundant biomass like carbohydrates and lignin is crucial, which is the basis for the development of bio-based polymer materials. Developing catalytic processes to depolymerize lignin into well-defined aromatic multifunctional monomers (A_2, B_3 types) for HER synthesis would be a monumental achievement.

Another problem for developing bio-based HERs is that the inherent rigidity of many bio-based aromatics leads to brittleness. Molecular design strategies, such as incorporating flexible aliphatic chains from fatty acids or isosorbide, or creating block-like architectures with rigid and flexible segments within the HER, will be essential to overcome this limitation and compete with petroleum-based analogs in toughness-demanding applications.

8.4.2 Optimization of Degradation and Recycling Strategies

The degradation conditions for most current recyclable HERs (e.g., strong acids or bases) are still relatively harsh. Future recyclable HERs will feature cleavable linkages (e.g., esters, acetals, silyl ethers, hexahydro-*s*-triazines) placed at strategic locations within the hyperbranched structure. Moreover, developing milder degradation methods, such as enzyme-catalyzed degradation or photo-triggered cleavage, is crucial for reducing energy consumption and environmental impact. The goal is to achieve complete depolymerization back to the original monomers or to well-defined oligomers under these triggers. Notably, future research should focus not only on the efficient recovery of monomers but also on their purification and repolymerization into new high-performance materials, establishing a true closed-loop recycling process.

A comprehensive cradle-to-grave or cradle-to-cradle life cycle assessment (LCA) must be conducted. This will provide transparent data on its true environmental impact compared to incumbent materials and guide researchers toward the most impactful innovations.

8.5 Industrialization Pathways and Standardization

8.5.1 Overcoming Scale-up Challenges

The engineering challenges of scaling up HER synthesis are nontrivial and require dedicated effort. The synthesis of high MW HERs can lead to dramatic viscosity increases and significant exotherms. Engineering solutions, such as the use of twin-screw extruders as continuous reactors or the development of novel high-shear mixing systems, will be necessary to manage heat and mass transfer effectively.

8.5.2 Lack of Specialized Processing Techniques and Equipment

While significant progress has been made in the proof-of-concept reprocessing and recycling of bio-based HERs, specialized techniques and equipment remain underdeveloped. Currently, recyclable HERs are predominantly processed using conventional thermoplastic equipment, with heated compression molding being the most widely reported method. This reliance on a single technique leaves alternatives like injection molding largely unexplored. Therefore, developing dedicated processing technologies is a critical next step not only for bio-based HERs but also for the entire vitrimer field.

8.6 Concluding Remarks

As a rising star in the epoxy resin family, HERs, with their unique structural advantages and tunable performance, offer a promising pathway to solving the long-standing bottlenecks of traditional thermosets: brittleness, high viscosity, and nonrecyclability. The future development of HERs necessitates close collaboration among chemists, materials scientists, chemical engineers, and computational theorists. Through multidisciplinary integration, we can collectively overcome the core challenges in synthesis, structure–property relationships, functional integration, and sustainability. It is foreseeable that as these challenges are progressively addressed, HERs will play an increasingly vital role in enabling next-generation technologies across various sectors, opening a new chapter in the era of high-performance and sustainable polymeric materials.

Index

f

g

h